Electrodynamics of Metamaterials

Electrodynamics of Metamaterials

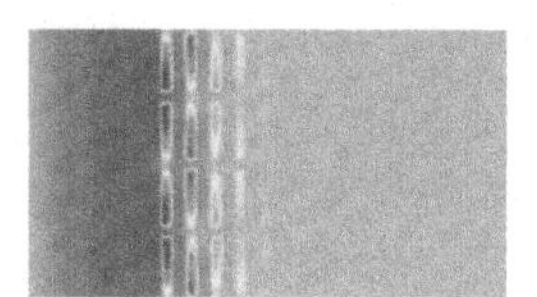

Andrey K Sarychev
Russian Academy of Sciences
Institute of Theoretical and Applied Electrodynamics

Vladimir M Shalaev
Purdue University
Birck Nanotechnology Center

World Scientific

NEW JERSEY • LONDON • SINGAPORE • BEIJING • SHANGHAI • HONG KONG • TAIPEI • CHENNAI

Published by

World Scientific Publishing Co. Pte. Ltd.

5 Toh Tuck Link, Singapore 596224

USA office: 27 Warren Street, Suite 401-402, Hackensack, NJ 07601

UK office: 57 Shelton Street, Covent Garden, London WC2H 9HE

British Library Cataloguing-in-Publication Data
A catalogue record for this book is available from the British Library.

ELECTRODYNAMICS OF METAMATERIALS

ISBN-13 978-981-02-4245-9
ISBN-10 981-02-4245-X

Printed in Singapore.

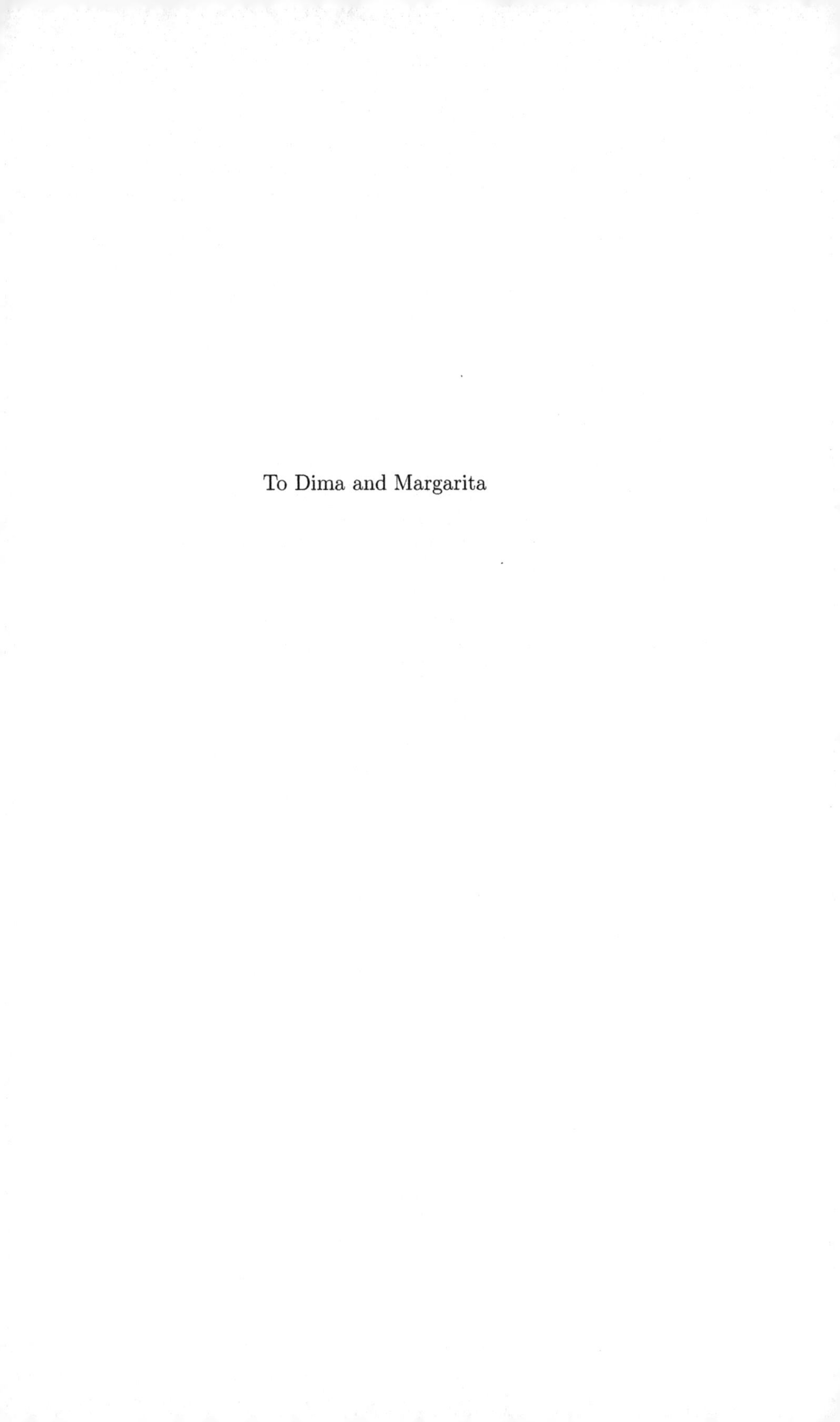

To Dima and Margarita

Preface

The current electronic techniques, as many believe, are running out of steam due to issues with RC-delay times, meaning that fundamentally new approaches are needed to increase data processing operating speeds to THz and higher frequencies. There is an undeniable and ever-increasing need for faster information processing and transport. The photon is the ultimate unit of information because it packages data in a signal of zero mass and has unmatched speed. The power of light is driving the photonic revolution, and information technologies, which were formerly entirely electronic, are increasingly enlisting light to communicate and provide intelligent control. Today we are at a crossroads in this technology. Recent advances in this emerging area now enable us to mount a systematic approach towards the goal of full system-level integration.

The mission that researchers are currently trying to accomplish is to fully integrate photonics with nanotechnology and to develop novel photonic devices for manipulating light on the nanoscale, including molecule sensing, biomedical imaging, and processing information with unparalleled operating speeds. To enable the mission, one can use the unique property of metal nanostructures to "focus" light on the nanoscale. Metal nanostructures supporting collective electron oscillations — plasmons — are referred to as plasmonic nanostructures, which act as optical nanoantennas by concentrating large electromagnetic energy on the nanoscale.

There is ample evidence that photonic devices can be reduced to the nanoscale using optical phenomena in the near-field, but there is also a scale mismatch between light at the microscale and devices and processes at the nanoscale that must be addressed first. Plasmonic nanostructures can serve as optical couplers across the nano–micro interface. They also have the unique ability to enhance local electromagnetic fields for a number of ultra-compact, sub-wavelength photonic devices. Nanophotonics is not only about very small photonic circuits and chips, but also about new ways of sculpting the flow of light with nanostructures and nanoparticles exhibiting fascinating optical properties never seen in the macro-world.

Plasmonic nanostructures utilizing surface plasmons (SPs) have been extensively investigated during the last decade and show a plethora of amazing effects and fascinating phenomena, such as extraordinary light transmission, giant field enhancement, SP nano-waveguides, and recently emerged metamaterials that are often based on plasmonic nanostructures. Metamaterials are expected to open a new gateway to unprecedented electromagnetic properties and functionality unattainable from naturally occurring materials. The structural units of metamaterials can be tailored in shape and size, their composition and morphology can be artificially tuned, and inclusions can be designed and placed at desired locations to achieve new functionality.

Light is in a sense "one-handed" when interacting with atoms of conventional materials. This is because out of the two field components of light, electric and magnetic, only the electric "hand" efficiently probes the atoms of a material, whereas the magnetic component remains relatively unused because the interaction of atoms with the magnetic field component of light is normally weak. Metamaterials, i.e., artificial materials with rationally designed properties, can enable the coupling of both the field components of light to meta-atoms, enabling entirely new optical properties and exciting applications with such "two-handed" light. Among the fascinating properties is a negative refractive index. The refractive index is one of the most fundamental characteristics of light propagation in materials. Metamaterials with negative refraction may lead to the development of a superlens capable of imaging objects and their fine structures that are much smaller than the wavelength of light. Other exciting applications of metamaterials include novel antennae with superior properties, optical nano-lithography and nano-circuits, and "meta-coatings" that can make objects invisible.

The word "meta" means "beyond" in Greek, and in this sense, the name "metamaterials" refers to "beyond conventional materials." Metamaterials are typically man-made and have properties not available in nature. What is so magical about this simple merging of "meta" and "materials" that has attracted so much attention from researchers and has resulted in exponential growth in the number of publications in this area?

The notion of metamaterials, which includes a wide range of engineered materials with pre-designed properties, has been used, for example, in the microwave community for some time. The idea of metamaterials has been quickly adopted in optics research, thanks to rapidly-developing nanofabrication and sub-wavelength imaging techniques. One of the most exciting opportunities for metamaterials is the development of "*left handed metamaterials*" (LHMs) with negative refractive index. These LHMs bring the concept of refractive index into a new domain of exploration and thus promise to create entirely new prospects for manipulating light, with revolutionary impacts on present-day optical technologies.

It is a rather unique opportunity for researchers to have a chance to reconsider and possibly even revise the interpretation of very basic laws. The notion of a negative refractive index is one such case. This is because the index of refraction enters into the basic formulae for optics. As a result, bringing the refractive index into a new domain of negative values has truly excited the imagination of researchers worldwide.

The refractive index gives the factor by which the phase velocity of light is decreased in a material as compared to vacuum. LHMs have a negative refractive index, so the phase velocity is directed against the flow of energy in a LHM . This is highly unusual from the standpoint of "conventional" optics. Also, at an interface between a positive and a negative index material, the refracted light is bent in the "wrong" way with respect to the normal. Furthermore, the wave-vector and vectors of the electric and magnetic fields form a left-handed system.

This book reviews the fundamentals of plasmonic structures and metamaterials based on such structures, along with their exciting applications for guiding and controlling light. Both random and geometrically ordered metamaterials are considered. Introductory Chapter 1 outlines the basic properties of surface plasmon resonances (SPRs) in metal particles and metal-dielectric composites along with the percolation model used for their description. Chapter 2 is focused on metal rods and their applications for LHMs . Chapters 3 and 4 describe the unique properties of metal-dielectric films, also referred to as semicontinuous metal films, and their important applications.

We also present there the general theory of the surface enhancement of the Raman signal and the theory of nonlinear optical phenomena in metal-dielectric metamaterials. At the end of Chapter 4 we discuss the analytical theory of the extraordinary optical transmittance (linear and nonlinear). Finally, Chapter 5 deals with electromagnetic properties of geometrically ordered metal-dielectric crystals.

The authors are grateful to their collaborators, Profs. Armstrong, Antonov, Aronzon, Bergman, Boccara, Brouers, Cao, Clerc, Ducourtieux, Dykhne, Feldmann, Gadenne, Golosovsky, Gresillon, Lagarkov, Markel, Matitsine, McPhedran, Pakhomov, Panina, von Plessen, Plyukhin, Podolskiy, Quelin, Rivoal, Rozanov, Safonov, Seal, Shvets, Tartakovsky, Vinogradov, Wei, Yarmilov, Ying, and Drs. Blacher, Bragg, Drachev, Genov, Goldenshtein, Kalachev, Karimov, Kildishev, Nelson, Poliakov, Seal, Shubin, Simonov, Smychkovich, Yagil, who did critical contributions without which this book would not be possible. Useful discussions with Profs. Aharony, Boyd, Bozhevolnyi, Efros, Gabitov, George, Grimes, Lakhtakia, Likalter, Maradudin, Moskovits, Narimanov, Nazarov, Noginov, Obukhov, Render, Pendry, Shklovskii, Sahimi, Sheng, Smith, Soukoulis, Stockman, Stroud, Tatarskii, Thio, Veselago, Weiglhofer, Weiner, and Yablonovitch are also highly appreciated. Special thanks go to Dr. Poliakov who did a lot of editorial work in preparing this book. Special thanks also go to Dr. Genov, who made important contributions to Sec. 3.3.3.

Contents

Chapter 1

Introduction

Current developments in optical technologies are being directed toward nanoscale devices with subwavelength dimensions, in which photons are manipulated using nano-scale optical phenomena. Although light is clearly the fastest means to send information to and from the nanoscale, there is a fundamental incompatibility between light at the microscale and devices and processes at the nanoscale. For most materials, light–matter interactions decrease as $(a/\lambda)^2$, where a is the structure size and λ is the wavelength. However, metals, which support surface plasmon modes, can concentrate electromagnetic (em) fields to a small fraction of a wavelength while enhancing local field strengths by several orders of magnitude. For this reason, plasmonic nanostructures can serve as optical couplers across the nano–micro interface: metal–dielectric and metal–semiconductor nanostructures can act as optical nanoantennas and enhance efficiency such that the optical cross-sections increase in magnitude from $\sim a^2$ to $\sim \lambda^2$.

As electronics shrinks to the nanoscale, photonics must follow to maintain the speed and parallelism necessary for nanoelectronics to achieve their full potential. Nanophotonics, united with nanoelectronics, is destined to be a vital technology in the global high-tech economy. Nanophotonics with plasmonic structures promises to create entirely new prospects for manipulating light, some of which may have revolutionary impact on present-day optical technologies. However, our understanding of the interplay between light and plasmonic nanostructures is still incomplete, and techniques to synthesize nanophotonic devices and circuits based on plasmonic materials are still relatively primitive. Full integration of light with nanoscale plasmonic devices and processes will require fundamental advances at this research interface.

In this book, we will describe the electrodynamics of metal-dielectric metamaterials, which form a large class of nanostructured composite materials. Nanostructured metal-dielectric composites exhibit fascinating optical properties at visible and near-infrared frequencies due to excitation of surface plasmon modes. The technology drift of adopting these properties for applications in photonics and chemical sensing have led to exponentially growing activity in the actual design of plasmonic materials with subwavelength dimensions. Strongly amplified electromagnetic fields can be generated both in a broad spectral range for disordered metal–dielectric composites and at selected frequencies for periodically ordered metal nanostructures. Periodicity of metamaterial geometry plays a crucial role in tuning the optical response: Such phenomena have been observed in numerous experimental and theoretical studies of enhanced plasmon effects in left-handed materials (also referred to as negative-index materials, NIMs), in surface-enhanced Raman scattering, extraordinary optical transmission and photonic band-gap structures, both in the visible and near-infrared wavelengths. Microstrucured analogues of nano-metamaterials also exhibit very interesting and unusual properties, particularly in the microwave band range, where such phenomena as artificial magnetism, negative refractive index, and plasmonic bandgap have been also observed.

1.1 Surface Plasmon Resonance

To give a reader a quick insight to the wonderland of metamaterials let us consider a relatively simple plasmonic system, such as two-dimensional (2D) arrays of metal nanoparticles embedded in a dielectric medium. We then provide numerical calculations and a simple analytical theory for electromagnetic (EM) field enhancements in such systems. The obtained numerical and analytical results are in good agreement, yielding values for the enhancement of surface-enhanced Raman scattering (SERS) signal as large as 10^{11} for arrays of silver or gold nano-disks. SERS enhancement factor is described by $G_R = \left\langle |E(\mathbf{r})/E_0|^4 \right\rangle$ equation, where $E(\mathbf{r})$ is the local electric field in Cartesian coordinates $\mathbf{r} = (x, y, z)$, E_0 is the amplitude of incident field. Such field enhancements for the nanostructured arrays are strongly dependent on the ratio of composing particle diameter a and inter-particle spacing δ, which determines both the intensity of local fields and the characteristic cross-section for sampling chemical and biomolecular analytes. Our numerical simulations further illustrate how the interplay between field enhancement and particle spacing can affect the design of array-based SERS sensors for tracing chemical analytes.

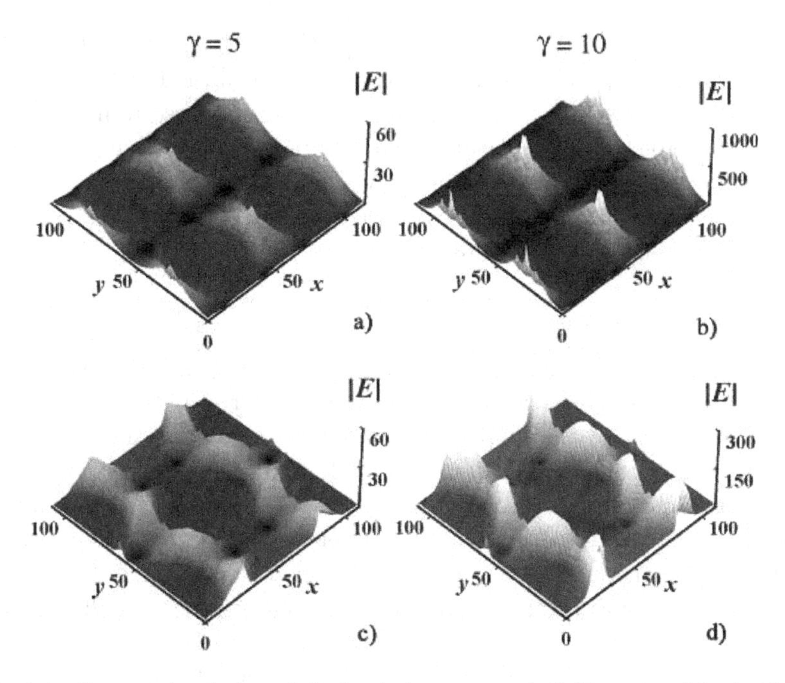

Fig. 1.1 Cross-sectional view of the local electromagnetic field produced by incident light, normal to the surface of periodic $2D$ arrays of gold nano-disks (λ = 647 nm, $E_0 = E_x$), embedded in a low-dielectric medium (susceptibility $\varepsilon_d = 1.5$). Two different lattice geometries are shown: a,b) $2D$ square lattice; c,d) $2D$ hexagonal lattice. Different diameter–spacing ratios γ are shown: a,c) $\gamma = 5$; and b,d) $\gamma = 10$. Note the change in scale for $|E_{loc}(r)/E_0|$ for arrays with different values of γ.

Calculations are performed for two-dimensional arrays of silver nano-disks arranged in square or hexagonal lattices in a dielectric medium of permittivity ε_d using periodic boundary conditions (see Fig. 1.1). Individual disk diameter a is much smaller than the wavelength of the incident light, $a \ll \lambda$. For the purpose of estimating local field factors $|E(\mathbf{r})/E_0|$, such arrays can be approximated by a planar system. By applying the charge conservation law $\operatorname{div} \mathbf{D} = 0$, which is then expressed in terms of a local electric potential $\phi(\mathbf{r})$, we obtain

$$\operatorname{div}\left[\varepsilon(\mathbf{r})\left(-\nabla\phi\left(\mathbf{r}\right)+\mathbf{E}_0\right)\right]=0, \tag{1.1}$$

where $\mathbf{E}_0$ is the incident electric field, and $\varepsilon(\mathbf{r})$ is the local permittivity. The permittivity takes values $\varepsilon(\mathbf{r}) = \varepsilon_d$ in the dielectric and $\varepsilon(\mathbf{r}) = \varepsilon'_m + i\varepsilon''_m = i4\pi\sigma_m/\omega$ in the metal, where it is complex valued. The metal permittivity is typically negative in the optical range of frequencies with a small imaginary part: $\varepsilon_m = \varepsilon'_m (1 - i\kappa)$, $\varepsilon'_m < 0$, $\kappa = \varepsilon''_m / |\varepsilon'_m| \ll 1$ (see e.g., [Johnson and Christy, 1972], [Palik, 1985], and [Kreibig and Volmer, 1995]; Eq. (3.1) and Table 3.3). Describing the continuity equation in these terms allows one to obtain the collective plasmon response, in a scale-invariant manner, under the quasistatic conditions [Sarychev and Shalaev, 2000]. In addition, $|E(\mathbf{r})/E_0|$ can be calculated as a continuous function of the packing density described by a single geometric parameter

$$\gamma = a/\delta, \tag{1.2}$$

where δ is the closest distance between the particles and γ is the constituent particle diameter-and-spacing ratio. The parameter γ is fundamentally important, because of the simple relationship $\varepsilon'_m \approx -\gamma\varepsilon_d$ obtained at the resonance conditions, which is derived directly from the Poisson's equation (1.1). It is important to note that the quasistatic approximation is valid for nanoparticle arrays with a periodicity below the diffraction limit, i.e., $a + \delta < \lambda$. Furthermore, for a $2D$ nanoparticle array where $\gamma \gg 1$, the spatial localization of the EM resonances between particles is well within the quasistatic limit.

Discretization of the equation for the local potential $\phi(\mathbf{r})$ under periodic boundary conditions yields the set of equations, which can be solved by the exact block elimination approach (see Ch.3.2.1 for details) or by commercially available simulation tools like COMSOL, etc. The enhancement factor G_R is obtained simply as the mean value of $G_{loc} = |E_{loc}(r)/E_0|^4$ within a unit cell of a periodic lattice. We note that these calculations are equally valid for both periodic arrays of nanowires and oblate metal nanodisks with high aspect ratio, whose electric fields and currents are both confined within the surface plane.

The intensities of the local and average EM field enhancements depend greatly on both incident wavelength and diameter–spacing ratio γ as seen in Figs. 1.1 – 1.3 for G_{loc} of gold nanodisk arrays. Nanodisk arrays made of gold nanoparticles with large diameter–spacing ratios ($\gamma \geq 30$) can produce G_{loc} values as large as 10^{10}, whereas the silver-based nanodisk arrays can produce G_{loc} values up to 10^{12}. Such large optical enhancements are due to the resonant modes of periodically ordered arrays. Arrays with a larger γ can produce the high field enhancement over a greater range of the excitation wavelengths, which has important practical applications for SERS and other spectroscopies. It is interesting to note that arrays of various metal structures have been used in microwave applications for a long time, for example, as frequency selective surfaces.[Mittra *et al.*, 1988].

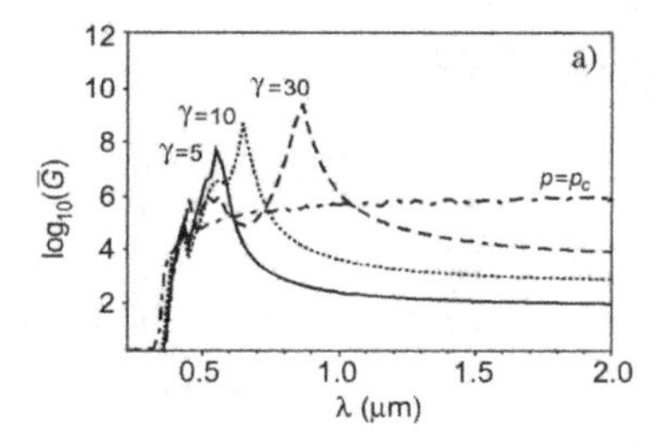

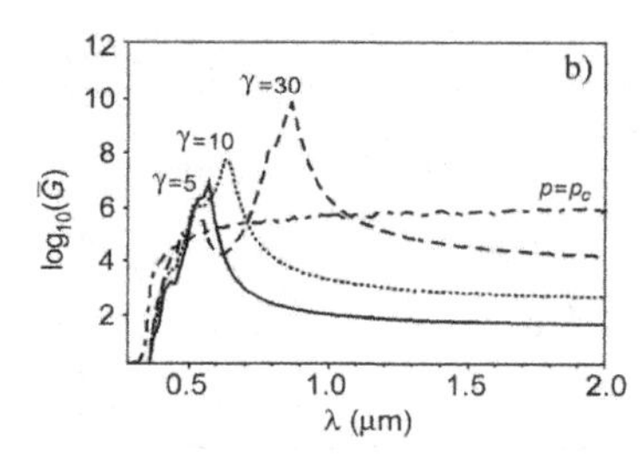

Fig. 1.2 Average EM enhancements ($\bar{G} \equiv G_R$) from *2D* arrays of gold nano-disks as a function of incident wavelength (λ) at fixed particle diameter–spacing ratios ($\gamma = 5$, *10*, and *30*): a) *2D* square lattice; b) *2D* hexagonal lattice. A plot of enhancement G_R from random metal–dielectric films at the percolation threshold ($p = p_c$, see Ch. 3) is included for comparison.

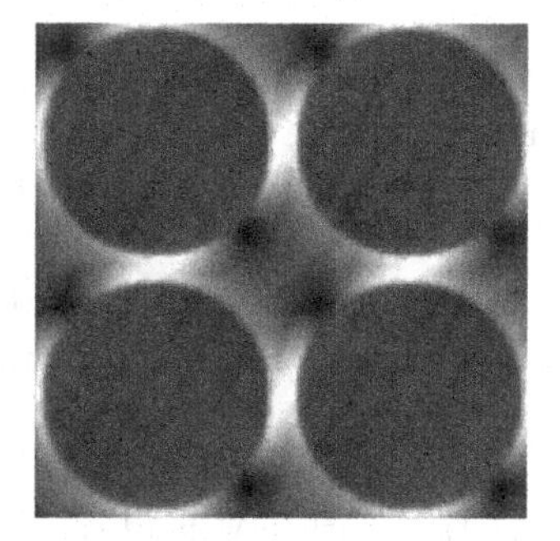

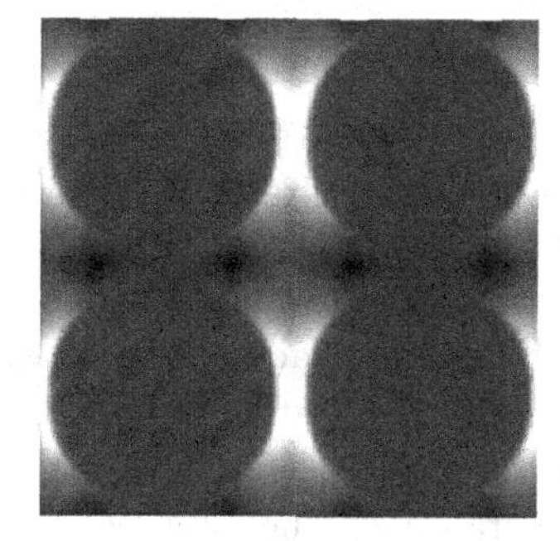

Fig. 1.3 Local EM field intensities at different polarization angles; left $\theta = 45°$, right $\theta = 0°$ ($E_0 = E_x$).

The intensities of the local fields that define G_R are highly dependent on their spatial distributions within a nanoparticle array as well as on the polarization of the incident light. In our calculations, values for E_{loc} were obtained at the fixed diameter–spacing ratios with the incident light being linearly polarized along the "x"-axis so that the local EM field enhancements are largest between the particles (see Fig. 1.1 for $\gamma = 5$ and 10). A change of E_0 direction, i.e., the change of the polarization angle, produces large shifts in the positions and intensities of the maximum local fields, with a reduction in the average EM enhancements (see Fig. 1.3). For example, the enhancement magnitudes produced by a square-lattice *2D* array decrease by 10^2 when θ changes from 0° to 45°. The polarization angle is an important parameter for single-molecule SERS and related spectroscopies, in which the analytes of interest are localized between nanoparticles

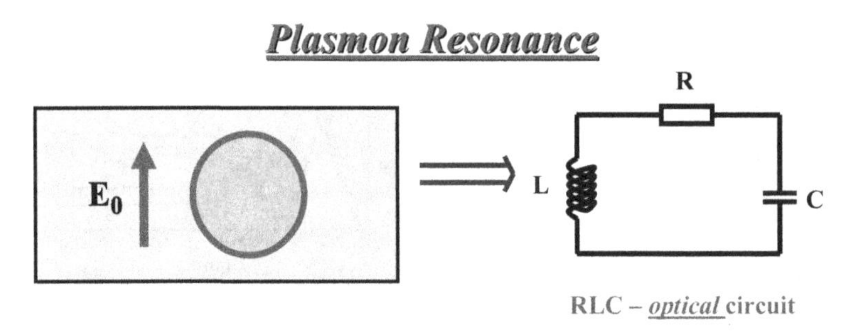

Fig. 1.4 *R-L-C* contour mapping of surface plasmon resonance of metal nanoparticle; inductance and resistance stand for metal particle, capacitor represents surrounding space.

[Kneipp *et al.*, 2002], [Wei, 2003], and [Xu *et al.*, 1999]. In this scenario, the Raman signals can be maximized by adjusting angle θ until the spatial position of the largest local field coincides with the exact position of the analyte molecule.

To confirm the validity of the numerical calculations, we also estimate the enhancement factor G_R via analytical calculations. Since the metal permittivity is negative in the optical frequency region and it is inversely proportional to the frequency squared we can model a metal particle as an inductance L. The interaction of a metal particle with EM field can be then presented as excitation of R-L-C contour, which is depicted in Fig. 1.4. Here, inductance L (with small losses described by resistance R) represents the metal particle while capacitance C represents the surrounding space. The resonance in R-L-C contour is analogous to the *surface plasmon resonance* in a single metal particle.

The EM coupling between metal nanoparticles can be conceptualized as arrays of R-L-C circuits across the inter-particle gap, with each element i representing a resonance defined by local spacing δ_i (see Fig. 1.5). The negative-valued permittivity of the metal ($\varepsilon'_m < 0$) and the dielectric ε_d are represented respectively by inductance R–L and capacitance C. The R-L-C model suggests that the collective plasmon resonances of *2D* nanoparticle arrays are shifted toward lower frequencies (high L and C) with increasing γ, in accordance with our numerical calculations, involving metal nanoparticle dimers [Hongxing *et al.*, 2000] and chain of particles [Stockman and Bergman, 2003]. In addition, the model agrees with the broadening of the plasmon bandwidth that accompanies the red-shifting of frequencies.

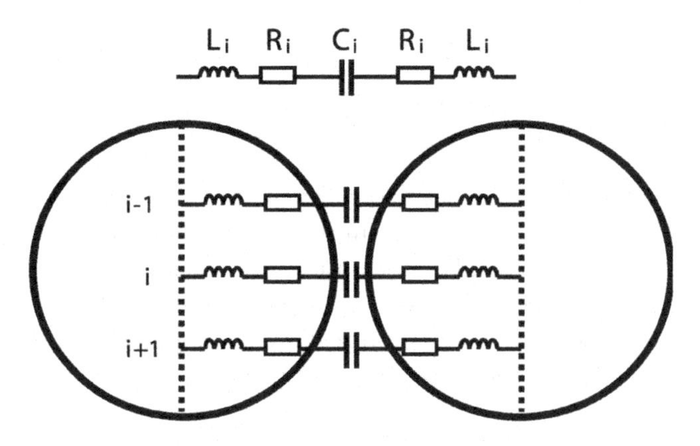

Fig. 1.5 Frequency-dependent plasmon response depicted as an array of *R-L-C* circuits.

The *R-L-C* model provides a theoretical framework for describing the variation of EM field enhancement as a function of γ and ε_m. We first consider the electric field distribution in a two-dimensional (*2D*) periodic array of infinite metal cylinders arranged in a square lattice. This is essentially a *2D* problem, since the local field depends only on the "x"-, "y"-coordinates in the plane perpendicular to the cylinder height. Therefore, the obtained solution is a good approximation for periodic arrays of metal nanodisks with high aspect ratio, since their electric fields and currents are essentially confined within the surface (disk) plane.

The most interesting effects appear when particle diameter $a = 2r$ is much larger than the interparticle spacing δ so that the parameter $\gamma \gg 1$ (see Eq. (1.2)) resulting in large local fields. We superimpose the Cartesian coordinate system onto the principal directions of the *2D* array. To estimate the local electric fields, we consider incident (macroscopic) electric fields E_0 polarized along the "x"-axis of the *2D* lattice and assume that the local field $E = E_x(y)$ has x component only, i.e. the electric field E is aligned with E_0. Using the central point between two particles as the origin, the electric field within the array can be expressed as

$$E_m(y) = E_0 \frac{x_0 \varepsilon_d}{x_d \varepsilon_m + x_m \varepsilon_d}, \tag{1.3}$$

$$E_d(y) = E_0 \frac{x_0 \varepsilon_m}{x_d \varepsilon_m + x_m \varepsilon_d}, \tag{1.4}$$

where ε_d is the dielectric permittivity between the particles, $x_0 = 2r + \delta$ is the distance between the centers, $x_m(y) = 2\sqrt{r^2 - y^2}$ ($|y| < r$), $x_m(y) = 0$ ($|y| > r$) and $x_d(y) = x_0 - x_m(y)$. We can now calculate the average dielectric displacement,

$$\langle D\rangle = \frac{1}{\Omega}\int_{-r-\delta/2}^{r+\delta/2} \left[\varepsilon_m E_m(y) + \varepsilon_d E_d(y)\right] dy, \tag{1.5}$$

where $\Omega = a^2(1+1/\gamma)^2$, and define the effective permittivity of the metal–dielectric composite as $\varepsilon_e = \langle D\rangle / E_0$. In the case of metals with low optical loss, the permittivity of metal at the optical frequencies can be expressed as $\varepsilon_m = \varepsilon_m'(1-i\kappa)$, where $\varepsilon_m' < 0$ and loss factor $\kappa \ll 1$. Performing the integration (1.5) and using the notations above, we obtain the equation for ε_e, which takes a simple form when $\gamma \gg 1$ and $\kappa \ll 1$:

$$\varepsilon_e = \varepsilon_d W\left(\frac{\pi}{\sqrt{2\kappa(\Delta - i)(W+1)}} - \frac{1+\pi/2}{1+W}\right), \tag{1.6}$$

where $W = |\varepsilon_m'|/\varepsilon_d$, and $\Delta = (W/\gamma - 1)/\kappa$. As seen from Eq. (1.6), the composite dielectric function reaches its maximum when $\Delta \to 0$, so that $\varepsilon_e^{MAX} = \varepsilon_d W(1+i)\pi/\sqrt{4\kappa(W+1)}$. When $\Delta > 1$, ε_e is mostly real and larger than ε_d, such that the array response is dominated by the interparticle capacitance. When $\Delta < 0$, the imaginary part of ε_e prevails, resulting in large losses and thus large fluctuations of the local field. It is worth pointing out that the effective absorption of the metal–dielectric composite, ε_e'', is proportional to $\sqrt{|\varepsilon_m|/\kappa}$ near the resonance and increases as κ goes to zero. A high effective absorption is necessary to produce giant fluctuations in the local fields between particles, and it determines the ability of *2D* arrays to accumulate and release EM energy.

The average enhancement for Raman scattering, which is given by

$$G_R \cong |E_0|^4 \Omega^{-1} \int_{-r-\delta/2}^{r+\delta/2} |E(y)|^4 dy, \tag{1.7}$$

can be calculated similar to the average electric displacement $\langle D\rangle$ (see Eq. (1.5)). In the case of the densely packed metal particles ($\gamma \gg 1$) and $W \gg 1$, we obtain

$$G_R \cong \frac{\pi}{8}\frac{W^{5/2}}{\kappa^{7/2}}\sqrt{\frac{4\Delta^2+9}{(\Delta^2+1)^{3/2}} - \frac{\Delta(4\Delta^4+15\Delta^2+15)}{(\Delta^2+1)^3}}. \tag{1.8}$$

The maximum EM enhancement, G_R^{MAX}, is achieved when $\Delta \to 0$. If we assume the Drude free-electron response (see Sec. 3.2) and substitute the equations for $\varepsilon_m' = 1 - (\omega_p/\omega)^2 \simeq -(\omega_p/\omega)^2$ and $\kappa = \omega_\tau/\omega$, with $\omega_p \gg \omega$

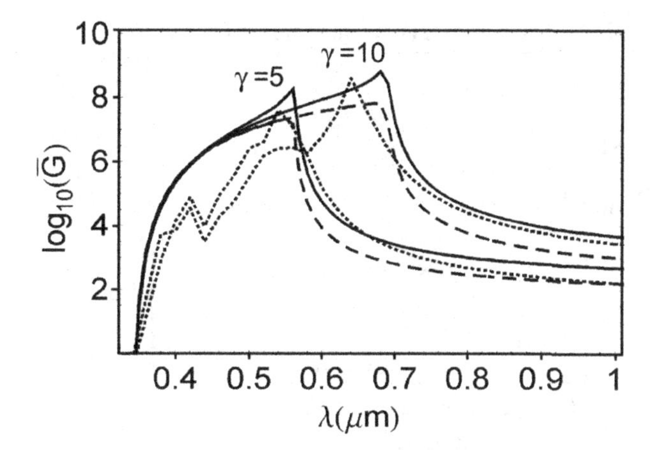

Fig. 1.6 Analytical solutions of average EM enhancements G_R as a function of λ for *2D* square-lattice arrays of *Au* nanodisks and nanospheres (solid and dashed curves, respectively), calculated for $\gamma = 5$ and 10. The corresponding numerical solution for the *2D* nanodisk array (dotted curve) is included for comparison.

and $\omega_\tau \ll \omega$, then parameter W can be approximated as $(\omega_p/\omega)^2/\varepsilon_d \gg 1$ while G_R^{MAX} can be estimated as

$$G_R^{MAX} \sim \frac{\pi}{\varepsilon_d^{5/2}} \left(\omega_p/\omega_{res}\right)^{3/2} \left(\omega_p/\omega_\tau\right)^{7/2}, \tag{1.9}$$

where we define the resonance frequency as $\omega_{res} \approx \omega_p(\gamma\varepsilon_d)^{-1/2}$. For the case of silver-particle nanodisk arrays, we use $\omega_p = 9.1$ eV and $\omega_\tau = 0.02$ eV ([Johnson and Christy, 1972], [Kreibig and Volmer, 1995]) to obtain the approximate G_R^{MAX} value of $2 \cdot 10^{11}\varepsilon_d^{5/2}\lambda_{res}^{3/2}$, where $\lambda_{res} = 2\pi c/\omega_{res}$ is the resonance wavelength expressed in micrometers.

The analytical expressions derived from the *R-L-C* model are in good agreement with the numerical calculations for *2D* nanodisk arrays at different values of γ (see Fig. 1.6). It should be noted, however, that the numerical calculations exhibit additional maxima at shorter wavelengths, which are not present in the analytical solution. These peaks are likely related to artifacts resulting from the discretization procedure. The numerical calculations and analytical theory [Sarychev and Shalaev, 2003] are in good agreement with the experimental results obtained by A. Wei for SERS in arrays of gold nanoparticles [Genov *et al.*, 2004].

We mention here two additional observations. First, for a given incident wavelength, G_R is essentially equivalent to G_R^{MAX} when γ is above the threshold value, which is clearly seen in Figs. 1.7 and 1.8. This is a

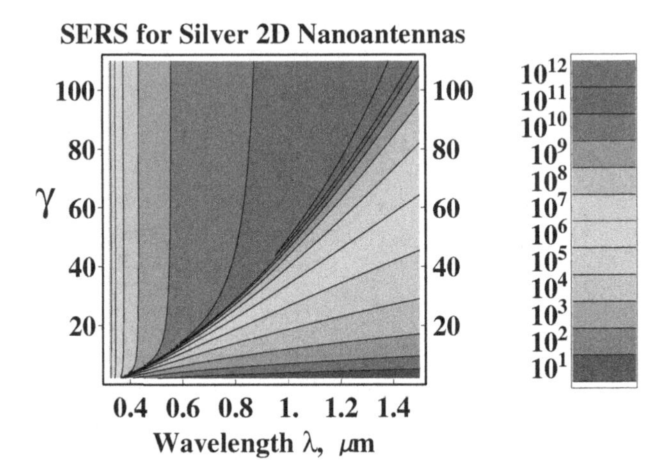

Fig. 1.7 Average EM enhancements G_R as a function of wavelength λ and diameter-spacing ratio γ for *2D* square-lattice array of silver nanodisks.

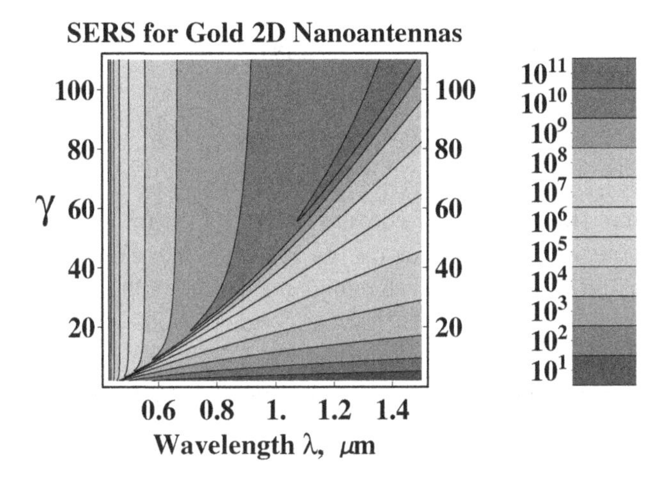

Fig. 1.8 Average EM enhancements G_R as a function of λ and γ for *2D* square-lattice array of gold nanodisks.

consequence of the trade-off between G_{loc}, with maximum intensities increasing rapidly with γ, and the total available field volume, which decreases with γ. The wide range of diameter–spacing ratios which can provide G_R^{MAX} offers the major practical benefit for designing *2D* nanoparticle arrays as SERS-based chemical sensors. However, it is also necessary to consider geometrical requirements for accessing high-field regions so that analyte molecules can be rapidly detected. Highly localized field enhancements have a lower probability of detecting exogenous analytes than those which are diffused over a larger cross-sectional area; therefore, the optimal diameter–spacing ratio should be close to the threshold value of γ, which produces G_R^{MAX} within the largest possible sampling volume.

Second, both theoretical and experimental G_R values for *2D* nanoparticle arrays [Genov *et al.*, 2004] remain approximately constant as the ratio a/λ increases (γ is constant) in contrast to that calculated and observed in an ensemble of isolated nanospheres for which G_R decreases as a/λ becomes larger. This indicates that size-dependent radiative damping, which is known to degrade field enhancement factors for isolated metal nanoparticles, does not have a strong effect on the SERS-based enhancement in *2D* nanoparticle arrays. The modest effect of particle size on G_R seen in experiments further supports the idea that the damping and retardation effects are of minor importance when the local EM fields are spatially confined to the small, subwavelength areas.

1.2 Percolation Threshold: Singularities in Metal-dielectric Composites

Here we consider the behavior of the electrical conductivity and the permittivity of a percolating composite near the percolation threshold. An example of such systems is a composite consisting of a disordered mixture of metallic and insulating particles. A reduction in the concentration p of the metallic (conducting) component reduces the static conductivity σ_e of the composite, so that it vanishes at some critical concentration p_c known as the *percolation threshold.* In other words, the composite undergoes a composition-dependent metal-insulator transition at the percolation threshold when $p = p_c$. The composite remains a dielectric below the percolation threshold. Yet, the presence of large metal clusters strongly influence its properties. The composite can behave as a metamaterial concentrating EM energy in a way as it is discussed below. The metal-insulator transition in the semicontinuous metal film is shown on Fig. 3.2.

The behavior of metal-dielectric composites near the percolation threshold has been the subject of numerous investigations (see for example, [Kirkpatrik, 1973], [Clerc *et al.*, 1990], [Bergman and Stroud, 1992], [Stauffer and Aharony, 1994], [Sarychev and Shalaev, 2000]). The results include expressions for the singular behavior near the threshold of basic quantities such as the conductivity, permittivity, and the probability for finding the infinite cluster of the conducting material; all of them are functions of the fraction of conducting material, p.

The dc conductivity σ_e is nonzero only above the threshold, decreasing as a power-law function of the detuning from the critical metal fraction

$$\sigma_e \simeq \sigma_m \Delta p^t, \tag{1.10}$$

where $\Delta p = (p - p_c)/p_c$ is the reduced metal concentration, σ_m is the metal conductivity, and t is a critical exponent. The recent computer simulations give $t \approx 2.1$ for three-dimensional materials ($D = 3$) and $t \approx 1.3$ for films ($D = 2$) (see [Bergman and Stroud, 1992], [Clerc *et al.*, 2000], [Xiong *et al.*, 2001], [Park *et al.*, 2004], [Keblinski and Cleri, 2004], [Byshkin and Turkin, 2005], and reference therein). Below p_c, there is a frequency-dependent ac conductivity, which increases as the threshold is approached. The dc dielectric constant ε_e diverges as p_c is approached from either side:

$$\varepsilon_e \simeq \varepsilon_d \left|\Delta p\right|^{-s} \tag{1.11}$$

where ε_d is the permittivity of the dielectric in the composite, and the critical exponent $s \approx 1.3$ for $D = 2$ and $s \approx 0.8$ for $D = 3$ [Efros and Shklovskii, 1976], [Bergman and Imry, 1977], [Grannan *et al.*, 1981], [Vinogradov *et al.*, 1984a], [Vinogradov *et al.*, 1984b], [Bergman and Stroud, 1992] [Yoshida *et al.*, 1993]. In close vicinity to the percolation threshold, both ε_e and σ_e exhibit a power-law dependence on ω. The frequency dependencies of σ_e and ε_e appear despite the fact that in the pure phases, σ_m and ε_d are frequency independent.

All these singular dependencies can be derived on the basis of a simple geometrical consideration. Near the percolation threshold the electric current flows through the random network of the conducting channels that span the entire system. One of such channels is shown in Fig. 3.2(c) schematically showing a model for a semicontinuous gold film. Skal and Shklovskii [Skal and Shklovskii, 1975] and De Gennes [de Gennes, 1976] had given a simple picture of the channel network which has proved to be very useful for the understanding of the conductivity behavior. The current-carrying metal channels can be idealized as long chains of metal granules with length l, which interconnect at the nodes of the average spacing $\xi_p < l$ as it is shown in Fig. 1.9.

Below we express all the spatial lengths in terms of the size a of the metal granules. Note that we neglect the dangling ends, i.e., metal clusters

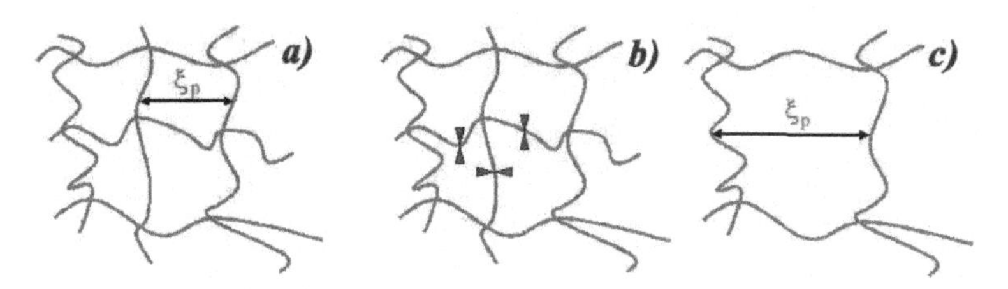

Fig. 1.9 (a) Conducting channels form macro-lattice in percolation metal-dielectric composite; macrobonds are chains of metal granules that carry the current; average distances between the macrobonds is the percolation correlation length ξ_p; (b)–(c) it is enough to remove some few metal granules from the macrobonds to increase ξ_p twice; on the scale much less than ξ_p the structure of the system does not change in transition from a) to (c).

are connecting to the channels only at one point since they do not carry the current. The conductance Σ of the metal chain, which is also known as a *macrobond*, is inversely proportional to its size

$$\Sigma \sim \sigma_m \xi_p^{-\zeta}, \tag{1.12}$$

where the critical exponent $\zeta \approx 1$ for $D = 2$ and $\zeta \approx 1.3$ for $D = 3$. The critical exponent ζ is larger than one when the arc length of the macrobond is larger than its geometrical size.

When the metal concentration decreases to the percolation threshold, the correlation length ξ_p increases, as it is illustrated in Fig. 1.9; the macro-lattice of the percolation channels become more and more rare and the conductivity vanishes. The behavior of ξ_p near the p_c is approximated as

$$\xi_p \sim |\Delta p|^{-\nu}, \tag{1.13}$$

where $\nu = 1.33$ for $D = 2$ and $\nu \approx 0.9$ for $D = 3$ (see [Bergman and Stroud, 1992], [Stauffer and Aharony, 1994], [Moukarzel and Duxbury, 1995], [Ballesteros *et al.*, 1999], [Newman and Ziff, 2001], [Tomita and Okabe, 2002], [Martins and Plascak, 2003], [Huinink *et al.*, 2003], [Nishiyama, 2006]). The percolation system is homogeneous on the spatial scale larger than the percolation correlation length ξ_p. Since the correlation length ξ_p gives the spatial scale of inhomogeneity, we estimate the effective conductivity as $\sigma_e \sim \Sigma/\xi_p^{D-2} \sim \sigma_m \xi_p^{-\zeta}/\xi_p^{D-2} \sim \sigma_m |\Delta p|^{\nu(\zeta+2-D)}$; the "macrobond" exponent ζ is expressed in terms of the critical exponents t and ν defined in Eqs. (1.10) and (1.13) as follows:

$$\zeta = t/\nu - 2 + D. \tag{1.14}$$

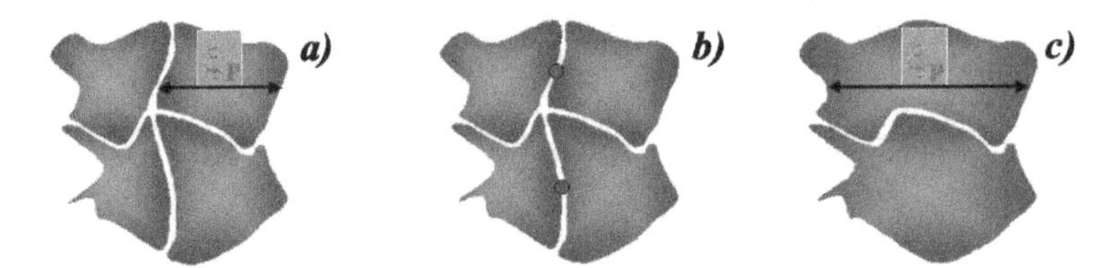

Fig. 1.10 (a) Large metal clusters are separated by dielectric gaps below the percolation threshold; (b)–(c) it is enough to add just two metal granules to increase twice the cluster size; on the scale much less than ξ_p, the structure of the system does not change in transition from (a) to (c).

There is no "infinite" percolation cluster below the percolation threshold. Yet, the system contains a lot of large metal clusters with the size increasing when we approach p_c from below the threshold. As a result, the effective permittivity ε_e diverges at p_c according to Eq. (1.11). The physical reason for the divergence of ε_e is the existence of many conducting clusters which stretch across the entire length of the system. Every channel of this type contributes to abnormally large capacitance, and all of them are connected in parallel as it is illustrated in Fig. 1.10. A typical size of largest clusters is still given by percolation correlation length ξ_p (Eq. (1.13)) since the size of the largest cluster is the inhomogeneity scale for the entire system below P_c [Stauffer and Aharony, 1994]. The capacitance C between the clusters in Fig. 1.10 increases with the correlation length as

$$C \sim \varepsilon_d \xi_p^{s_1}, \tag{1.15}$$

where ε_d is permittivity of the dielectric component and s_1 is "intercluster" exponent. The effective dielectric constant is expressed in terms of intercluster capacitance as $\varepsilon_e \sim C/\xi_p^{D-2} \sim \varepsilon_d \xi_p^{s_1+2-D}$. By comparing this estimate with Eq. (1.11), we obtain that "intercluster" exponent s_1 is equal to

$$s_1 = s/\nu - 2 + D. \tag{1.16}$$

This picture also suggests that strong nonlinearity can occur in the dielectric response when ε_e is large due to the large electric fields in the thin dielectric barriers (see Sec. 3.6). Likewise, quantum-mechanical tunneling through the barriers may become important near p_c [Sarychev and Brouers, 1994].

We are now ready to consider the high-frequency properties of a percolation composite in the critical region near the percolation threshold. According to the scaling theory, the system looks the same on the scales

$l < \xi_p$ regardless of the metal concentration p. It is impossible to determine whether the system is below or above p_c dealing with the fragment of sizes $l < \xi_p$. This assumption, which is the core of the modern theory of critical phenomena [Kadanoff *et al.*, 1967], [Stanley, 1981], is confirmed by all known experiments and computer simulations [Stauffer and Aharony, 1994]. Thus, we assume that the conductance Σ of the finite cluster of size ξ_p (see Fig. 1.10) is still given by Eq. (1.12). The high frequency conductivity $\sigma_e(\omega)$ of the composite depends on the intracluster conductance Σ and intercluster reactive conductance $-i\omega C$, that is $\sigma_e(\omega, \Delta p) = F(\Sigma, -i\omega C)/\xi_p^{D-2}$. Since we consider a linear system this equation can be rewritten as $\sigma_e(\omega, \Delta p) = \Sigma f_1(i\omega C/\Sigma)/\xi_p^{D-2}$. By substituting here Eqs. (1.12) and (1.15) we obtain

$$\sigma_e(\omega, \Delta p) = \sigma_m \Delta p^t f\left(\frac{i\varepsilon_d\,\omega}{\sigma_m}\Delta p^{-(t+s)}\right), \tag{1.17}$$

where $f(z)$ is an analytical function.

For fixed Δp and frequencies approaching zero, the effective conductivity can be expand in series of ω. We obtain that $f(z) = A_+ - B_+ z + C_+ z^2 + \ldots$ for $\Delta p > 0$ and $f(z) = -B_- z + C_- z^2 + \ldots$ for $\Delta p < 0$, where the coefficients $A_+,\ B_+,\ C_+$ and $(A_- = 0),\ B_-, C_-$ are different so that $z = 0$ is a branch point. It follows from Eq. (1.17) that above the percolation threshold

$$\sigma_e(\omega, \Delta p) = A_+ \sigma_m \Delta p^t - iB_+ \varepsilon_d\,\omega \Delta p^{-s}; \tag{1.18}$$

below the percolation threshold, $\sigma_e(\omega, \Delta p) \sim -i\varepsilon_d\,\omega\,|\Delta p|^{-s}$ and the effective permittivity is estimated as $\varepsilon_e(\omega, \Delta p) = i4\pi\sigma_e/\omega \sim \varepsilon_d\,|\Delta p|^{-s}$, which is in agreement with Eq. (1.11). Exactly at the percolation threshold $(\Delta p \to 0)$, the effective conductivity σ_e does not vanish when the frequency ω is finite. Therefore, the function $f(z)$ in Eq. (1.17) is approximated as $f(z) \sim z^{t/(t+s)}$ for $|z| \to \infty$ so the conductivity is described as

$$\sigma_e(\omega, p_c) \sim \sigma_m \left(\frac{\varepsilon_d\,\omega}{\sigma_m}\right)^{t/(t+s)} \tag{1.19}$$

and the effective dielectric permittivity is given by

$$\varepsilon_e(\omega, p_c) \sim \varepsilon_d \left(\frac{\varepsilon_d\,\omega}{\sigma_m}\right)^{-s/(t+s)}. \tag{1.20}$$

Note that they both depend on frequency despite the fact that σ_m and ε_d are frequency independent. This dependence is due to the ramified $R - C$ circuits existing in the composite near the percolation threshold.

We consider now a composite made of two dielectrics with permittivities ε_d and ε_m. If we assume that both $\varepsilon_m, \varepsilon_d > 0$ and $|\varepsilon_m| \gg \varepsilon_d$; then all the above scaling formalism can be applied to this problem, and we obtain for $p = p_c$ that the effective dielectric constant is equal to

$$\varepsilon_e(\omega, p_c) = A\varepsilon_d \left(\frac{\varepsilon_d}{\varepsilon_m}\right)^{-s/(t+s)}, \tag{1.21}$$

where the coefficient $A > 0$. We suppose now that $\varepsilon_m < 0$ and $|\varepsilon_m| \gg \varepsilon_d$; then, we can still use Eq. (1.21) obtaining

$$\varepsilon_e(\omega, p_c) = A\varepsilon_d \left[\cos\left(\frac{\pi s}{t+s}\right) + i \sin\left(\frac{\pi s}{t+s}\right)\right] \left(\frac{\varepsilon_d}{|\varepsilon_m|}\right)^{-s/(t+s)}. \tag{1.22}$$

This implies that the composite, where both components are lossless ($\mathrm{Im}\,\varepsilon_d = 0$ and $\mathrm{Im}\,\varepsilon_m = 0$), has the effective loss at the percolation threshold ($\mathrm{Im}\,\varepsilon_e > 0$). This puzzling result becomes clear when we realize that the EM energy can accumulate in the composite. The effects of the energy accumulation in metamaterials and the corresponding gigantic local fields will be discussed in the following chapters.

Equations (1.20) – (1.22) give properties of a percolation composite not only at $p = p_c$ but also in some vicinity of the percolation threshold, where the scaling parameter $|z| = \omega\,|\varepsilon_d\,\Delta p^{-s}|\,/\,|\sigma_m \Delta p^t| > 1$ (see Eq. (1.17)). In the opposite case $|z| \ll 1$ the crossover occurs to the "static" behavior, given by Eqs. (1.10) and (1.11).

The scaling for the percolation conductivity was suggested in the pioneering papers by Straley, Efros, Shklovskii, Bergman, and Imry [Straley, 1976], [Straley, 1977], [Efros and Shklovskii, 1976], [Bergman and Imry, 1977]. The scaling also follows from the real space renormalization group developed for percolation composites in [Stinchcombe and Watson, 1976], [Sarychev, 1977], [Reynolds *et al.*, 1977], and [Bernasconi, 1978]. Before any scaling was introduced, A.M. Dykhne proved [Dykhne, 1971] the exact result $\varepsilon_e = \sqrt{\varepsilon_d \varepsilon_m}$ for $p = p_c$ for two dimension self-dual percolating systems. Equation (1.21) reproduces this important result since for $D = 2$ the critical exponents are equal to $s = t = 1.3$. It also follows from Eq. (1.17) that the loss tangent δ, which is defined as $\tan\delta = \mathrm{Im}[\varepsilon_e]/\,\mathrm{Re}[\varepsilon_e] = -\,\mathrm{Re}[\sigma_e]/\,\mathrm{Im}[\sigma_e]$, is a function of the scaling parameter z only. The loss tangent obtained in computer simulations in [Vinogradov *et al.*, 1988] is shown in Fig. 1.11. We see that the data obtained for various metal concentrations and frequencies collapse to a single curve as expected from the scaling theory.

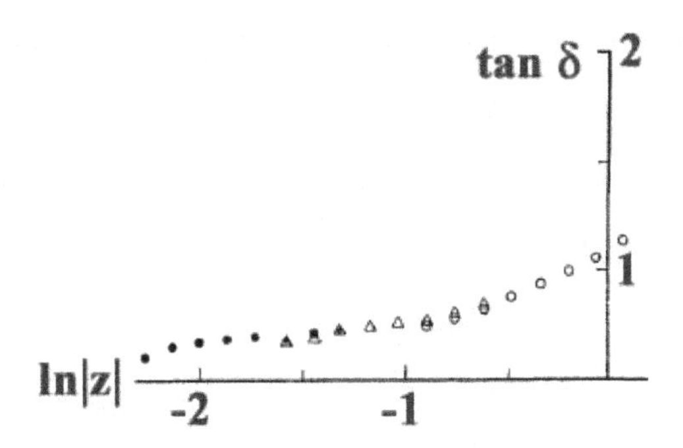

Fig. 1.11 Loss tangent $\delta = \mathrm{Im}\left[\varepsilon_e(\Delta p,\omega)\right]/\mathrm{Re}\left[\varepsilon_e(\Delta p,\omega)\right]$ for three dimensional metal-dielectric percolation composite as a function of the scaling parameter $z = \Delta p\left(\omega\varepsilon_d/4\pi\sigma_m\right)^{1/(s+t)}$. Circles, triangles, and filled circles on the curve correspond to $\Delta p = 0.32, 0.16$, and 0.08, respectively.

The scaling can be also used to derive the explicit equation (so-called parametric equation) for the effective conductivity. The form and the order of the parametric equation both depend on the critical exponents s and t . In the three-dimensional case, we can approximate the exponents as $s \simeq 1$ and $t \simeq 2$. Then, the parametric equation reduces to the simple cubic equation

$$h + \alpha x^2 \Delta p = \beta x^3 \qquad (1.23)$$

where $x = \sqrt{\sigma_e\left(\Delta p,\omega\right)/\sigma_m}$ and $h = \left(-i\omega\varepsilon_d/\left(4\pi\sigma_m\right)\right)^{1/3}$; α and β are numerical coefficients on the order of one. We assume that $\Delta p \ll 1$ and $|h| \ll 1$; then it is easy to check that Eq. (1.23) reproduces the asymptotic behavior for $\Delta p > 0$, $\Delta p < 0$ and $\Delta p = 0$ given by Eqs. (1.10), (1.11), and (1.19), respectively. Indeed, at the percolation threshold ($\Delta p = 0$) Eq. (1.23) gives $x = (h/b)^{1/3}$, that is $\sigma_e \sim \sigma_m\left(\varepsilon_d\omega/\sigma_m\right)^{2/3} \sim \sigma_m\left(\varepsilon_d\omega/\sigma_m\right)^{t/(t+s)}$ (recall that we approximate the exponents as $s = 1$ and $t = 2$). Above the threshold ($\Delta p > 0$) when $\Delta p \gg |h|^{1/3}$, Eq. (1.23) gives $x \sim \Delta p$, $\sigma_e \sim \sigma_m \Delta p^2$. Below the threshold, when $\Delta p < 0$, $|\Delta p| \gg |h|^{1/3}$, we obtain $x \sim \sqrt{h/|\Delta p|}$ and, therefore, $\sigma_e \sim h/|\Delta p|$. Thus we obtain the conductivity and permittivity of metal-dielectric composites can be found form the solution to the cubic equation (1.23) that holds in the entire range of concentrations and frequencies. In two dimensional case, where the critical exponents $t = s \approx 4/3$, the parametric equation takes more complicated form than Eq. (1.23). Yet, it is still algebraic equation that can be easily solved. The parametric equations for percolating systems were first suggested in [Vinogradov *et al.*, 1988].

Chapter 2

Conducting Stick Composites and Left Handed Metamaterials

2.1 Metamaterial

In this chapter, we consider a special class of metal nanocomposites that contain elongated conducting inclusions, or "sticks", that embedded in a dielectric host (Fig. 1.1). The sticks are assumed to be randomly distributed and oriented. We will calculate and analyze the macroscopic dielectric and magnetic response of these conducting composites. The metal-dielectric composites, where conducting inclusions have a very elongated shape, pose an interest because these systems particularly well describe physical structures occurring both in nature and technology. For example,there are many porous rocks in nature such as sandstones or similar geological formations that have channel-like or sheet-like porous structure. In spite of the great importance, there is no universal concept for rock conductivity, permeability, and dielectric susceptibility (see discussion in [Hilfer, 1992; Hilfer, 1993; Nettelblab and Niklasson, 1994]). Considering the stick-like inclusions as a model for the rock pores, one can reproduce the structure of porous rocks by fitting the shape and concentration of the model sticks [Balberg, 1987]. Then it is possible to calculate the rock conductivity, permeability, and dielectric susceptibility by the methods that will be discussed here. The analysis of conducting stick composites is also important for industrial applications. Ceramic and plastic materials reinforced by carbon or metallic fibers are becoming increasingly attractive as engineering materials. The physical–chemical and mechanical properties of such materials became the subject of great interest (see [Theocaris, 1987], [Monette *et al.*, 1994], [Chang *et al.*, 2006], and references therein). The dielectric properties of the materials may play an important role in their characterization and diagnostics [Bergman and Stroud, 1992].

The results discussed in this chapter can be also applied for characterization of semicontinuous metal films. The semicontinuous films, used to fabricate selective surfaces for solar photo thermal energy conversion, are usually prepared by thermal evaporation or spattering of the metal on an insulating substrate. The semicontinuous films have been studied for example by an image analysis (See Ref. [Blacher *et al.* 1993].) The authors have observed that in some films, elementary conducting cells had elongated shapes. It is therefore important to establish a correct picture how the film's dielectric properties vary with the shape of conducting inclusions. We will also show that dielectric and magnetic properties of the conducting stick composites are unusual and very interesting in their own regard. These properties demonstrate how important is to take into account the actual shape of metal grains as well as non-quasistatic, i.e. retardation, effects when a composite interacts with electromagnetic field.

Properties of conducting stick composites at high frequencies had been first investigated in the Institute of Theoretical and Applied Electrodynamics (Moscow) in the end of eighties. It was probably first example of what is now called "metamaterials", namely, materials that bulk properties are defined by electromagnetic excitations in the material. Microwave properties of the conducting sticks composites had been so unusual that the research team, to which one of the authors (A. Sarychev) belonged at that time, was not able to obtain reliable data for the bulk dielectric susceptibility for some extensive period of time. We were forced then to develop a new experimental technique to investigate the composites where conducting inclusion was on the order of the wavelength [Kalachev *et al.*, 1991]. Still, even after successful characterization, it was very difficult to accept that the effective dielectric constant of solid, microwave material could be negative. Now the conducting stick composites with such unusual properties are routinely used to manufacture various artificial dielectrics and magnetics, both in microwave frequency region and in optics.

The geometric properties of composites with conducting sticks have been studied experimentally [see, e.g. [Lagarkov *et al.*, 1998; Lagarkov *et al.*, 1997a; Grimes *et al.*, 2000; Grimes *et al.*, 2001]], by Monte Carlo simulation [Balberg *et al.*, 1983; Balberg *et al.*, 1984; Balberg, 1985; Balberg, 1991] and analytically [Lagarkov *et al.*, 1998; Theocaris, 1987; Zuev and Sidorenko, 1985a; Lagarkov and Sarychev, 1996]. It was found that the percolation threshold p_c was inversely proportional to the stick aspect ratio $p_c \sim b/a$, where b is the radius of a stick and $2a$ is its length. The same estimation $p_c \sim b/a$ was obtained by [Balberg, 1986] from an

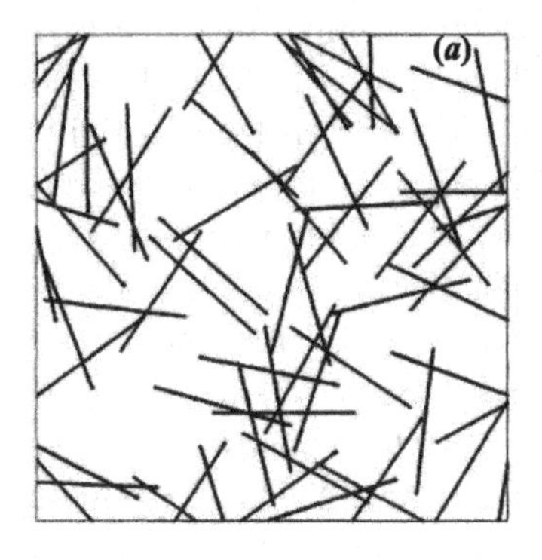

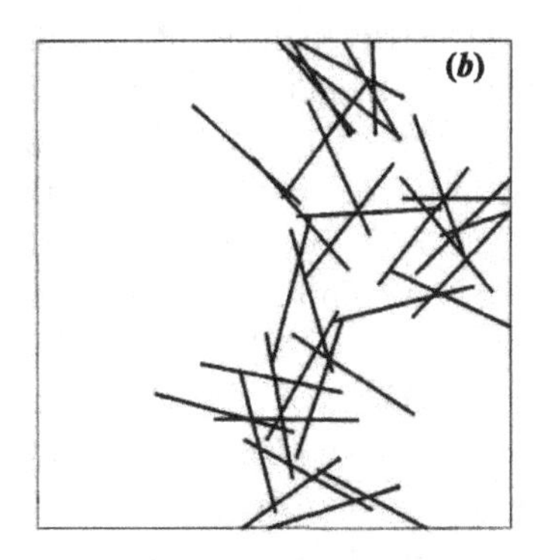

Fig. 2.1 (a) Conducting stick composite; (b) Backbone of the "infinite" cluster that spans from a top to a bottom.

excluded volume explanation of Archie's law for the porous rock permeability. This important result may be explained as follows: A conducting stick intersects on average with N other sticks in the composite. In a limit $N \ll 1$, the sticks are separated from each other and the probability to percolate through the system of the conducting sticks is equal to zero. In another limit $N \gg 1$, the sticks form a carpet as it is shown in Fig. 2.1, and the probability to percolate is equal to one. Consequently, there is a critical number of intersections, N_c, that correspond to the percolation threshold. An important finding of Refs. [Balberg, 1991; Balberg *et al.*, 1983; Balberg *et al.*, 1984; Balberg, 1985; Zuev and Sidorenko, 1985b; Zuev and Sidorenko, 1985a] is that the critical number N_c is about one and it does not depend on the stick aspect ratio a/b in the limit $a \gg b$. The averaged number of intersections, N, can be calculated using elementary theory of probability. It appears to be proportional to the aspect ratio and to the volume concentration of the conducting sticks: $N = (4/\pi)\, p\, (a/b)$ for $D = 2$ and $N = 2p\,(a/b)$ for $D = 3$ (see, e.g., [Nettelblab and Niklasson, 1994]). Therefore, the percolation threshold is $p_c = N_{c2}(\pi/4)\, b/a \sim b/a$ for $D = 2$ and $p_c = N_{c3}b/(2a) \sim b/a$ for $D = 3$, where N_{c2} and N_{c3} are critical numbers of the stick intersections for the two- and three-dimensional composites, respectively. It follows from this result that the percolation threshold p_c may be very small for composites with elongated conducting inclusions. Moreover, p_c tends to approach to zero when the aspect ratio a/b goes to infinity. A small value of the percolation threshold is one of the distinguishing features of the conducting stick composites. Another one is the anomalous dielectric response of such composites.

The dielectric response of the metal-dielectric composites has attracted the attention of many researchers for a long time. The problem of interest was the calculation of the macroscopic, effective permittivity ε_e and magnetic permeability μ_e of the composites in terms of the dielectric and magnetic responses of its constituents. The most results were obtained for the composites with spherical conducting (see, e.g., [Bergman and Stroud, 1992; Niklasson and Granquist, 1984; Clerc *et al.*, 1990; Stauffer and Aharony, 1994; Shin and Yeung, 1988; Tsui and Yeung, 1988; Shin *et al.*, 1990] and Ch. 1.2). The attempts to extend models to nonspherical inclusions were restricted, almost entirely, to mean-field approximations. There are two different mean-field approximations considered in the literature: Bruggeman effective-medium theory (EMT) [Bergman and Stroud, 1992; Bruggeman, 1935; Landauer, 1978] and the symmetric Maxwell-Gannet approximation introduced by P. Sheng (MGS) [Sheng, 1980; Gibson and Buhrman, 1983; Niklasson and Granquist, 1984]]. Both the EMT and MGS theories reproduce, at least qualitatively, the behavior of the effective parameters of the composites in the entire range of the conducting component concentration p. It was found that the last one gives better agreement with experimental data for the optical properties of the composites [Niklasson and Granquist, 1984]. Numerous approximations, described in the literature, usually obtain the effective parameters of the composites in the limit of small concentrations p or in the case when the properties of the constituents slightly differ. A list of such approximations can be found in Refs. [Shin and Yeung, 1988; Tsui and Yeung, 1988; Shin *et al.*, 1990]. We restrict our further consideration exclusively to EMT and MGS theories, which both give nontrivial values of p_c that are independent of the conductivities of the composite components. However, we would like to point out that our approach is close to some extent to that proposed by [Tsui and Yeung, 1988].

Both the EMT and MGS theories have been developed originally to describe the properties of composites containing spherical conducting grains. Different generalizations of the EMT have been suggested for composites with randomly oriented prolate conducting inclusions [Stroud, 1975; Granqvist and Hunderi, 1977; Granqvist and Hunderi, 1978]. It is easy to show [Brouers, 1986] that all these approaches conclude that a percolation threshold is proportional to the depolarization factor of an inclusion in the direction of its major axis $p_c \sim g_{\parallel}$. For very elongated inclusions, the depolarization factor takes form $g_{\parallel} \sim (b/a)^2 \ln(a/b)$, where $2a$ and b

are the stick length and radius, respectively (see, e.g., [Landau *et al.*, 1984] Sec. 4). Based on that, the EMT gives $p_c \sim (a/b)^2$, which is in obvious disagreement with the results of Refs. [Balberg, 1991; Balberg *et al.*, 1983; Balberg *et al.*, 1984; Balberg, 1985; Zuev and Sidorenko, 1985b; Zuev and Sidorenko, 1985a]. The percolation threshold given by the MGS theory for the randomly oriented sticks depends on the typical shape of the dielectric regions that are assumed in the theory [Niklasson and Granquist, 1984]. Since the sticks are randomly oriented, it is quite natural to assume that the dielectric regions have a spherical shape on average [Blacher *et al.*, 1993; Granqvist and Hunderi, 1978]. Then the percolation threshold given by MGS theory, $p_c = 0.46$, is independent of the stick shape [Niklasson and Granquist, 1984]. An alternative approach uses dielectric regions of the same shape as conducting sticks [Bergman and Stroud, 1992; Niklasson and Granquist, 1984; Stroud, 1975]. The MGS theory gives $p_c \sim \sqrt{b/a}$ [Brouers, 1986], which also disagrees with results for the percolation threshold, $p_c \sim b/a$, obtained in Refs. [Lagarkov *et al.*, 1998; Balberg, 1991; Balberg *et al.*, 1983; Balberg *et al.*, 1984; Balberg, 1985; Zuev and Sidorenko, 1985b; Zuev and Sidorenko, 1985a].

The percolation threshold is very important parameter for characterizing metal-dielectric composites since it determines the concentration of inclusions for which dramatic changes in the dielectric properties occur. A shown above discrepancy between the values of the percolation threshold leads to variation between the dielectric data. Consider, for example, the composites with the stick aspect ratio $a/b = 10^2$, which have been investigated experimentally [Kolesnikov *et al.*, 1991; Lagarkov *et al.*, 1998]. The observed values of the percolation threshold $p_c \sim b/a$ are many tens times larger than $p_c \sim (b/a)^2$ predicted by EMT and many times smaller than $p_c = 0.46$ or $p_c \sim \sqrt{b/a}$ given by MGS theory. Therefore, the accuracy of existing theories in describing the conducting stick composites is in question.

The percolation threshold is not a single problem with the conducting stick composites. The effective dielectric constant ε_e of the composite with aligned conducting spheroids has been considered in details [Barrera *et al.*, 1993]. Suppose that the prolate spheroids with semiaxes a and b ($a > b$) are aligned along "z" axis. The simple scale transformation $x = (b/a)^{1/3}x^*$, $y = (b/a)^{1/3}y^*$, $z = (a/b)^{1/3}\, z^*$ reduces the composite to the system of an isotropic conducting spheres distributed in an anisotropic host [Shklovskii and Efros, 1984]. Therefore, the percolation threshold for the composites with aligned sticks coincides with that of the spherical particles. Nevertheless, it appears that the dielectric response of aligned stick composites is quite different from that of composites with spherical inclusions even for the small concentrations considered in the literature [Barrera *et al.*, 1993]. This difference is due to long-range correlations in the interaction of the sticks. It is shown in Ref. [Barrera *et al.*, 1993] that long-range correlations are a distinct feature of the sticks; while for spherical particles, they are negligible [Felderhof, 1989]. Another important result of studies is the possibility of internal manifold mode excitation [Barrera *et al.*, 1993]. This observation correlates with the results discussed in this chapter. The high-frequency dielectric properties of the aligned stick composites have been considered in Refs.[Sarychev and Smychkovich, 1990; Panina *et al.*, 1990; Lagarkov *et al.*, 1998; Lagarkov *et al.*, 1997a; Lagarkov and Sarychev, 1996; Lagarkov *et al.*, 1992; Makhnovskiy *et al.*, 2001] beyond the usual quasistatic approximation. Bellow, we follow theory first presented by [Lagarkov and Sarychev, 1996].

Let us consider the interaction of randomly oriented conducting sticks excited by the constant external electric field E_0. To estimate the dipole moment of a conducting stick, we consider a perfect conductor, which is aligned with the external field E_0. The stick is polarized in the field and acquires the induced electric charges ($+Q$ and $-Q$) at its ends. The electric field at the center of the stick is $E_{in} \sim E_0 - 2Q/a^2$. By taking into account the fact that the internal electric field (from the induced charges, $+Q$ and $-Q$) compensates the external field E_0, $E_{in} = 0$, in the perfect conducting stick, we fund the value of $Q \sim E_0 a^2$. The dipole moment of the stick is thus given by $\mathcal{P} \sim Qa \sim a^3 E_0$ (Ref. [Landau *et al.*, 1984] Sec. 3). This estimate holds for a stick with the finite conductivity σ_m for case $\sigma_m \gg \omega$, where ω is the frequency of the external field E_0. Then the effective dielectric constant of the conducting stick composite is $\varepsilon_e \sim a^3 N \sim a^3 p/(ab^2)$, where N is the number of the sticks in the unit volume. We can rewrite the last expression for the dielectric constant in terms of the percolation threshold $p_c \sim b/a$, namely, $\varepsilon_e \sim (a/b)\ (p/p_c)$. Throughout this Chapter, we will discuss the composites with high aspect ratio ($a/b \gg 1$, i.e., elongated inclusions), resulting in the dielectric susceptibility ε_e that is large even for concentrations p, which are far below the percolation threshold p_c. We can estimate the number N_{int} of sticks inside the scale range of the dipole interactions a as $N_{int} \sim a^3 N \sim \varepsilon_e$. Therefore, one can expect strong dipolar interaction between the sticks as soon as limit $\varepsilon_e \gg 1$ is met, which occurs for sufficiently small stick concentrations $p < p_c \ll 1$.

The effective dielectric susceptibility may be introduced for the composite samples whose size are much larger than the stick length $2a$. However, to calculate the effective parameters it is necessary to start with scales smaller than the stick radius b. The aspect ratio $a/b \gg 1$ may be considered as a minimal dimensionless correlation length of the problem. When the concentration p increases, the correlation length further increases. As we have pointed out above, the sticks usually strongly interact. Therefore, the conducting stick composite is a system with strong long-range interactions. As a result, well-developed methods of the percolation theory similar to the renormalization group theory in real space or mean field approximations cannot be directly applied to the system.

Computer simulations of the dielectric properties of the conducting stick composites are also difficult. To calculate the effective dielectric susceptibility, we have to know the electromagnetic field distribution over the entire system. The field distribution are usually obtained by solving of finite-difference Laplace or Maxwell equations. In a crude approximation, on can use the stick radius b as an elementary length. Then the equations must be solved on a three-dimensional ($3D$) lattice whose number of elementary sites is much larger than $(a/b)^3$. If, for example, the stick aspect ratio is $a/b \sim 10^3$, then the minimal number of such sites is on the order of 10^9, implying that the solution of a finite difference equation on such lattice is a difficult problem even for the most powerful computers.

The results of first experimental attempts [Kolesnikov *et al.*, 1991; Lagarkov *et al.*, 1997a; Lagarkov *et al.*, 1998; Grimes *et al.*, 2000; Grimes *et al.*, 2001] to investigate the dielectric properties of conducting stick composites resulted in evident disagreement with a standard percolation theory [Bergman and Stroud, 1992; Clerc *et al.*, 1990; Stauffer and Aharony, 1994]. The percolation theory predicts that the dielectric constant exhibits a power-law behavior $\varepsilon_e \sim (\omega/\sigma_m)^{s/(t+s)}$ in a small critical region around p_c defined as $|\Delta p| \ll (\omega/\sigma_m)^{1/(t+s)}$, where the critical exponents are equal to $t \cong 2.0$, $s \cong 0.8$ for $D = 3$; σ_m is the conductivity of the conducting inclusions. Out of this critical region, there should be no dispersion at all (for details, see the discussion in Ch. 1.2.) In contrast to the predictions, the dispersion behavior of the dielectric function ε_e was observed in the microwave range 10^9-10^{10} Hz [Kolesnikov *et al.*, 1991; Lagarkov *et al.*, 1997a; Lagarkov *et al.*, 1998] and in radio frequency range [Grimes *et al.*, 2000; Grimes *et al.*, 2001] for all investigated concentrations of carbon fibers and metal microwires. The dispersion behavior for the fibers with conductivity $\sigma_m \sim 10^{14}$ s^{-1} was similar to the Debye relaxation behavior

$\varepsilon_e(\omega) \sim 1/(1 - i\omega\tau)$ [Debye, 1912], where the relaxation time τ depends on the stick length and its conductivity. The frequency dependence of the dielectric susceptibility ε_e has a resonance form for the composites with metal microwires ($\sigma_m \simeq 10^{17}$ s^{-1}). The real part of the dielectric function $\varepsilon_e(\omega)$ drops to zero at some resonance frequency ω_1 and then acquires negative values for the frequencies $\omega > \omega_1$. The power-law behavior predicted by the percolation theory had not been observed in these experiments.

Below we present a comprehensive theoretical study of the dielectric and magnetic response of conducting stick composites. We will study in details the interaction of a conducting stick with the external electric field. Then we will develop an effective medium theory for the conducting stick composites. Our approach will be based on the well-known Bruggeman effective medium theory (EMT), which also borrows an idea from MGS theory: The local environment of an inclusion may be different for different inclusions. We will show that the dielectric constant is a nonlocal quantity for the composites. It depends on the spatial scale for scales less than the stick length. We will consider composite on these geometrical scales and derive equations for the composite dielectric susceptibility.

The theory presented in this chapter gives the percolation threshold $p_c \sim b/a$, in agreement with the results of Refs. [Balberg, 1991; Balberg *et al.*, 1983; Zuev and Sidorenko, 1985b; Zuev and Sidorenko, 1985a; Lagarkov *et al.*, 1998]. It also reproduces the dispersion behavior obtained in the experiments [Kolesnikov *et al.*, 1991; Lagarkov *et al.*, 1997a; Lagarkov *et al.*, 1998; Grimes *et al.*, 2000; Grimes *et al.*, 2001]. For the quasistatic case, when the skin effect in the sticks is negligible, we will obtain the relaxation behavior of the effective dielectric constant $\varepsilon_e \sim 1/(1-i\omega\tau)$, for a wide concentration range: below and above the percolation threshold. We will actually determine the stick relaxation time τ, which as it turns out depends on the stick shape and the stick conductivity. For frequencies $\omega \sim 1/\tau$, the power-law dispersion behavior predicted by the percolation theory is observed in the vicinity of the percolation threshold. We will extend our consideration to the non-quasistatic case when it is a strong skin effect in the conducting sticks. Then the relaxation behavior of the effective dielectric function $\varepsilon_e(\omega)$ changes to the resonance dependence. The real part of $\varepsilon_e'(\omega)$ accepts negative values when the skin effect is strong. We will also discuss the wave localization and giant local field fluctuations for the conducting stick composites in optical spectral range.

We will also consider the magnetic response of conducting stick composites and show that a composite composed of pairs of conducting sticks has effective magnetic response even when all its constituents have no natural magnetism. The effective magnetic permeability demonstrates resonance behavior and acquires negative values near the resonance frequency. This means that the conducting stick composites can be used to manufacture metamaterial where both dielectric constant and magnetic permeability are negative. Such materials, which have negative refraction index, are commonly called left-handed materials and are in focus of modern electrodynamics since they have very exciting and engineering properties including a possibility to manufacture a superlens with the resolution above the diffraction limit. What is remarkable though is that the conducting stick composites can have left handed properties not only in microwave but also in an optical region. They are relatively easy for fabrication and can be considered as very promising materials for obtaining optical materials with negative refraction. Indeed first optical metamaterial with negative refractive index was obtained from gold nanosticks [Shalaev *et al.*, 2005a], [Shalaev *et al.*, 2006], [Kildishev *et al.*, 2006]. Thus the investigation started twenty years ago in the Institute of Theoretical and Applied Electrodynamics led us to the negative index metamaterials.

2.2 Conductivity and Dielectric Constant: Effective Medium Theory

Let us consider a system of randomly oriented conducting prolate spheroids with semiaxes $a \gg b$. Such shape offers the amenable geometry for an analysis and is a good model for the sticks. The prolate spheroids are randomly embedded in a dielectric matrix characterized by a dielectric constant ε_d. We neglect the direct "hard-core" interaction assuming the spheroidal–type sticks could penetrate each other. Our objective is the calculation of the effective complex dielectric constant $\varepsilon_e = \varepsilon_e' + i\varepsilon_e''$ or the effective complex conductivity $\sigma_e = -i\omega\varepsilon_e/4\pi$ for the stick composite. To find the effective parameters, one has to know the distribution of the electric field $\mathbf{E}(\mathbf{r})$ and current density $\mathbf{j}(\mathbf{r})$ in the system when an external field $\mathbf{E}_0$ is applied. The effective complex conductivity σ_e is determined by the definition

$$\langle \mathbf{j}(\mathbf{r}) \rangle = \sigma_e \langle \mathbf{E}(\mathbf{r}) \rangle \,, \tag{2.1}$$

where $\langle \ldots \rangle$ denotes an average over the system volume. In a real composite, both the current $\mathbf{j}(\mathbf{r})$ and field $\mathbf{E}(\mathbf{r})$ are highly inhomogeneous and statistically random so that it will be very difficult to calculate them precisely. The typical correlation length of the field and the electric current fluctuations are about the stick length $2a$. Therefore the effective conductivity σ_e can be defined only for the scales l, which are larger than the stick length $2a$. On the other hand, the field and current inside of a stick are both determined on scales corresponding to the stick radius $b \ll a$. Then the conductivity σ_e and other effective parameters are determined by the field distribution in a volume larger than a^3, while the field fluctuations are important within a volume $b^3 \gg a^3$. As a result, the effective parameters of the composites essentially depend on the aspect ratio a/b, that is, on the shape of the conducting inclusions. In a peculiar situation like the one above, the standard methods of percolation theory, discussed in Ch. 1.2, cannot be applied and we has to develop new approaches to find the effective parameters. Let us illustrate the latter by an example from effective-medium theory.

The method widely used to calculate the effective properties of a composite is a self-consistent approach known as effective medium theory (EMT) [Bergman and Stroud, 1992; Bruggeman, 1935; Landauer, 1978]. EMT has the virtue of relative mathematical and conceptual simplicity, and it is a method that provides quick insight into the effective properties of metal-dielectric composites. In EMT, one makes two approximations: *(a)* the metal grains as well as dielectric are embedded in the same homogeneous effective medium that will be determined self-consistently and *(b)* the metal grains as well as dielectric grains are taken to have the same shape. For conducting stick composites, these approximations mean that the metal and dielectric grains are assumed to have the same shape as prolate spheroids. The internal field inside a spheroid embedded in the effective medium with conductivity σ_e can be easily calculated ([Landau *et al.*, 1984] Sec. 8).

The internal fields in the conducting and dielectric spheroids averaged over all the orientations are equal, respectively, to [Panina *et al.*, 1990]

$$\mathbf{E}_m = \frac{1}{3}\mathbf{E}_{m\parallel} + \frac{2}{3}\mathbf{E}_{m\perp}, \tag{2.2}$$

$$\mathbf{E}_{m\parallel} = \frac{\sigma_e}{\sigma_e + g_\parallel(\sigma_m - \sigma_e)}\mathbf{E}_0, \tag{2.3}$$

$$\mathbf{E}_{m\perp} = \frac{\sigma_e}{\sigma_e + g_\perp(\sigma_m - \sigma_e)}\mathbf{E}_0, \tag{2.4}$$

$$\mathbf{E}_d = \frac{1}{3}\mathbf{E}_{d\parallel} + \frac{2}{3}\mathbf{E}_{d\perp}, \tag{2.5}$$

$$\mathbf{E}_{d\parallel} = \frac{\sigma_e}{\sigma_e + g_\parallel(\sigma_d - \sigma_e)}\mathbf{E}_0, \tag{2.6}$$

$$\mathbf{E}_{d\perp} = \frac{\sigma_e}{\sigma_e + g_\perp(\sigma_d - \sigma_e)}\mathbf{E}_0, \tag{2.7}$$

where σ_m, is the conductivity of the sticks, $\sigma_d = -i\omega\varepsilon_d/4\pi$ is the dielectric host conductivity, and $g_\parallel$ and $g_\perp$are the spheroid depolarization factors in the direction of the major axis and in the transverse direction, respectively. Note that for very elongated inclusions, $g_\parallel \sim (b/a)^2 \ll 1$ and $g_\perp \approx 1/2$ ([Landau *et al.*, 1984], Sec. 4).

The currents in the conducting and dielectric spheroids averaged over all the orientations are equal to

$$\mathbf{j}_m = \sigma_m\mathbf{E}_m, \quad \mathbf{j}_d = \sigma_d\mathbf{E}_d, \tag{2.8}$$

respectively. To find the effective conductivity σ_e, we substitute Eqs. (2.2)–(2.7) and (2.8) in Eq. (2.1) and take into account that the conducting spheroids occupy the volume fraction p within the system. Thus, we obtain the following equation to determine the effective conductivity:

$$\frac{p}{3}\left[\frac{\sigma_m - \sigma_e}{\sigma_e + g_\parallel(\sigma_m - \sigma_e)} + \frac{2(\sigma_m - \sigma_e)}{\sigma_e + g_\perp(\sigma_m - \sigma_e)}\right] + \frac{1-p}{3}\left[\frac{\sigma_d - \sigma_e}{\sigma_e + g_\parallel(\sigma_d - \sigma_e)} + \frac{2(\sigma_d - \sigma_e)}{\sigma_e + g_\perp(\sigma_d - \sigma_e)}\right] = 0. \tag{2.9}$$

Numerical factor 3 in denominator occurs due to averaging over all the possible directions of the spheroid. Eq. (2.9) has a transparent physical meaning. Namely, the effective conductivity σ_e is chosen in such a way that the averaged scattered field vanishes [Antonov *et al.*, 1990], [Brouers, 1986]. For spherical particles ($g_\parallel = g_\perp = 1/3$), Eq. (2.9) coincides with the usual EMT equation

$$(1-p)\frac{\sigma_d - \sigma_e}{\sigma_d + 2\,\sigma_e} + p\frac{\sigma_m - \sigma_e}{2\,\sigma_e + \sigma_m} = 0 \tag{2.10}$$

and gives the percolation threshold p_c, equal to $p_c = 1/3$. Indeed, we put $\sigma_d \to 0$ in Eq. (2.10) obtaining $\sigma_e = \sigma_m(3p-1)/2$ for $p > p_c = 1/3$ and $\sigma_e = 0$ for $p < p_c$. In the case of elongated conducting particles, for which $g_\perp = 1/2$, Eq. (2.9) defines the percolation threshold

$$p_c = (5 - 3g_\parallel)g_\parallel/(1 + 9g_\parallel). \tag{2.11}$$

We now consider composites formed by conducting particles with high aspect ratio $a/b \gg 1$. Then the depolarization factor is small, $g_{\parallel} \approx (b/a)^2 \ln(a/b) \ll 1$ and the percolation threshold given by Eq. (2.11) equals to $p_c = 5(b/a)^2 \ln(a/b) \ll b/a$. However, this result is in strict contradiction with theoretical result $p_c \sim b/a$ obtained for the stick composites [Balberg, 1991; Balberg *et al.*, 1983; Balberg *et al.*, 1984; Balberg, 1985; Zuev and Sidorenko, 1985b; Zuev and Sidorenko, 1985a; Lagarkov *et al.*, 1998] and in the experiment by [Kolesnikov *et al.*, 1991; Lagarkov *et al.*, 1997a; Lagarkov *et al.*, 1998]. Therefore, Eq. (2.9), derived from EMT, cannot be used for the actual calculation of the effective parameters of the composites.

To understand the reason for this discrepancy, let us examine the basic approximations of EMT. It is obvious that the earlier assumption *(b)* stating that "the metal grains as well as dielectric ones have the same shape" cannot be correct for stick composites. Indeed, the dielectric in such composites cannot be considered as an aggregate of individual grains. It fills all the space between randomly oriented conducting sticks, and the averaged shape of the dielectric regions is rather spherical than a that of a prolate spheroid. For spheres, the electric field in the dielectric regions is given by

$$\mathbf{E}_d = \frac{3\sigma_e}{2\sigma_e + \sigma_d}\mathbf{E}_0, \tag{2.12}$$

where we have assumed that dielectric regions are surrounded by the "effective medium" with conductivity σ_e.

Since the dielectric regions have a spherical shape, which is different form the shape of the conducting sticks, there is no reason to assume that local environments of dielectric regions and conducting sticks are the same. It means that the first approximation *(a)* of the standard EMT, "the metal grains as well as dielectric ones are both embedded in the same homogenous effective medium," should be also revised. At this point, we follow the ideas of the symmetric Maxwell-Garnet approximation by [Sheng, 1980].

Let us consider a small domain of the composite within sizes $l \sim b \ll a$. The probability that the domain contains a conducting stick estimates as $p(l) \sim l^3 N(p) \sim b^3 N(p)$, where $N(p) \sim p/\left(ab^2\right)$ is the stick concentration. The probability $p(l)$ is small even for the concentrations, corresponding to the percolation threshold, where p estimates as $p(l) \sim b^3 N(p_c) \sim (b/a)^2 \ll 1$. Since there is an infinitesimal chance to find a conduction stick in such a small domain, the conductivity of the whole domain is equal to σ_d. On the other hand, we can still prescribe bulk effective properties to the domain, when a characteristic size l is much larger than the stick length $2a$. This implies that the local conductivity $\sigma(l)$ depends on the scale of consideration: For $l \longrightarrow 0$, the conductivity is equal to $\sigma(l) = \sigma_d$ and when $l > a$, it recovers its bulk value $\sigma(l) = \sigma_e$. To satisfy this model, we accept the simplest assumption:

$$\sigma(l) = \sigma_d + (\sigma_e - \sigma_d)l/a, \quad l < a,$$
$$\sigma(l) = \sigma_e, \quad l > a, \tag{2.13}$$

so that a conducting stick is surrounded in the composite by the medium with conductivity Eq. (2.13).

To find the internal field E_m in a conducting stick embedded in such "effective medium," note that Eqs. (2.2)–(2.4) for the field is obtained under very general conditions (see [Landau *et al.*, 1984] Sec. 8) and it holds for the scale-dependent stick environment. The internal field was calculated for this case in [Lagarkov and Sarychev, 1996] where it was shown that E_m is still given by Eqs. (2.3) and (2.4). However, the depolarization factors $g_\parallel$ and $g_\perp$ in the limit $a \gg b$ now take the form

$$g_\parallel = \frac{b^2\sigma_e}{a^2\sigma_d}\ln\left(1+\frac{a\sigma_d}{b\sigma_e}\right) = \frac{b^2\varepsilon_e}{a^2\varepsilon_d}\ln\left(1+\frac{a\varepsilon_d}{b\varepsilon_e}\right), \quad g_\perp = \frac{1}{2}. \tag{2.14}$$

Note that the depolarization factor $g_\parallel$ reduces to the usual expression $g_\parallel = (b/a)^2 \ln(a/b)$ in the dilute case when $\sigma_e \cong \sigma_d$ (see derivation of Eq. (2.65) in Sec. 2.3.1).

Now we would like to summarize the two main assumptions of our Effective-Medium Theory for Stick Metamaterial (EMTSM):

(a)* Each conducting stick is embedded in the effective medium with conductivity $\sigma(l)$ that depends on the scale l by means of Eq. (2.13). The conductivity $\sigma(l)$ equals to the effective conductivity σ_e for $l > a$. The value of σ_e will be determined in a self-consistent manner.

(b)* The dielectric regions are to have a spherical shape and embedded in the effective medium with conductivity σ_e. To find the effective conductivity σ_e, we substitute Eqs. (2.2a), (2.12), and (2.8) into Eq. (2.1) to determine the effective conductivity for the EMTSM theory:

$$\frac{p}{3}\left[\frac{\sigma_m-\sigma_e}{\sigma_e+g_{\parallel}(\sigma_m-\sigma_e)}+\frac{2(\sigma_m-\sigma_e)}{\sigma_e+g_{\perp}(\sigma_m-\sigma_e)}\right] + (1-p)\frac{3(\sigma_d-\sigma_e)}{2\sigma_e+\sigma_d}=0, \tag{2.15}$$

where the polarization factors $g_{\parallel}$ and $g_{\perp}$are given by Eq. (2.14).

Equation (2.15) can be simplified for the case of the high-aspect-ratio conducting elongated inclusions ($\sigma_m \gg \omega$, $a \gg b$). The following simplifications are adopted

(1) Since we are interested in the effective conductivity for the stick concentration p, which is below and in a vicinity of the percolation threshold $p_c \sim b/a \ll l$, we neglect the concentration p in the first term but keep it in the second term of Eq. (2.15).

(2) Since the concentration $p \ll 1$, the effective conductivity σ_e is much smaller than the stick conductivity $\sigma_e \ll \sigma_m$, we neglect σ_e in comparison with σ_m in all the terms of Eq. (2.15).

(3) We neglect the stick polarizability in a direction perpendicular to the external field since it is much smaller than the polarizability in the direction of the field. That is, we can neglect the second term in the square brackets of Eq. (2.15).

After simplifications (1)-(3), Eq. (2.15) takes the form

$$\frac{1}{3}p\frac{\sigma_m}{\sigma_e}\frac{1}{1+\dfrac{b^2\sigma_m}{a^2\sigma_d}\ln\left(1+\dfrac{a\sigma_d}{b\sigma_e}\right)}+3\frac{\sigma_d-\sigma_e}{2\sigma_e+\sigma_d}=0. \tag{2.16}$$

This is the basic EMTSM equation for the effective parameters of the metamaterials containing elongated conducting inclusions. It allows analyze easily the behavior of σ_e, especially in some extreme cases considered below.

In the static limit $\omega \to 0$, $\sigma_d \to 0$, the composite conductivity σ_e decreases with decreasing the stick concentration p and vanishes at the percolation threshold value of

$$p_c = \frac{9}{2}\frac{b}{a}. \tag{2.17}$$

Note that this result for the percolation threshold p_c is in the best agreement with $p_c \sim b/a$ result,obtained in Refs. [Balberg, 1991; Balberg *et al.*, 1983; Balberg *et al.*, 1984; Balberg, 1985; Zuev and Sidorenko, 1985b; Zuev and Sidorenko, 1985a; Lagarkov *et al.*, 1998] and in the experiment [Kolesnikov *et al.*, 1991; Lagarkov *et al.*, 1997a; Lagarkov *et al.*, 1998]. The static conductivity vanishes near the percolation threshold as

$$\sigma_e(\Delta p) = \frac{b}{a}\sigma_m \Delta p^t, \tag{2.18}$$

where $\Delta p = (p - p_c)/p_c$ is the reduced concentration of the conducting sticks, $t = 1$. It follows from Eq. (2.18) that $\sigma_e \ll \sigma_m$ for concentrations $\Delta p < 1$, which is in an agreement with the assumption (1) made before deriving Eq. (2.16). To understand this result, let us consider sticks that form the backbone of an infinite cluster. Each stick that belongs to the backbone connects a finite number of other backbones sticks [see Fig. 1(b)]. Therefore, the joint cluster's backbone consists of segments whose length is proportional to the stick length $2a$. Since the conductance of an individual segment is proportional to b/a, we immediately obtain this factor in Eq. (2.18).

The EMTSM discussed above has disadvantages common for mean-field theories. For example, it gives the "conductivity" critical exponent $t = 1$ in Eq. (2.18) instead of $t = 2.0$, given by the percolation theory discussed in Ch. 1.2 (see [Bergman and Stroud, 1992]). In spite of that, we believe that the EMTSM correctly reproduces the main features of the metal-dielectric composites, at least qualitatively. Thus, the factor b/a has to appear in a "true" scaling equation for the static conductivity $\sigma_e(\Delta p)$ simply because the conductance of an elementary bond in the backbone is proportional to this factor. We, therefore, suggest that the scaling equation for $\sigma_e(\Delta p)$ has the form (2.18) with a critical exponent t is equal to $t = 2.0$.

Our theory also predicts the static dielectric constant $\varepsilon_e(\Delta p)$ diverges when p_c is approached from either side of the percolation threshold,

$$\varepsilon_e = \frac{1}{2}\varepsilon_d \frac{a}{b}|\Delta p|^{-s}, \tag{2.19}$$

where the critical exponent s is equal to $s = 1$. The "dielectric" critical exponent $s = 1$ in Eq. (2.19) is not far away from the known value $s = 0.8$ [Bergman and Stroud, 1992]. Therefore, we believe that Eq. (2.19) may be used for a quantitative estimation of $\varepsilon_e(\Delta p)$ near p_c. Note that the factor $a/b \gg 1$ again appears in Eq. (2.19) due to the large polarizability of conducting sticks discussed above. As a result, the dielectric constant $\varepsilon_e(\Delta p)$ achieves very large values in the critical region Δp near the percolation threshold p_c as shown in Fig. 2 (the dashed line in Fig. 2 indicates the position of p_c). Actually, the concentrations region, where $\varepsilon_e(\Delta p) \gg 1$,largely exceeds the usual condition of the percolation theory $|\Delta p| \ll 1$.

The dielectric susceptibility $\varepsilon_e = \varepsilon_e' + i\varepsilon_e'' = 4i\pi\sigma_e/\omega$ does not diverge at the percolation threshold for finite frequencies ω, but its real part ε_e' still has a maximum at the percolation threshold as it follows from the solution of Eq. (2.16). Let us consider the scaling properties of the effective permittivity $\varepsilon_e(\Delta p, \omega)$ in the critical region $\Delta p \ll 1$, $|\varepsilon_m| \gg 1$ near the percolation threshold. The critical behavior of $\varepsilon_e(\Delta p, \omega)$ for ordinary metal dielectric composites is described by a scaling expression, which is equivalent to Eq. (1.17)

$$\varepsilon_e\left(\Delta p, \omega\right) = \varepsilon_d \Delta p^{-s} f_2\left[\Delta p\left(\varepsilon_m/\varepsilon_d\right)^{1/(t+s)}\right], \tag{2.20}$$

where the scaling function $f_2(z)$ has the following asymptotic behavior

$$f_2(z) \cong \left\{\begin{matrix} (-1)^{-s}, & |z| \gg 1,\ \Delta p < 0 \\ z^s, & |z| \ll 1 \\ z^{t+s} + 1, & |z| \gg 1,\ \Delta p > 0 \end{matrix}\right\}. \tag{2.21}$$

This scaling behavior has been obtained in experiments and numerical simulations (see, for example, [Bergman and Stroud, 1992; Grannan *et al.*, 1981; Vinogradov *et al.*, 1988; Vinogradov *et al.*, 1984a]), where the composites with approximately spherical conducting grains were studied. For $\omega \to 0$ and $\Delta p > 0$, Eqs. (2.20) and (2.21) give an asymptotic behavior both for the effective dielectric susceptibility $\varepsilon_e \sim \varepsilon_d |\Delta p|^{-s}$ and effective conductivity $\sigma_e \sim \sigma_m (\Delta p)^t$, somewhat different from the asymptotic behavior of the same quantities for the conducting stick composites given by Eqs. (2.18) and (2.19). To incorporate Eqs. (2.18) and (2.19) into the scaling model, we rewrite the scaling Eq. (2.20) in the form

$$\varepsilon_e\left(\Delta p, \omega\right) = \frac{a}{b} \varepsilon_d \Delta p^{-s} f_2\left[\Delta p\left(1/\omega^*\right)^{1/(t+s)}\right], \tag{2.22}$$

where the scaling function $f_2(z)$ is still given by Eq. (2.21), while the dimensionless frequency $\omega^* = (a/b)^2 \left(\varepsilon_d/\varepsilon_m\right)$ is large in an absolute value due to the large stick's aspect ratio a/b. Equation (2.22) can be considered as a generalization of Eq. (2.20) to the conducting stick composites. This equation has a scaling form and gives the asymptotic (2.18) and (2.19).

From Eqs. (2.21) and (2.22), it follows that when $|\Delta p|$ is small in the critical region

$$|\Delta p| < \left|\frac{a^2 \omega}{b^2 \sigma_m}\right|^{1/(t+s)}, \tag{2.23}$$

the complex dielectric function exhibits a power-law dependence

$$\varepsilon_e' \sim \varepsilon_e'' \sim \left(\frac{a}{b}\right)^{(t-s)/(t+s)} \left|\frac{\omega}{\sigma_m}\right|^{-s/(t+s)} . \tag{2.24}$$

Note that the concentration range where the dielectric constant has the scaling dependence (2.24) increases by a factor $(a/b)^{2/(t+s)}$. Consider, for example, a hypothetical composite prepared from industrial carbon fibers with a thickness of about 1 μm that are cut into sticks with length of $\sim$ 1 cm. The typical conductivity of such carbon fibers is about $\sigma_m \sim 10^{14}$ s^{-1}. Suppose that we measure the dielectric susceptibility in the vicinity $|\Delta p| < 1$ of the percolation threshold. Then the power-law behavior given by Eq. (2.24) should be observed for the frequencies above $\nu = \omega/2\pi > 100$ kHz, in an agreement with the experimental results (see e.g. [Lagarkov *et al.*, 1998], [Grimes *et al.*, 2000], [Grimes *et al.*, 2001]). One can compare this frequency band with the band for the dispersion, $\nu > 10$ THz, given by the usual scaling expressions (2.20) and (1.17). Thus, a conducting stick composite is a convenient reference system to verify quantitatively the predictions of the percolation theory. Note that the loss due to the electron tunneling occurring between metal clusters near the percolation threshold (see e.g., [Sarychev and Brouers, 1994]) are much suppressed in the composites for the reason that the metal concentration corresponding to the percolation threshold is smaller by factor $b/a \ll 1$ than that one in the usual metal-dielectric composites.

Consider now the behavior of the dielectric function ε_e for concentrations p below the percolation threshold outside of the critical region. We

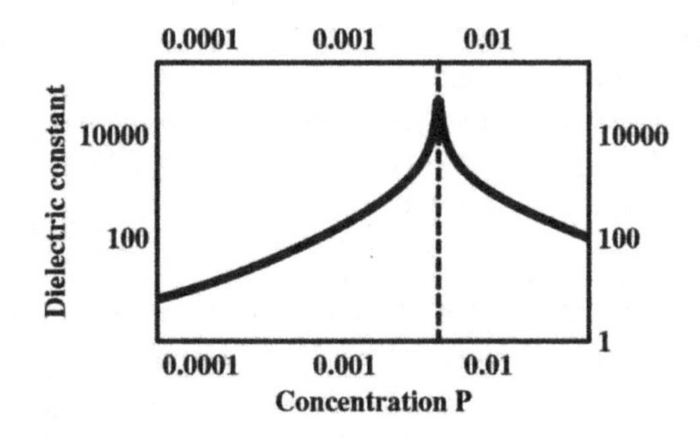

Fig. 2.2 Static dielectric constant of the conducting stick composite as a function of the stick concentration p; the stick aspect ratio and the dielectric susceptibility of the host are chosen to be $a/b = 1000$ and $\varepsilon_d = 2$, respectively. The dashed line signifies the position of the percolation threshold p_c.

will still assume that $|\varepsilon_e|/\varepsilon_d \gg 1$ (see Fig. 2.2.) Then Eq. (2.16) takes the form

$$\varepsilon_e(p,\omega) = \varepsilon_d \frac{2}{9} p \frac{a^2}{b^2 \ln(a/b)} \frac{1}{1 - i\omega^*}, \tag{2.25}$$

where we have reintroduced the dimensionless frequency (slightly different from the above):

$$\omega^* = \left(\frac{a}{b}\right)^2 \frac{\varepsilon_d \omega}{4\pi\sigma_m \ln(a/b)}, \tag{2.26}$$

and have assumed for simplicity that metal conductivity σ_m is real. It is interesting to note that the newly defined dimensionless frequency ω^* coincides [up to a factor $1/\ln(a/b)$] with the dimensionless frequency found in the scaling Eq. (2.22), obtained in the different concentration region.

The dimensionless frequency ω^* is an important parameter that determines the dependence $\varepsilon_e(p,\omega)$ for the entire range of concentrations. One can rewrite Eq. (2.26) for the dimensionless frequency as $\omega^* = \omega\tau_R$, where τ_R is the characteristic stick relaxation time, meaning that the electric charges, induced by an external field and are distributed over the stick, will relax to their stationary distribution state within the time-frame τ_R. For spherical particles, the relaxation time coincides with Maxwell's characteristic time $\tau_{Max} = 1/(4\pi\sigma_m)$. However, for the conducting sticks, the relaxation time $\tau_R \sim (a/b)^2 \tau_{Max}$ may be many orders of magnitude larger. This is the origin of the relaxation behavior of the effective dielectric constant obtained in the experiments of several groups [Kolesnikov *et al.*, 1991; Lagarkov *et al.*, 1997a; Lagarkov *et al.*, 1998; Grimes *et al.*, 2000; Grimes *et al.*, 2001].

The theoretical predictions agrees with experiment, where composites with carbon nanotubes as conducting inclusions were investigated [Grimes *et al.*, 2000; Grimes *et al.*, 2001]. In Fig. 2.3, we observe that the dielectric function indeed has the relaxation behavior in a wide frequency range: from 75 MHz to almost 2 GHz.

To obtain the behavior of the effective dielectric susceptibility in an entire range of the concentrations, the EMTSM equation (2.16) is solved numerically. The calculated real $\varepsilon_e'(\Delta p)$ and imaginary $\varepsilon_e''(\Delta p)$ parts of the dielectric function are presented in Fig. 2.4 for different values of the dimensionless frequency ω^*. The dependence $\varepsilon_e'(\Delta p)$ has "percolation like" behavior with a sharp peak at p_c when $\omega^* \ll 1$. The peak in $\varepsilon_e'(\Delta p)$ growths wider, and shifts to the larger concentration far away from the

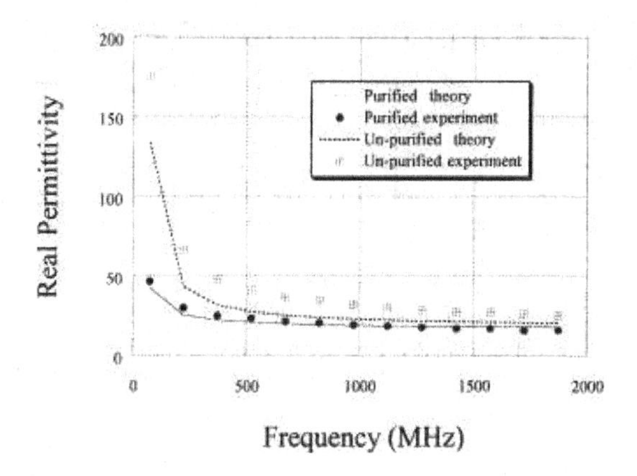

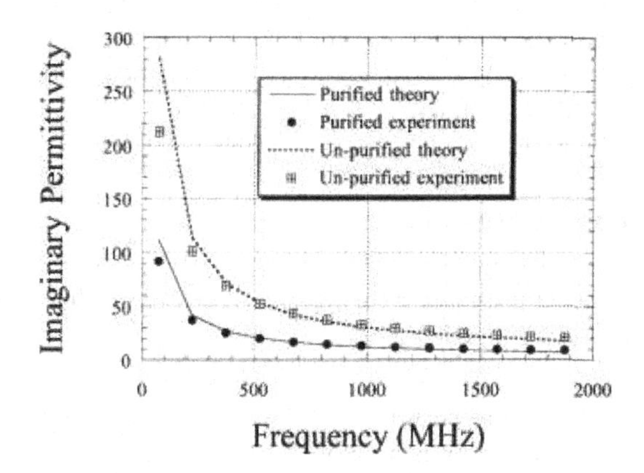

Fig. 2.3 Comparison of the real and imaginary susceptibility spectra of polystyrene loaded with unpurified and graphitized (catalyst particles removed) carbon multiwall nanotubes. Symbols denote the quantities measured experimentally; lines represent the theoretical simulations based on Eq. (2.16).

percolation threshold when the frequency ω^* increases. Since the dielectric function has no critical behavior at the percolation threshold when $\omega^* > 1$, the EMTSM equation (2.16) gives quantitatively correct results for all stick concentrations including the percolation threshold. For $\omega^* > 1$ the effective dielectric constant has a relaxation dispersion $\varepsilon_e(\omega) \sim 1/(1 - i\omega\tau_R)$ in the entire range of concentrations p while the relaxation time τ_R somewhat decreases with increasing p above the percolation threshold p_c.

As it follows from this section discussion, the dispersive behavior of stick composites has richer physics than that of ordinary metal-dielectric

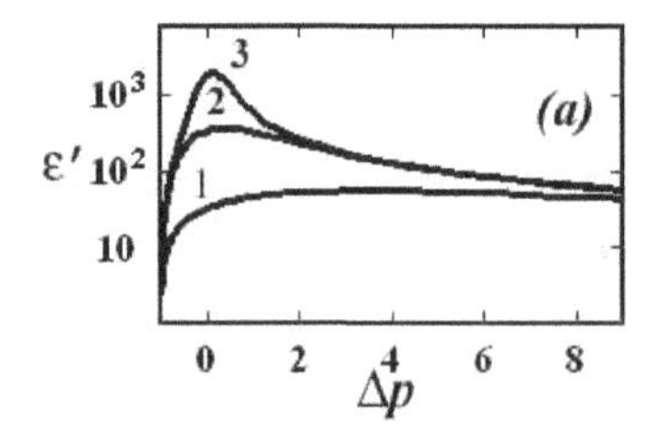

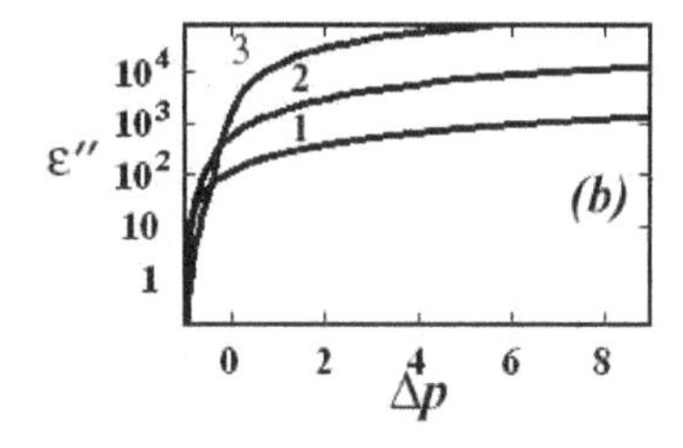

Fig. 2.4 Dependence of real ε' and imaginary ε'' parts of the effective dielectric function ε_e'' on the dimensionless stick concentration $\Delta p = (p - p_c)/p$ for different values of the dimensionless frequency $\omega^* = (a/b)^2 \varepsilon_d \omega / (4\pi |\sigma_m| \ln(a/b))$. The aspect ratio is $a/b = 1000$, $\varepsilon_d = 2$.

composites with spherical particle inclusions. We thus can conclude that the stick composites represent a special class of metamaterials that properties essentially depend on the local geometry of the system. As we will see in the next section, this dependence becomes much stronger for high frequencies when non-quasistatic (retardation) effects are important.

2.3 High-frequency Response

Composite materials containing conducting sticks dispersed in a dielectric matrix have unusual properties at high frequencies. When frequency ω increases, the wavelength $\lambda = 2\pi c/\omega$ of an external electromagnetic field becomes comparable with the stick length $2a$. In this case, one might think that the sticks act as an array of independent micro-antennas and the external EM wave should be scattered in all directions. Yet, the composite materials have well defined dielectric and magnetic properties at these frequencies, and we will demonstrate that. The description based on the "effective-medium" theory is possible because a thin conducting stick interacts with the external field like an elementary dipole. Therefore, we can still use the effective dielectric constant ε_e or effective conductivity $\sigma_e = -i\omega\varepsilon_e/4\pi$ to describe the interaction of stick composites with the external electromagnetic wave. We note, however, that the formation of large stick clusters near the percolation threshold may result in scattering.

Since conducting stick composites do have effective parameters for all concentrations p outside the percolation threshold, we can use the percolation theory to calculate their effective conductivity σ_e. However, the theory has to be generalized to take into account the nonquasistatic effects. The problem of effective parameters of composites beyond the quasistatic limit has been considered in [Vinogradov *et al.*, 1989b], [Lagarkov *et al.*, 1992], [Rousselle *et al.*, 1993], [Lagarkov and Sarychev, 1996], [Sarychev *et al.*, 2000b], [Sarychev and Shalaev, 2000]. It was shown that the mean-field approach can be extended to find the effective dielectric constant and magnetic permeability at high frequencies. Results of these conclusions can be briefly summarized as it follows: One first finds the polarizability α_m for an individual particle in the composite illuminated by the external electromagnetic wave (the particle is embedded into the "effective medium" with dielectric some constant ε_e). Then, the effective dielectric susceptibility ε_e is determined by the self-consistent requirement that the averaged polarizability of all the particles vanishes. Thus, for the non-quasistatic case, the problem is reduced to the calculation of the polarizability of an elongated conducting inclusion.

2.3.1 *Scattering of electromagnetic wave by conducting stick*

The diffraction of electromagnetic waves on a conducting stick is the classical problem of electrodynamics. A rather tedious theory for this process is presented in several textbooks [Vainshtein, 1988], [Hallen, 1962]. We show below that the problem can be solved analytically in the case of very elongated sticks with the large aspect ratio: $\ln(a/b) \gg 1$. We will restrict our consideration to wavelengths $\lambda > 2a$ that still include a main, a half-wave, resonance at $\lambda \simeq 4a$.

We consider a conducting stick of length $2a$ and radius b, illuminated by an electromagnetic wave. We suppose that the electric field in that wave is directed along the stick and the stick is embedded into a medium with permittivity $\varepsilon_d = 1$. The external electric field excites the current $I(z)$ in the stick, and z is the main axis (directed along the stick), referenced from its midpoint. The dependence $I(z)$ is nontrivial when the wavelength λ is of the order of the stick length or smaller. There is also a nontrivial charge distribution $Q(z)$ along the stick in this case. The charge distribution $Q(z)$ determines the polarizability of the stick. To find $I(z)$ and $Q(z)$, we introduce the potential $U(z)$ of the charges $Q(z)$ distributed over the stick surface. From the equation for the electric charge conservation we obtain the following formula

$$\frac{dI(z)}{dz} = i\omega Q(z), \tag{2.27}$$

which relates the charge per unite length $Q(z)$ and the current $I(z)$. Note that the electric charges q are induced by the external field E_0 and they can be expressed as the divergence of the polarization vector, $q = -4\pi \operatorname{div} \mathbf{P}$. Then, the polarization $\mathbf{P}$ can be included in the definition of the electric displacement $\mathbf{D}$, $\operatorname{div} \mathbf{D} = 0$. However, during calculation of a high frequency field in the conducting stick, it is convenient to explicitly consider charges generated by the external field.

To find an equation for the current $I(z)$ we treat a conducting stick as a prolate conducting spheroid with semiaxes a and b. The direction of the major axis is thought to coincide with the direction of the electric field $\mathbf{E}_0 \exp(-i\omega t)$ of the incident wave. The electric potential of the charge q is given by the solution to the Maxwell's equations (see, e.g., [Hallen, 1962], p. 377)

$$U(z) = \frac{1}{\varepsilon_d} \oint \frac{q(\mathbf{r}') \exp(ik|\mathbf{r} - \mathbf{r}'|)}{|\mathbf{r} - \mathbf{r}'|} ds' \cong \frac{1}{\varepsilon_d} \int_{-a}^{a} \frac{Q(z') \exp(ik|z - z'|)}{\sqrt{(z - z')^2 + \rho(z)^2}} dz', \tag{2.28}$$

where the integration is initially performed over the surface of the stick, $\mathbf{r}$ and $\mathbf{r}'$ are the points on the rode's surface, z and z' are the corresponding coordinates along the stick, $\rho(z) = b\sqrt{1 - z^2/a^2}$ is the radius of the cross section, and $k = \omega/c$ is the wave vector of the external field. In order to arrive to the second integral in Eq. (2.28), we have neglected terms of the order of $\rho(z)/a < b/a \ll 1$.

We can divide the second integral in Eq. (2.28) into two parts by setting $Q(z') \exp(ik|z - z'|) = Q(z) + [Q(z') \exp(ik|z - z'|) - Q(z)]$, i.e.,

$$U(z) \cong \frac{Q(z)}{\varepsilon_d} \int_{-a}^{a} \frac{dz'}{\sqrt{(z - z')^2 + \rho(z)^2}} + \int_{-a}^{a} \frac{Q(z') \exp(ik|z - z'|) - Q(z)}{\varepsilon_d |z - z'|} dz'. \tag{2.29}$$

The first integral in Eq. (2.29) is given by

$$\frac{1}{C_0} = \int_{-a}^{a} \frac{dz'}{\varepsilon_d \sqrt{(z - z')^2 + \rho(z)^2}} \cong \frac{2}{\varepsilon_d} \ln\left(2\frac{a}{b}\right) + \frac{1}{2\varepsilon_d} \left(\frac{b}{a}\right)^2 \frac{a^2 + z^2}{(a^2 - z^2)} + \ldots \cong \frac{2}{\varepsilon_d} \ln\left(\frac{a}{b}\right), \tag{2.30}$$

where we have neglected small terms on the order of $1/\ln(a/b)$ in last equality. This result does not hold near the ends of the stick ($z \to \pm a$), however, these regions are non-important in calculating the polarizability. The second integral in Eq. (2.29) has no singularity at $z = z'$ and, therefore its value proportional to $\sim Q(z)$, which is an odd function of the coordinate z. We assume for simplicity that $Q(z)$ is proportional to z and in this approximation

$$\begin{aligned}\frac{1}{C_1} &= \int_{-a}^{a} \frac{Q(z')\exp(ik|z-z'|) - Q(z)}{\varepsilon_d |z-z'|} dz' \\ &= 2\frac{Q(z)}{\varepsilon_d}\left[-e^{i\,a\,k} + \mathrm{Ei}(a\,k)\right], \end{aligned} \tag{2.31}$$

where the function $\mathrm{Ei}(x)$ is defined as

$$\mathrm{Ei}(x) = \int_0^x \left[\exp(it) - 1\right]/t\,dt. \tag{2.32}$$

By substituting Eqs. (2.30) and (2.31) in Eq. (2.29), we obtain

$$U(z) = Q(z)/C, \tag{2.33}$$

where the capacitance C is given by

$$\begin{aligned}\frac{1}{C} &= \frac{1}{C_0} + \frac{1}{C_1} = \frac{2}{\varepsilon_d}\left[\ln(2a/b) - e^{i\,a\,k} + \mathrm{Ei}(a\,k)\right] \\ &\simeq \frac{2}{\varepsilon_d}\ln(a/b) + i\frac{2}{\varepsilon_d}(\mathrm{Si}(ak) - \sin(ak)). \end{aligned} \tag{2.34}$$

Function Si is defined as $Si(x) = \int_0^x \sin(t)/t\,dt$ and we have neglected the corrections on the order of (or smaller) than $1/\ln(a/b)$. The capacitance C takes its usual value $C = 1/\left[2\ln(a/b)\right]$ in the quasistatic limit $ka \to 0$. But the retardation effects result in the additional term in Eq. (2.34), which is smaller in comparison with the leading logarithmic term, yet equally important. We preserve the radiation term since it manifests the radiative loss. This result is obtained within the logarithmic accuracy: Its relative error is on the order of $1/\ln(a/b)$, and the ratio a/b (and its logarithm) is large.

By substituting Eq. (2.33) into Eq. (2.27), we obtain the following equation

$$\frac{dI(z)}{dz} = i\omega C U(z), \tag{2.35}$$

which relates the current $I(z)$ and the surface potential $U(z)$. The electric current $I(z)$ and electric field $E(z)$ on the stick surface are related by the usual Ohm's law

$$E(z) = RI(z), \tag{2.36}$$

where R is the impedance per unit length. Since the stick is excited by the external field $E_0 \exp(-i\omega t)$, which is parallel to its axis, the electric field $E(z)$ equals to

$$E(z) = E_0 - \frac{dU(z)}{dz} + ikA_z(z). \tag{2.37}$$

We consider now the vector potential $A_z(z)$ induced by the current $I(z)$ flowing in the stick and obtain $A_z(z)$ by the same procedure as was used to estimate the potential U. Thus, with the same logarithmic accuracy we find

$$\begin{aligned} A_z(z) &= \frac{1}{c}\int_{-a}^{a} \frac{I(z')\exp(ik|z-z'|)}{\sqrt{(z-z')^2+\rho(z)^2}}dz' \\ &\simeq \frac{2}{c}I(z)\ln\left(\frac{2a}{b}\right) + \frac{1}{c}\int_{-a}^{a} \frac{I(z')\exp(ik|z-z'|)-I(z)}{|z-z'|}dz', \end{aligned} \tag{2.38}$$

where c is the speed of light. To estimate the second integral in Eq. (2.38) we approximate the current $I(z)$, which is an even function of z, as $I(z) = I(0)\left[1-(z/a)^2\right]$. Note that $I(z) = 0$ for $z = \pm a$, which is in an agreement with the boundary conditions for the electric current in the stick. Thus we obtain for $z \ll a$ that

$$\frac{1}{c}\int_{-a}^{a} \frac{I(z')\exp(ik|z-z'|)-I(z)}{|z-z'|} \simeq \frac{1}{c}I(z)\left[-1+l(ka)\right], \tag{2.39}$$

where the function $l(x)$ is given by

$$l(x) = \left[2+2e^{ix}(ix-1)+x^2\right]x^{-2} + 2\,\mathrm{Ei}(x). \tag{2.40}$$

By substituting Eq. (2.39) in Eq. (2.38), we obtain the following relation between the vector potential and the current

$$A_z(z) = \frac{L}{c}I(z), \tag{2.41}$$

where L is the inductance per unit length,

$$L \simeq 2\ln\left(\frac{2a}{b}\right) - 1 + l\,(ka) \simeq L_0 + iL_1$$
$$= 2\ln(a/b) + 2i\left[(ka\cos(ka) - \sin(ka))/(ka)^2 + \mathrm{Si}\,(ka)\right], \qquad (2.42)$$

where we have again neglected corrections to the large integral $1/\ln(a/b) \ll 1$, but preserved the radiation terms responsible for loss as they play an important role in the electromagnetic response of a conducting stick.

By substituting Eqs. (2.37) and (2.41) in Eq. (2.36), we obtain the following form of the Ohm's law:

$$-\frac{dU(z)}{dz} = \left(R - i\frac{\omega L}{c^2}\right) I(z) - E_0. \qquad (2.43)$$

To obtain a closed equation for the current $I(z)$, we differentiate Eq. (2.35) with respect to z and substitute the result into Eq. (2.43) for $dU(z)/dz$. We obtain

$$\frac{d^2 I(z)}{dz^2} + i\omega C\left[\left(R - i\frac{\omega L}{c^2}\right) I(z) - E_0\right] = 0, \qquad (2.44)$$

with the boundary conditions requiring the current vanish at the ends of the stick,

$$I(-a) = 0, \quad I(a) = 0. \qquad (2.45)$$

A solution for Eq. (2.44) gives the current distribution $I(z)$ in a conducting stick irradiated by an electromagnetic wave. Based on that, we can calculate the charge distribution and the polarizability of the stick.

As mentioned, we consider the conducting stick as a prolate spheroid with semiaxes such that $a \gg b$. To determine the impedance R in Eq. (2.44) we recall that the cross-section area of a spheroid at some coordinate z equals to $\pi b^2[1 - (z/a)^2]$; thus we have the following expression for the impedance:

$$R = \frac{1}{\pi b^2[1 - (z/a)^2]\sigma_m^*}, \qquad (2.46)$$

where σ_m^* is the renormalized stick conductivity, calculated by taking into account the skin effect. In calculations, we have assumed that the conductivity σ_m change due to the skin effect occurs in the same way as the conductivity of a long wire of radius b (see, for example, Ref. [Landau *et al.*, 1984] Sec. 61):

$$\sigma_m^* = \sigma_m f(\Delta), \quad f(\Delta) = \frac{2}{\Delta}\frac{J_1(\Delta)}{J_0(\Delta)}, \tag{2.47}$$

where J_0 and J_1 are the Bessel functions of the zeroth and the first order, respectively, and the parameter Δ equals to

$$\Delta = bk\sqrt{\varepsilon_m} \equiv bk_m. \tag{2.48}$$

When the skin effect is weak, i.e., $|\Delta| \ll 1$, the function $f(\Delta) = 1$ and the renormalized conductivity σ_m^* equals to the stick conductivity $\sigma_m^* = \sigma_m$. In the opposite case of a strong skin effect ($|\Delta| \gg 1$), the current I flows within a thin skin layer at the surface of the stick and $\sigma_m^* \ll \sigma_m$ as it follows from Eq. (2.48).

For further consideration, it is convenient to rewrite Eqs. (2.44) and (2.45) in terms of the dimensionless coordinate $z_1 = z/a$ and dimensionless current

$$I_1 = I/\left(\sigma_m^* \pi b^2 E_0\right). \tag{2.49}$$

We also introduce the dimensionless relaxation parameter

$$\gamma = 2\,i\frac{b^2\,\pi\,\sigma_m^*}{a^2\,C\,\omega} = \varepsilon_m^* g_{\shortparallel} + \varepsilon_m^* \left(\frac{b}{a}\right)^2 (\mathrm{Si}(ak) - \sin(ak)), \tag{2.50}$$

where $\varepsilon_m^* = i4\pi\sigma_m^*/\omega$ is the renormalized dielectric constant for metal, and

$$g_{\shortparallel} = (b/a)^2 \ln(a/b) \tag{2.51}$$

is the depolarization factor for a prolate ellipsoid (see, e.g., [Landau *et al.*, 1984], Sec. 4). We also introduce a dimensionless frequency

$$\Omega(ak) = ak\sqrt{LC} \simeq ak\left(1 + i\frac{ak\cos(ak) + ((ak)^2 - 1)\sin(ak)}{2(ak)^2 \ln(a/b)}\right), \tag{2.52}$$

and still assume that $\ln(a/b) \gg 1$. By substituting parameters γ and Ω in Eqs. (2.44) and (2.45), we obtain the following differential equation for the dimensionless current

$$\frac{d^2 I_1(z_1)}{dz_1^2} + \left[-\frac{2}{\gamma(1 - z_1^2)} + \Omega^2\right] I(z_1) + \frac{2}{\gamma} = 0, \tag{2.53}$$

$$I_1(-1) = I_1(1) = 0.$$

To understand the physical meaning of Eq. (2.53) let us consider first the quasistatic case when the skin effect is negligible and $ka \ll 1$. Then, $|\Omega^2\gamma| \sim |\Delta| \ll 1$ as it follows from Eqs. (2.50) and (2.53). Therefore, we can neglect the second term in the square brackets in Eq. (2.53) and find the current,

$$I_1(z_1) = \frac{1 - z_1^2}{1 + \gamma}, \tag{2.54}$$

and the electric field E_m inside a conducting stick (see Eqs. 2.36 and 2.46),

$$E_m = \frac{1}{1+\gamma} E_0 = \frac{1}{1 + \varepsilon_m g_{\parallel}} E_0. \tag{2.55}$$

As anticipated, the electric field E_m is uniform and coincides with the quasistatic internal field in the case of a prolate conducting spheroid.

In the opposite case of a strong skin effect the product $\Omega^2\gamma \sim \Delta$ is much larger than unity: $|\Omega^2\gamma| \gg 1$. Therefore, we can neglect the first term in the square brackets in Eq. (2.53) and find that

$$I_1(z_1) = \frac{2}{\Omega^2\gamma}\left[\frac{\cos(\Omega z_1)}{\cos(\Omega)} - 1\right]. \tag{2.56}$$

As it follows from the equation, the current has maxima when $\cos(\Omega) \approx 0$ that corresponds to the well-known antenna resonance [Vainshtein, 1988], [Hallen, 1962] at $\lambda = 4a$. In a general case of arbitrary Ω and γ, the solution for Eq. (2.53) can neither be obtained in a close-from solution nor be expressed as a finite set of known special functions [Stratton, 1935]. Still, this equation can be solved numerically. Its numerical integration shows that for relatively large wavelengths ($\lambda > 2a$), the solution can be approximated by an equation

$$I_1(z_1) = \frac{1 - z_1^2}{1 + \gamma\cos(\Omega)}, \tag{2.57}$$

which is an interpolation for Eqs. (2.54) and (2.56).

When the current I is known, we can calculate the specific polarizability P_m of a conducting stick: $\alpha_m = (VE_0)^{-1}(\varepsilon_m - 1)\int E_m\, dV$, where $V = 4\pi ab^2/3$ is the stick volume. Assuming that $|\varepsilon_m| \gg 1$, we obtain

$$\alpha_m = \frac{\varepsilon_m^*}{1 + \gamma\cos(\Omega)}. \tag{2.58}$$

Above, we have assumed that the stick is aligned with the electric field of the incident electromagnetic wave. Stick composites can be formed by randomly oriented rods. In this case, we have to modify Eq. (2.58) for the stick polarizability. We consider a conducting stick directed along the unit vector $\mathbf{n}$ for the case when the stick is irradiated by an electromagnetic wave

$$\mathbf{E} = \mathbf{E}_0 \exp\left[i\left(\mathbf{k}\cdot\mathbf{r}\right)\right], \tag{2.59}$$

with $\mathbf{k}$ being a wave vector, defined inside the composite media.

The major contribution in generating current I within a stick with high aspect ratio comes from a component of the electric field, that is parallel to the stick

$$\mathbf{E}_{\parallel}\left(z\right) = \mathbf{n}\left(\mathbf{n}\cdot\mathbf{E}_0\right) \exp\left[i\left(\mathbf{k}\cdot\mathbf{n}\right)z\right], \tag{2.60}$$

where z is the coordinate along the stick.

The field $\mathbf{E}_{\parallel}$ averaged over the stick orientations is aligned with the external field $\mathbf{E}_0$ and has the following magnitude

$$E_0^*\left(z\right) = \frac{E_0}{\left(kz\right)^2}\left[\frac{\sin\left(kz\right)}{kz} - \cos\left(kz\right)\right]. \tag{2.61}$$

The current in the stick is a linear function of the filed $\mathbf{E}_{\parallel}$. Since the average field $\mathbf{E}_{\parallel}$ is aligned with $\mathbf{E}_0$, the current averaged over the stick orientations is also parallel to the external field $\mathbf{E}_0$.

To obtain the current $\langle\langle I(z)\rangle\rangle$ averaged over all the stick orientations and the average stick polarizability $\langle\langle P_m\rangle\rangle$, we substitute the field $E_0^*\left(z\right)$ given by Eq. (2.61) into Eq. (2.44) for the field E_0. Hereafter, the sign $\langle\langle\ldots\rangle\rangle$ denotes the average over all the stick orientations. Then, the current $\langle\langle I\rangle\rangle$, polarizability $\langle\langle P_m\rangle\rangle$, and the effective dielectric permittivity depend on frequency ω and, in addition, on the wave vector k. This means that a conducting-stick composite is a medium with spatial dispersion. This result is easy to understand, if we recall that a characteristic scale of inhomogeneity is the stick length $2a$, which can be on the order of or larger than the excitation wavelength. Therefore, it is not surprising that the interaction of an electromagnetic wave with such composite has a nonlocal character and, therefore, the spatial dispersion is important. We can expect that additional waves can be excited in the composite in the presence of strong spatial dispersion.

Bellow we consider wavelengths such that $\lambda > 2a$; therefore, we can expand $E_0^*(z)$ in a series as

$$E_0^*(z) = \frac{E_0}{3}\left(1 - \frac{(kz)^2}{10}\right). \tag{2.62}$$

When considering the dielectric properties, we restrict ourselves to the first term. Since the average field is given by $E_0^*(z) = E_0/3$, then the average current is equal to $\langle\langle I\rangle\rangle = I/3$, where the current I is defined by Eq. (2.57). As a result, the stick polarizability averaged over all the orientations is equal to $\langle\langle P_m\rangle\rangle = P_m/3$, and P_m is given by Eq. (2.58).

2.3.2 *High-frequency effective dielectric function*

To calculate the effective properties of a conducting stick composite, whose sticks randomly distributed in a dielectric matrix, we use the approach developed in Sec. 2.2. We consider the polarizability of a conducting stick immersed in a scale-dependent effective medium with permittivity $\varepsilon_e(l)$ given by Eq. (2.13). The scale dependence of the local dielectric constant results in a modification of the stick capacitance C_0 in Eq. (2.34), namely, it is now given by the integral

$$\begin{aligned}\frac{1}{C_0} &= \int_{-a}^{a} \frac{dz'}{\varepsilon_e(|z-z'|)\sqrt{(z-z')^2 + \rho(z)^2}} \\ &\cong \int\limits_{\substack{|z-z'|>b \\ |z'|<a}} \frac{dz'}{\varepsilon_e(|z-z'|)\,|z-z'|} \cong \frac{2}{\varepsilon_d}\ln\left(1 + \frac{\varepsilon_d\, a}{\varepsilon_e\, b}\right),\end{aligned} \tag{2.63}$$

where ε_d and ε_e are host and bulk dielectric susceptibilities (see Eq. 2.13), respectively. The stick inductance L_0 in Eq. (2.42) does not depend on the dielectric susceptibility and still equals to $L_0 = 2\ln(a/b)$. We replace ka by $\sqrt{\varepsilon_e}ka$ in retardation terms for C_1 and L_1, effectively making a transformation $C_1 \longrightarrow \varepsilon_e C_1\left(\sqrt{\varepsilon_e}ka\right)$ and $L_1 \longrightarrow L_1\left(\sqrt{\varepsilon_e}ka\right)$ as we assume that the radiating electromagnetic field forms around a stick on the scales $l < a$. The current distribution along the stick is still governed by Eq. (2.53), with the parameters γ and Ω that now depend on the effective dielectric constant ε_e. For example, the scale dependence of the local dielectric constant in Eq. (2.13) results in a modification of the parameter γ to

$$\gamma = \frac{\varepsilon_m^*}{\varepsilon_d}\left[g_{\shortparallel} + \left(\frac{b}{a}\right)^2 (\mathrm{Si}(\sqrt{\varepsilon_d}ka) - \sin(\sqrt{\varepsilon_d}ka))\right], \tag{2.64}$$

where the depolarization factor $g_{\shortparallel}$ is given now by the following equation

$$g_{\shortparallel} = \left(\frac{b}{a}\right)^2 \ln\left(1 + \frac{\varepsilon_d\, a}{\varepsilon_e\, b}\right). \tag{2.65}$$

The effective susceptibility ε_e is determined from the condition that the polarizability averaged over all the inclusions vanishes. Since the sticks are randomly oriented, the dielectric regions of the composite have spherical shapes on average, as it was assumed in Sec. 2.2. The specific polarizability of a dielectric region is given then by a well-known quasistatic equation (see, for example, [Bergman and Stroud, 1992])

$$\alpha_d = \frac{3\,(\varepsilon_d - \varepsilon_e)}{2\varepsilon_e + \varepsilon_d}. \tag{2.66}$$

The polarizability of a conducting stick embedded into the effective medium (2.13) is obtained from Eq. (2.58), by replacing in the numerator ε_m^* with $\varepsilon_m^*/\varepsilon_e$. Then the condition that the average polarizability should vanish gives the following equation:

$$\begin{aligned}&\langle\langle 4\pi\alpha_m\rangle\rangle + (1-p)\,4\pi\alpha_d \\ &\quad = \frac{1}{3}p\frac{\varepsilon_m^*}{\varepsilon_e}\frac{1}{1+\gamma\cos\Omega} + 3\frac{\varepsilon_d - \varepsilon_e}{2\varepsilon_e + \varepsilon_d} = 0,\end{aligned} \tag{2.67}$$

where p is the volume concentration of the conducting sticks and the sign $\langle\langle\cdots\rangle\rangle$, as before, denotes the average over the orientations. Equation (2.67) gives the high frequency permittivity $\varepsilon_e\,(p,\omega)$ of the conducting stick composites.

To understand the high frequency properties of the composite we consider the solution of Eq. (2.67) for the stick concentrations p below the percolation threshold $(b/a)^2 \ll p \ll b/a$. Assuming that $\varepsilon_d \ll |\varepsilon_e| < a/b$, we obtain the explicit equation for the effective dielectric permittivity

$$\begin{aligned}\varepsilon_e &\simeq \varepsilon_d\frac{2}{9}p\frac{a^2}{b^2}\frac{1}{\Phi\cos\left[\Omega\left(\sqrt{\varepsilon_d}ka\right)\right] + \varepsilon_d/\varepsilon_m^*}, \\ \Phi &= \left[\ln(a/b) + i\,(\mathrm{Si}(\sqrt{\varepsilon_d}ka) - \sin(\sqrt{\varepsilon_d}ka))\right],\end{aligned} \tag{2.68}$$

where function $\Omega\,(x)$ is defined in Eq. (2.52).

We consider then the effective dielectric permittivity ε_e for perfect metals ($|\varepsilon_m| \to \infty$). In this case, the electromagnetic wave does not penetrate the stick and, as it follows from Eq. (2.47), the renormalized conductivity ε_m^* tends to approach the infinity. We neglect the second term in the denominator of Eq. (2.68) and obtain that the effective permittivity ε_e has maximum when $\operatorname{Re}\Omega = \pi/2$, which corresponds approximately to the wavelength $\lambda_r = 4a/\sqrt{\varepsilon_d}$.

Now we consider the behavior of the effective dielectric function near the resonance frequency $\omega_r = 2\pi c/\lambda_r$. By expanding the denominator of Eq. (2.68) in a power series of $\omega - \omega_r$ and taking into account that $\varepsilon_m^* \to \infty$ and $\ln(a/b) \gg 1$, we obtain

$$\varepsilon_e \simeq \varepsilon_d p \frac{4}{9\pi \ln(a/b)} \frac{a^2}{b^2} \frac{1}{(\omega_r - \omega)/\omega_r - i\Gamma}, \tag{2.69}$$

where the loss factor $\Gamma = \left(\pi^2 - 4\right) / \left[\pi^2 \ln(a/b)\right] \ll 1$.

It is interesting to point out that the effective dielectric function is independent of the metal conductivity ε_m, as it should be for the limiting case of a purely reflective metal. At the resonance frequency $\omega = \omega_r$, the real part of ε_e changes the sign and becomes negative when $\omega > \omega_r$. The imaginary part of permittivity ε_e has the resonance maximum of

$$\varepsilon_e''(\omega_r) \simeq \varepsilon_d \frac{4\pi}{9\left(\pi^2 - 4\right)} p \frac{a^2}{b^2} \tag{2.70}$$

and does not depend on the stick conductivity. As we can see, the imaginary part of the effective dielectric function does not vanish for the composites formed by perfectly conducting sticks. The presence of the effective loss, in this case, is due to the excitation of the internal modes in the composite. When neither the metal nor the dielectric have losses, the amplitudes of these modes continuously increase in time. In the real life, there are always some loss in the metal composites as well as in the dielectric host. Therefore, the internal fields stabilize at some large values and one can anticipate the existence of giant local fields in the conducting composites.

Microwave metamaterials with negative dielectric permittivity were first obtained in [Kolesnikov *et al.*, 1991], [Lagarkov *et al.*, 1997a], [Lagarkov *et al.*, 1998]. Fig.1.5 presents experimental and theoretical results for the dielectric function of composites containing very thin aluminum microwires in a microwave region [Lagarkov *et al.*, 1998]. In such metamaterials, the real part of ε_e is negative for the frequencies above the resonance one.

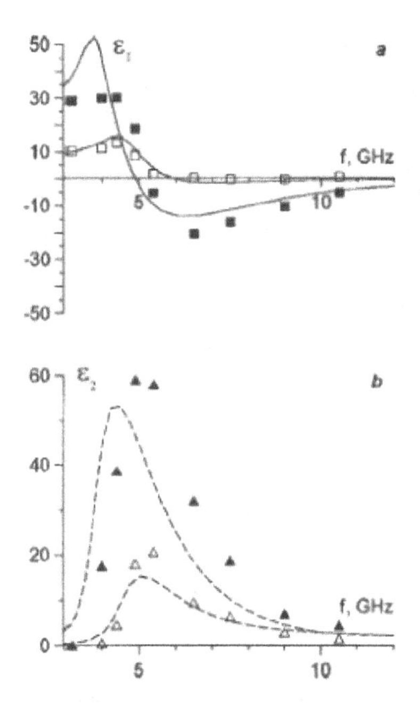

Fig. 2.5 Real (a) and imaginary (b) parts of the dielectric permittivity for a composite filled with aluminum-coated fibers of 20 mm long (thickness $\sim 1\,\mu m$). The fiber volume concentration is 0.01% and 0.03%. Data points indicate experimental data with the lines describing theoretical results.

2.4 Giant Enhancements of Local Electric Fields

We consider now the field distributions in very thin ($\sim$ 10 μm), but relatively long ($\sim$ 1 μm) metal sticks that we will refer as nanowires. The problem of finding field distribution around such small metal formations is complicated: It cannot be solved analytically, or in a closed form. Below, we describe a numerical model based on the discrete dipole approximation (DDA) [Podolskiy *et al.*, 2002a], [Podolskiy *et al.*, 2003].

The DDA was first introduced by Purcell and Pennypacker [Purcell and Pennypacker, 1973]. In the DDA approximation, an object is represented by a large amount of small spherical particles of same radius R as shown in Fig. 2.6. Each particle is placed in a node of a cubic lattice with period a. We denote the position of an individual particle by $\mathbf{r}_i$. We assume that the particle radius R is much smaller than the wavelength λ so that the interactions of the particle with an external electromagnetic field and other particles are well described in a framework of particles' dipoles $\mathbf{d}_i$. The

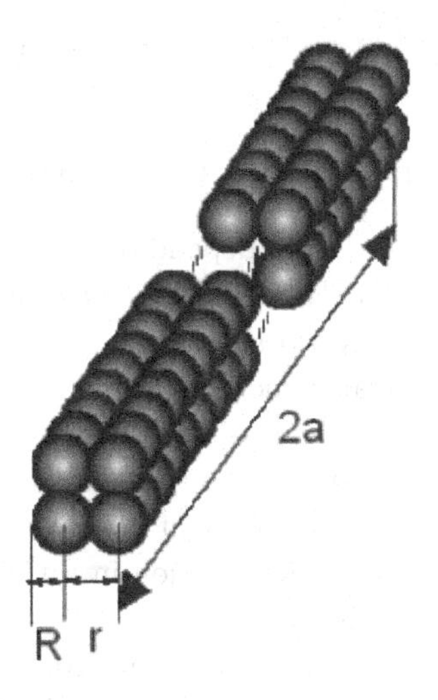

Fig. 2.6 Example of a long stick simulated by chains of spheres.

particle interacts with an incident field $\mathbf{E}_0$ and secondary, scattered fields (from all other particles). Therefore, the dipole moments of particles are coupled to the incident field and to each other and can be found by solving of the following coupled – dipole equations (CDEs):

$$\mathbf{d}_i = \alpha \left[\mathbf{E}_0(\mathbf{r}_i) + \sum_{j \neq i} \hat{G}(\mathbf{r}_i - \mathbf{r}_j)\, \mathbf{d}_j \right], \tag{2.71}$$

α is the polarizability of a particle, $\mathbf{E}_0(\mathbf{r}_i)$ is the incident field at site $\mathbf{r}_i$, and $\hat{G}(\mathbf{r}_i - \mathbf{r}_j)$ is a free-space dyadic Green's function. Operation $\hat{G}(\mathbf{r}_i - \mathbf{r}_j)\,\mathbf{d}_j$ gives essentially the scattered field produced by dipole $\mathbf{d}_j$ at location $\mathbf{r}_i$. In a matrix form, the Green's function is defined as

$$\begin{aligned} G_{\alpha\beta} &= k^3 \left[A(kr)\, \delta_{\alpha\beta} + B(kr)\, r_\alpha r_\beta \right], \\ A(x) &= \left[x^{-1} + i x^{-2} - x^{-3} \right] \exp(ix), \\ B(x) &= \left[-x^{-1} - 3 i x^{-2} + 3 x^{-3} \right] \exp(ix), \end{aligned} \tag{2.72}$$

with $\hat{G}\mathbf{d} \equiv G_{\alpha\beta} d_\beta$. Note that the Greek indices represent Cartesian components and the summation over the repeated indices is implied. The polarizability α of an individual dipole is given by Lorentz-Lorenz formula with the radiative correction introduced by Draine [Draine, 1988]:

$$\alpha_{LL} = R^3 \frac{\varepsilon_m - 1}{\varepsilon_m + 1}, \qquad \alpha = \frac{\alpha_{LL}}{1 - i\,(2/3)\,(kR)^3\,\alpha_{LL}}. \tag{2.73}$$

The results of calculations for (2.71) depend on the intersection ratio r/R, i.e., the ratio of the distance between neighboring particles and its radius (see Fig. 2.6). We choose the ratio as $r/R \approx 1.66$ to reproduce the quasistatic polarizability of elongated metal ellipsoid.

In numerical simulations performed in [Podolskiy *et al.*, 2002a], [Podolskiy *et al.*, 2003], a single nanostick was represented by four parallel chains of spherical particles to take into account the skin effect (see Fig. 2.6). Specifically, we consider the field distribution in a vicinity of a conducting stick with roughly $2b = 30$ nm thickness and $2a \approx 15$ μm long, illuminated by a plane wave with the wavelength of 540 nm. Our results shown in Fig. 2.7 clearly identify the interference pattern between irradiation and the plasmon-polariton wave excited on the metal surface. Similar interference patterns were observed in experiments by [Moskovits, 2004] and computer simulations [Aizpurua *et al.*, 2003]. Note that the electromagnetic field is concentrated around the nanowire surface, which suggests the possibility to use nanowires as nano waveguides.

Simulations for shorter sticks ($2a = 480$ nm) presented in Fig. 2.8 also show the existence of sharp plasmon resonances when the wavelength of the light is a multiple of surface plasmon (half)-wavelengths. The enhancement of the local field intensity in the resonance can reach the magnitude of 10^3. The spatial area, where the field concentrates is highly localized around the nanowire and can be as small as 100 nm. This plasmon resonance is narrowband, with the spectral width of about 50 nm in a single nanowire, which corresponds to the discussed above Eqs. (2.68) and (2.69).

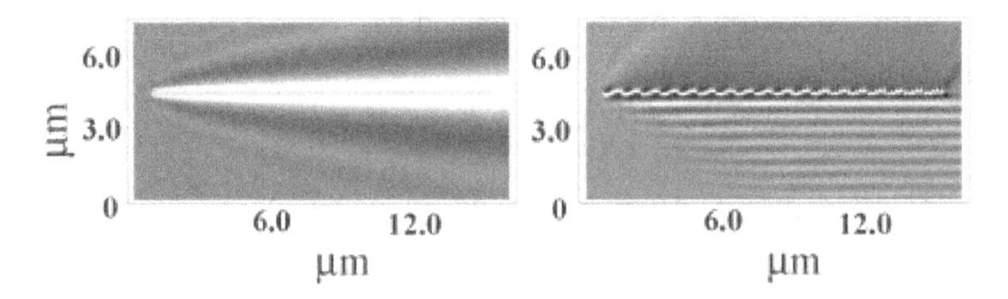

Fig. 2.7 EM field distribution for a long needle. The wavelength of incident light is 540 nm; the angle between the wave vectors of incident light and the needle is (a) 0°, (b) 30°.

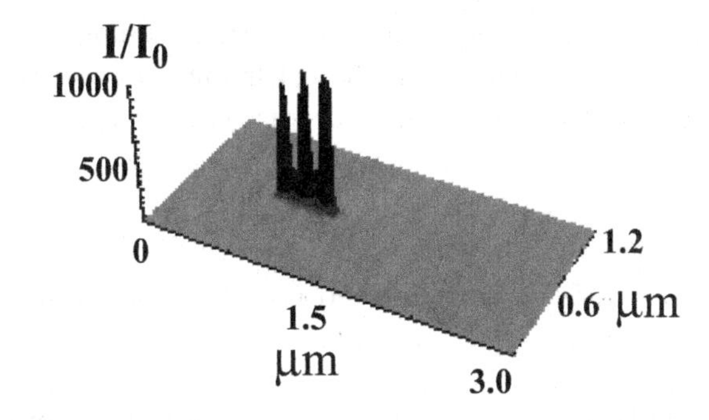

Fig. 2.8 The intensity distribution of the electric field at surface plasmon-polariton resonance in a silver nanowire excited by a plane electromagnetic wave. The angle between the nanowire and the wavevector of the incident wave is 30° degree. The wavevector and **E** vector of the incident radiation are in the plane of the figure; the needle length is 480 nm.

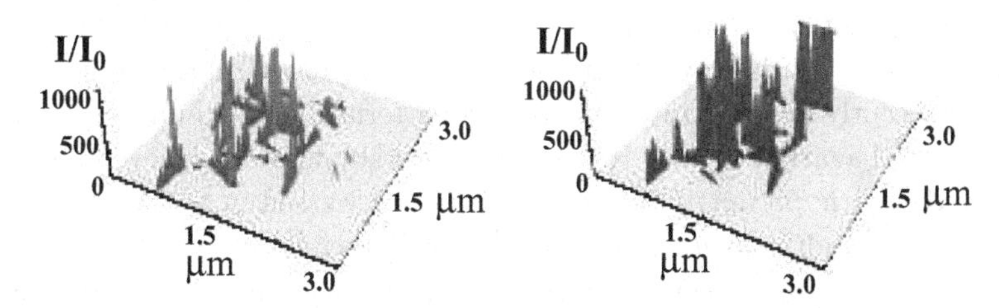

Fig. 2.9 Field distribution in nanowire percolation silver composite for the incident wavelength $\lambda = 550$ nm (left) and $\lambda = 750$ nm (right). In both figures, the normal incidence is implied (field $(E||x)$ is parallel to the film surface).

We simulate such composites by distributing randomly identical metal nanowires over a dielectric surface. In these simulations, the length of an individual nanowire is $2a = 480$ nm, while its diameter is 30 nm. Figure 2.9 shows the intensity $|E|^2$ of the local electric field at wavelengths $\lambda = 540$ and 750 nm. Our simulations emphasize the existence of localized plasmon modes in the composites. Similar to localization of quasistatic plasmon modes discussed in the next chapter, the localization of plasmon-polaritons bounded in the metal nanowires leads to large enhancement of local optical fields. The simulations also suggest that the local intensity enhancement factor can reach up to 10^3. Our simulations also show that plasmon modes cover a broad spectral range. The incident field at a given wavelength excites small resonant parts of the percolation system resulting in large enhancement of the local fields in these elements. In our case, such resonating elements can be single nanowires or groups of nanowires. Different clusters of wires resonate at different frequencies and all together they cover a broad spectral range where the stick composites have plasmon modes.

2.5 Optical Magnetism, Left-handed Optical Materials and Superresolution

Natural magnetism is not possible in optics. Relaxation times of paramagnetic and ferromagnetic processes are long in comparison with the optical period and collective magnetic responses become small at large frequencies. With no collective effects, the magnetic susceptibility, proportional to $\sim v^2/c^2 \sim \beta^2 \sim 10^{-4}$, where v is the velocity of electron in atom, is very small ($\beta = e^2/\hbar c \cong 1/137$ [Landau *et al.*, 1984]). To extend the range of electromagnetic properties of materials, the development of artificial (or meta-) materials has been invoked and promoted. For example, it has been recently demonstrated that meta-materials may exhibit artificial magnetism ([Lagarkov *et al.*, 1992] and [Lagarkov and Sarychev, 1996]), negative dielectric permittivity $\varepsilon < 0$ as it was discussed in Sec. 2.3.2 (see, for example [Pendry *et al.*, 1996] and [Lagarkov *et al.*, 1992], [Lagarkov and Sarychev, 1996]); negative magnetic permeability $\mu < 0$ ([Lagarkov *et al.*, 1997b], [Pendry *et al.*, 1999], [Wiltshire *et al.*, 2003]); end even both $\varepsilon, \mu < 0$ ([Veselago, 1968] [Pendry, 2000]).

In a paper published in 1964 [Veselago, 1968], Victor Veselago predicted that electromagnetic plane waves in a medium having simultaneously negative permittivity and permeability would propagate in a direction opposite to that of the flow of energy. This result follows not from the wave equation, which remains unchanged in the absence of sources, but rather from the individual Maxwell curl equations. The Maxwell equation $\text{curl}\,\mathbf{E} = i\,(\omega/c)\,\mathbf{B}$ for the electric field provides an unambiguous "right-hand" rule between the directions of the electric field $\mathbf{E}$, the magnetic induction $\mathbf{B}$, and the direction of the propagation vector $\mathbf{k}$, namely $(\mathbf{k}\cdot[\mathbf{E}\times\mathbf{B}]) = (\omega/c)\,B^2 > 0$. The direction of energy flow, however, is given by $\mathbf{S} = c\,[\mathbf{E}\times\mathbf{H}]\,/\,(4\pi)$. Where the permeability μ is negative, the direction of propagation $\mathbf{k}$ is reversed with respect to the direction of energy flow $\mathbf{S}$, the vectors $\mathbf{E}$, $\mathbf{H}$, and $\mathbf{k}$ forming a left-handed system; thus such material is referred as "left handed materials" (LHM).

It follows from the energy conservation that LHM, where both permittivity $\varepsilon < 0$ and permeability $\mu < 0$, have a negative refractive index n. While there are many examples of systems that can exhibit reversal of phase and group velocities the designation of negative refractive index is unique to LHMs. An isotropic negative index leads to the important property of LHM that it exactly reverses the propagation paths of rays; thus, LHM have the potential to form highly efficient low reflectance surfaces by exactly cancelling the scattering properties of other materials. Detailed consideration of negative refraction index n is given by [Smith and Kroll, 2000]. The negative refraction in optics was convincingly demonstrated by Lezec, Dionne, and Atwater in April 2007 using the plasmonic waveguides [Lezec *et al.*, 2007]

LHM with negative refractive index will focus light even when it is in the form of a parallel-sided slab of material [Veselago, 1968]. The focusing action of such a slab is sketched in Fig. 2.10, assuming that the slab permittivity ε and permeability μ are both equal to $\varepsilon = \mu = -1$, so that the refractive index $n = \sqrt{\varepsilon\mu} = -1$. The rays obey Snell's laws of refraction at the surface because light inside the medium makes a negative angle with the surface normal. A simple ray diagram reveals the double focusing effect. The double-negative case of $\varepsilon < 0$ and $\mu < 0$ (often referred as negative refractive regime) is particularly interesting because of the possibility of making a "perfect" lens with sub-wavelength spatial resolution [Pendry, 2000]. To illustrate the idea of such superlens suppose that the source of

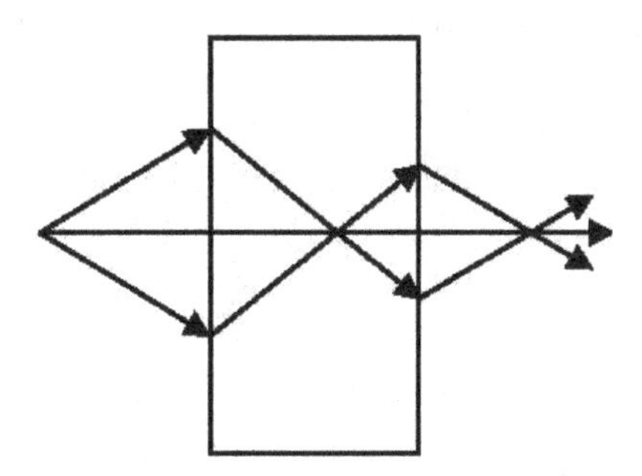

Fig. 2.10 Passage of rays of light through a plate of thickness d made of a left-handed material. The source is paced at $z = -h - d/2$. Negative refractive index bends light to a negative angle with the surface normal. Light formerly diverging from a point source is set in reverse and converges back to a point at $z = h - d/2$ inside the slab; released from the medium the light reaches second focus at $z = 3d/2 - h$ in front of the superlens.

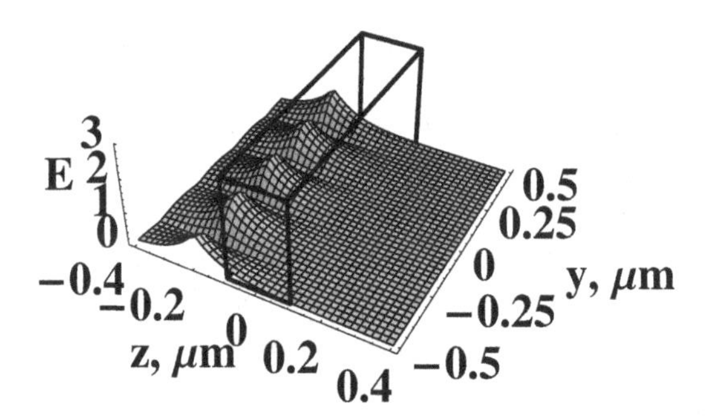

Fig. 2.11 Absolute value of electric field in vicinity of object; wavelength $\lambda = 1.5$ μm, electric field in the object plane $E\,(z = -0.2\ \mu m) = \cos\,(x/a)$, $a = 0.5\ \mu m < \lambda$; the field exponentially decays with increasing distance from the object.

the electric field

$$\mathbf{E}_0 = \{0, E_0\,\cos(q\,x), 0\}, \tag{2.74}$$

where $q > k$ ($k = \omega/c$), is located in the plane $\{x, y, -h - d/2\}$ at the distance $h < d$ behind the slab (see Fig. 2.11). We assume first that the permittivity and permeability of the slab coincide with the surrounding space, i.e., $\varepsilon = \mu = 1$. Then em field exponentially decays with increasing the distance from the object as it is shown in Fig. 2.11. We cannot see the details of the structure even at the distance of 100 nm from the object.

On the next figure we show the electric field in the presence of the superlens that is LHM slab, where $\varepsilon = \mu = -1$. The field still equals to $\mathbf{E}_0 = \{0, E_0\,\cos(q\,x), 0\}$ in the object plane. Electric field still decays around the object, however the LHM increases the field and creates the images in the focus planes.

The Maxwell equations take form curl $\mathbf{E} = ik\mathbf{H}$ and curl $\mathbf{H} = -ik\mathbf{H}$ outside LHM and curl $\mathbf{E} = -ik\mathbf{H}$ and curl $\mathbf{H} = ik\mathbf{H}$ inside the slab. We can check by direct substitution that the electric field

$$E_y\,(z) = e^{\kappa(z+d/2+h)}\,E_0\,\cos(q\,x), \qquad z \leq -h - \frac{d}{2}, \tag{2.75a}$$

$$E_y\,(z) = e^{-\kappa(z+d/2+h)}\,E_0\,\cos(q\,x), \qquad -h - \frac{d}{2} \leq z \leq -\frac{d}{2}, \tag{2.75b}$$

$$E_y\,(z) = e^{\kappa(z+d/2-h)}\,E_0\,\cos(q\,x), \qquad -\frac{d}{2} \leq z \leq \frac{d}{2}, \tag{2.75c}$$

$$E_y\,(z) = e^{-\kappa(z-3d/2+h)}\,E_0\,\cos(q\,x), \qquad z \leq \frac{d}{2}, \tag{2.75d}$$

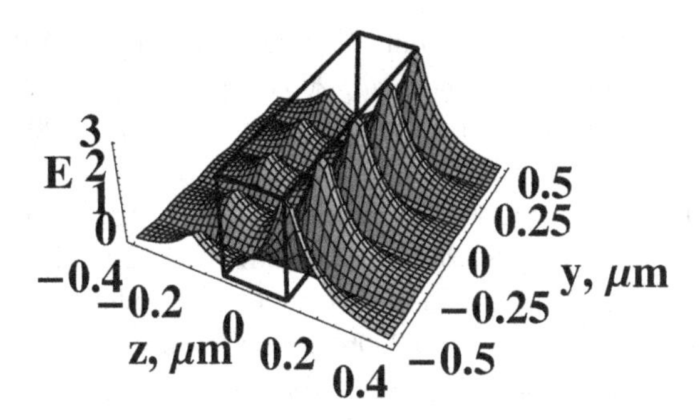

Fig. 2.12 Absolute value of electric field; electric field in the object plane $E\,(z=-0.2\ \mu m) = \cos(x/a)$, $a = 0.5\ \mu m$, wavelength $\lambda = 1.5\ \mu m$; superlens ($\varepsilon = \mu = -1$) with thickness $d = 0.2\ \mu m$ is placed between $z = -0.1\ \mu m$ and $z = 0.1\ \mu m$ before the source; the superlens "increases" evanescent modes and makes the object visible; superlens gives perfect image in the plane $z = 0$ inside LHM and another perfect image in the plane $z = 0.2\ \mu m$ in front of the superlens.

where $\kappa = \sqrt{q^2 - k^2}$, satisfies the Maxwell equations. The electric field equals to the object field in the image plane $z = -d/2 + h$ inside the slab, where the rays first intersect in Fig. 2.10. Electric field again equals to the object field in the outside image plane $z = 3d/2 - h$, which corresponds to the second ray intersection in Fig. 2.10. Therefore the superlens completely restores the source field in the image planes. Any spatial object like a point source, two subwavelength sources, virus, DNA, etc., can be expanded in Fourier spectrum, each Fourier component completely restores in the image planes. The image reproduces the object without any distortion and loss of any information. The superlens gives the super resolution and in this sense it is perfect. The further development of the superlens includes, for example, cylindrical and spherical superlenses that give the magnification of the object [Pendry and Ramakrishna, 2002], [Pendry, 2003], [Smith and Schurig, 2003] and employment of gain medium for compensation of the loss in metal [Sarychev and Tartakovsky, 2006], [Sarychev and Tartakovsky, 2007]. The first far field image of the subwavelength object was obtained by Liu, Lee, Xiong, Sun, and Zhang in March 2007 [Liu *et al.*, 2007] using idea of hyper lens by [Jacob *et al.*, 2006], [Salandrino and Engheta, 2006], and [Govyadinov and Podolskiy, 2006].

The super resolution and other unusual and sometimes counter-intuitive properties of LHMs make them prime candidates for applications in resonators, waveguides, in other microwave and optical elements [Podolskiy *et al.*, 2003], [Houck *et al.*, 2003], [Parazzoli *et al.*, 2003], and

[Alu and Engheta, 2004]. For recent reference on left-handed materials (LHMs) see, for example, [Aydin *et al.*, 2004], [O'Brien *et al.*, 2004], [Depine and Lakhtakia, 2004], [Foteinopoulou and Soukoulis, 2003] [Koschny *et al.*, 2004], [Lagarkov and Kissel, 2004], [Koschny *et al.*, 2004], [Pendry and Smith, 2004], [Pendry, 2004a], [Pokrovsky and Efros, 2003], [Pendry, 2004b], [Sarychev and Shalaev, 2004], [Sarychev *et al.*, 2003], [Smith *et al.*, 2004a], [Smith *et al.*, 2004b], [Linde *et al.*, 2004], [Shalaev *et al.*, 2005a], [Zhang *et al.*, 2005], [Shalaev *et al.*, 2006], and references therein. However, this field develops so fast that it is literally impossible to suggest a complete and fully applicable list of references so that our choice is rather subjective.

Negative refraction has been convincingly demonstrated in the microwave regime [Smith *et al.*, 2000], [Houck *et al.*, 2003], [Parazzoli *et al.*, 2003], and [Lagarkov and Kissel, 2004]. For microwave LHMs , artificial magnetic elements (providing $\mu < 0$) are the split-ring resonators or the "swiss roll" structures. In the microwave part of the spectrum, metals can be considered as the perfect conductors since the characteristic skin depth is much smaller than the characteristic size of metal features. The strong magnetic response is achieved by operating in the vicinity of the LC- resonance of the split ring [Pendry *et al.*, 1999], [Wiltshire *et al.*, 2003]. The same technique of obtaining negative magnetic permittivity using was recently extended to mid-IR [Linde *et al.*, 2004], [Katsarakis *et al.*, 2004] by scaling down the dimensions of the split rings. In the mid-IR part of the spectrum, metals can also be approximated as perfect conductors as in the case of the microwave frequencies, though, to a lesser degree. Therefore, the frequencies of the LC resonances are determined entirely by the split ring geometry and by an overall size but not by the electromagnetic properties of the metal. In accordance with this, the ring response is resonantly enhanced at some ratio of the radiation wavelength and the characteristic structure size. This allows us to refer to the LC resonances of perfectly conducting metals as geometric LC (GLC) resonances.

The situation drastically changes in the optical part of the spectrum, where thin (sub-wavelength) metal features behave very differently as the nanostructure sizes become less than the skin depth. For example, the *electrical* surface plasmon resonance (SPR) occurs in the optical and near-IR parts of the spectrum due to collective electron oscillations in metal structures. Many important plasmon-enhanced optical phenomena and applications of metal nano-composites are based on the electrical SPR (see, for example, [Sarychev and Shalaev, 2000]). The plasmonic nature of the electromagnetic response in metals for optical/mid-IR frequencies is the main reason why the original methodology of GLC resonances in microwave/mid-IR spectral range is not extendable to larger frequencies.

For the optical range, LHMs with a negative refractive-index have been for the first time demonstrated experimentally in by [Zhang *et al.*, 2005] and in Refs. [Shalaev *et al.*, 2005b], [Shalaev *et al.*, 2005a], and [Shalaev *et al.*, 2006], where the authors have observed the real part of the refractive index $n = -0.3$ at $\lambda = 1.5\,\mu m$. In their report, the authors experimentally verified an early theoretical prediction for negative refraction in an array of parallel metal nanosticks [Sarychev and Shalaev, 2003], [Podolskiy *et al.*, 2002a]. However, the loss become progressively important with increasing frequency towards the optical band. Moreover, the elementary cell of resulting structure is on the order of the wavelength. Making a true LHM requires the cell size to be less than $\lambda/2$. Therefore, miniaturization of the cell is of a great importance, and can be accomplished, for example, by utilizing plasmonic effects.

It is also necessary to take into account that dielectric permittivity $\varepsilon = \varepsilon' + i\varepsilon''$ and magnetic permeability $\mu = \mu' + i\mu''$ are always complex values. As an example, the structure that exhibits the negative real dielectric permittivity $\varepsilon' \approx -0.7$ and negative real of magnetic permeability $\mu' \approx -0.3$ at $\lambda \approx 0.5\,\mu\text{m}$ has been investigated in details in Ref. [Grigorenko *et al.*, 2005]. However, rather large resonance imaginary components ($\mu'' \approx 1.0$) have not allowed the observation of the negative refraction.

Thus the demonstration of a negative index meta-material in the regime where plasmonic effects are important remains elusive. Plasmonic effects must be correctly accounted for while designing a meta-material with optical magnetism. Below we show that specially arranged metal nanoparticles can support, along with the electrical SPR, *magnetic plasmon resonance* (*MPR*). The *MPR*'s resonance frequency ω_r can be independent of the absolute characteristic structure size a and $\lambda/a \equiv 2\pi c/\omega a$. The only defining parameters are the plasmonic permittivity $\varepsilon_m(\omega)$ and the structure geometry. Such structures act as optical nanoantennas at the optical frequencies by concentrating large electric and magnetic energies on the nanoscale. The magnetic response is characterized by the magnetic polarizability α_M with the resonant behavior similar to the electric SPR polarizability α_E: The real part of α_M changes the sign near the resonance and becomes negative for $\omega > \omega_r$, as required for the negative index meta-materials. Later in this chapter, we will show that the electrostatic resonances must replace (or strongly modify) *GLC* resonances in the optical/mid-IR range once a strong magnetic response is desired.

The idea of using electrostatic resonances for inducing optical magnetism is relatively recent. For example, electrostatic resonances of periodic plasmonic nanostructures have been recently employed to induce magnetic properties due to close proximity of adjacent nanowires [Shvets, 2003] , [Shvets and Urzhumov, 2004]. Higher multipole electrostatic resonances were shown to hybridize in such a way as to induce magnetic moments in individual nanowires [Shvets and Urzhumov, 2005]. Strong electrostatic resonances of regularly shaped nanoparticles (including nanospheres and nanowires) occur in an interval of $-2 < \varepsilon_m' < -1$. The resistive damping characterized by the ratio $\varepsilon_m''/\varepsilon_m'$ (ε_m is the dielectric permittivity of a metal) is fairly strong for the frequencies corresponding to $|\varepsilon_m'| \sim 1$. However, $\varepsilon_m''/\varepsilon_m'$ is known to decrease for $|\varepsilon_m| \gg 1$. Therefore, there is a considerable incentive to design nanostructures, which exhibit such resonances that $\varepsilon_m' \ll -1$. Such horseshoe-shaped structures, first suggested in Ref. [Sarychev and Shalaev, 2004] are described below. Spectrally, these structures support strong magnetic moments at the frequencies larger than the microwave/mid-IR frequencies supported by the traditional split-ring resonators (see [Pendry *et al.*, 1999], [Wiltshire *et al.*, 2003], [Linde *et al.*, 2004], and [Katsarakis *et al.*, 2004] for details). Conceptually, the horseshoe-shaped structures described here are distinct from the earlier attempts of low-frequency structures because the former are not relying on the *GLC* resonance for producing a strong magnetic response. Again, the plasmonic properties of the metal become very important here: when the sizes are small and the operational frequencies are high.

Below we present a three-dimensional theory of the magnetic resonance in a plasmonic structure and related two-dimensional numerical simulations. Specifically, we will develop a comprehensive theory and perform numerical simulations for negative index meta-materials based on horseshoe-shaped structures. The possibility of optical magnetism in such structures was first theoretically predicted in [Sarychev and Shalaev, 2004] and recently experimentally verified by [Enkrich *et al.*, 2005] and [Dolling *et al.*, 2005]. Closely related split-ring resonator structures were also shown to possess optical magnetism [Yen *et al.*, 2004]. We will also demonstrate that such horseshoe-shaped structures may have negative dielectric permittivity, in parallel with negative magnetic permeability, and thus can be used for building a metamaterial with a negative-refractive index.

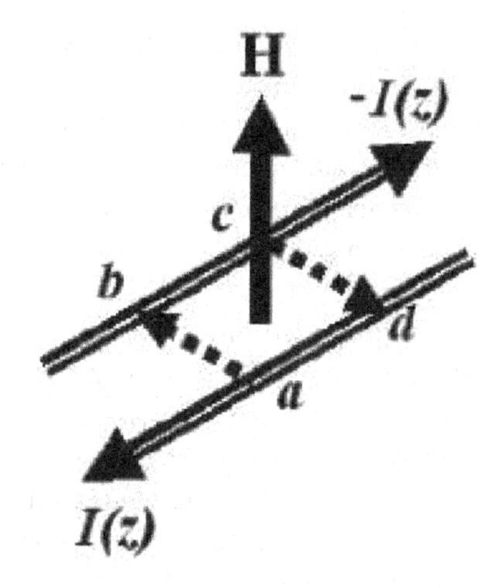

Fig. 2.13 Currents in two parallel metal wires excited by external magnetic field **H**. Displacement currents, "closing"the circuit, are shown by dashed lines.

2.5.1 *Analytical theory of magnetic plasmon resonances*

First, we consider a pair of parallel metallic rods. The external magnetic field excites the electric current in the pair of the rods as shown in Fig. 2.13. The magnetic moment associated with the circular current flowing in the rods results in the magnetic response of the system. Suppose that an external magnetic field $\mathbf{H} = \{0,\ H_0 \exp(-i\omega t),\ 0\}$ is applied perpendicular to the plane of the pair (We suppose that magnetic field is along y axis and the rods are in $\{x, z\}$ plane). The circular current $I(z)$ excited by the time-varying magnetic field flows in the opposite directions in the nanowires as shown in Fig. 2.13. The displacement currents flowing between the nanowires close the circuit. We then introduce the "potential drop" $U(z) = \int_a^b \mathbf{E} d\mathbf{l}$ between the pair where the integration is performed along the line $\{a(z), b(z)\}$. To find the current $I(z)$, we integrate the Faraday's Law $\text{curl}\,\mathbf{E} = ik(\mathbf{H}_0 + \mathbf{H}_{in})$ over the contour $\{a, b, c, d\}$ in Fig. 2.13, where $k = \omega/c$ is the wave vector, and $\mathbf{H}_{in} = \text{curl}\,\mathbf{A}$ is the magnetic field induced by the current. In our derivations, we assume that the nanowire length $2a$ is much larger than the distance d between the nanowires and the radius b of a nanowire. We also assume that $kd \ll 1$. Under these assumptions, the vector potential $\mathbf{A}$ is primarily directed along the nanowires (z direction). Then the integral form of the Faraday's Law yields

$$(IR - i2kA_z + dU/dz)\Delta z = -ikH_0 d\Delta z, \tag{2.76}$$

where $R \simeq 2/(\sigma\pi b^2) \approx 8i/(\varepsilon b^2\omega)$ is the pair impedance, $\varepsilon = i4\pi\sigma/\omega \ll -1$ is the metal complex permittivity, and $\pm IR/2$ are the electric fields on the surface of the nanowires. Note that the wire resistivity is explicitly taken into account. This distinguishes our calculations from earlier works on the resonances of conducting split-ring resonators [Marques *et al.*, 2002] and conducting stick composites [Lagarkov and Sarychev, 1996] because the plasmonic wire resonances are fully accounted in our consideration.

Electric field $\mathbf{E}$ can be presented in terms of the vector potential $\mathbf{A}$ and electric potential ϕ, $\mathbf{E} = -\nabla\phi + ik\mathbf{A}$. For the standard Lorentz gauge, the electric potential ϕ equals to $\phi(\mathbf{r}_1) = \int \exp(ikr_{12})\, q(\mathbf{r}_2)/r_{12}\, d\mathbf{r}_2$ while the vector potential is $\mathbf{A}(\mathbf{r}_1) = c^{-1}\int \exp(ikr_{12})\, \mathbf{j}(\mathbf{r}_2)/r_{12}\, d\mathbf{r}_2$, where $r_{12} = |\mathbf{r}_1 - \mathbf{r}_2|$, q and $\mathbf{j}$ are the corresponding charge and the current density. In the case of the two long wires, the currents flow inside the wires. Correspondingly, the vector potential $\mathbf{A}$ has the only component: in "z"-direction: $\mathbf{A} = \{0, 0, A_z\}$. Since the vector potential $\mathbf{A}$ is perpendicular to the line $\{a(z), b(z)\}$, the potential drop U in Eq. (2.76) equals to $U(z) = \int_a^b \mathbf{E}d\mathbf{l} = \phi_a - \phi_b$, where ϕ_a and ϕ_b are the electric potentials at the points $a(z)$ and $b(z)$, respectively. Now, consider the excitation of an antisymmetric mode when the currents in the wires are the same but run in the opposite directions (see Fig. 2.13), resulting in the electric charges per unite length being $Q(z) = Q_a = -Q_b$. If we assume that the diameter b of the wire is much smaller than the inter-wire distance d, and the wire length is much bigger: $2a \gg d$, then the potential drop $U(z)$ between the pair can be written as $U(z) = Q(z)/C$, where the interwire capacitance C is independent of the coordinate z and is estimated in Sec. 2.5.3 ($C \simeq [4\ln(d/b)]^{-1}$).

The vector potential $A_z(z)$ is proportional to the electric current $A_z(z) = (L/c)\, I(z)/2$, where the wire pair inductance is estimated as $L \simeq 4\ln(d/b)$ (see Sec. 2.5.3) Note that both C and L are purely geometric factors that do not depend on the plasmonic nature of the metal rods. The entire $LC-$ product can be estimated as $LC \simeq 1$. Yet, in the above consideration, we have never assumed that the wires were made of the perfect metal or from a metal with real conductivity (i.e., imaginary permittivity). Moreover, the nanowire pair (two nanoantennas) has most interesting behavior when metal dielectric constant is real and negative. The reason is the plasmonic nature of the metal, which is accounted below.

First, we substitute $U(z)$ and $A_z(z)$ into Eq. (2.76), taking into account the charge conservation law $dI/dz = i\omega Q(z)$, and obtain a differential equation for the current:

$$\frac{d^2 I(z)}{d z^2} = -g^2 I(z) - \frac{C d\omega^2}{c} H_0, \tag{2.77}$$

where $-a < z < a$, $I(-a) = I(a) = 0$, and the parameter g is given by

$$g^2 = k^2 \left[LC - 8C \left((kb)^2 \varepsilon_m \right)^{-1} \right]. \tag{2.78}$$

The two-wire antenna is resonantly excited when $G = ga = N\pi/2$, where N is an integer number. Note that the material properties of the metal enter the resonant parameter G through the dielectric permittivity ε_m. In the context of the wire pair, the earlier discussed GLC resonances [Lagarkov and Sarychev, 1996], [Pendry *et al.*, 1999], [Podolskiy *et al.*, 2002a], [Sarychev and Shalaev, 2003], [Marques *et al.*, 2002], [Panina *et al.*, 2002], [Linde *et al.*, 2004] correspond to the wire thickness,which is much larger than the characteristic skin depth $(k^2 |\varepsilon_m|)^{-1/2}$. This approximation, typically valid for microwave and mid-IR frequencies, yields $g = k/\sqrt{LC} \cong k$ and the resonant condition $ka = \pi/2$, which is also known as the antenna resonance.

Let us now consider the opposite ("electrostatic", as explained below) limit of $\left| 8C \left[(kb)^2 \varepsilon_m \right]^{-1} \right| \gg 1$. In the electrostatic regime, G depends only on the metal permittivity and the aspect ratio:

$$G^2 \simeq -2(a/b)^2 \ln(d/b)/\varepsilon_m\,. \tag{2.79}$$

Note it does not depend on the wavelength and absolute wire length. Sharp resonance in Eq. (2.77) requires that G^2 be positive, possibly with a very small imaginary part. Indeed, ε_m is negative (with a smaller imaginary part) for IR/visible frequencies and typical low-loss metals (Ag, Au, etc.) Metal dielectric constant ε_m can be approximated by the Drude formula:

$$\varepsilon_m(\omega) \cong \varepsilon_b - (\omega_p/\omega)^2 / (1 - i\omega_\tau/\omega), \tag{2.80}$$

where ε_b is a "polarization" constant, ω_p is the plasma frequency, and $\omega_\tau = 1/\tau$ is the relaxation rate. For considered here silver nanoantennas, $\varepsilon_b \approx 5$, $\omega_p \approx 9.1$ and $\omega_\tau \approx 0.02$ [Johnson and Christy, 1972], [Kreibig and Volmer, 1995]. For example, the silver dielectric constant estimates as $\varepsilon'_m \approx -120$ with $\varepsilon''_m / |\varepsilon_m| \approx 0.025$ at $\lambda = 1.5\,\mu m$.

We consider now the electric field in the system of two conducting rods, still assuming the electrostatic limit when the propagation constant G is given by Eq. (2.79). The electric charge $Q(z)$ and the current $I(z)$ ($Q(z) = (i\omega)^{-1} dI(z)/dz$) are given by the solution to Eq. (2.77)

$$Q(z) = Q_0 \frac{\sin(Gz/a)}{\cos G}, \tag{2.81}$$

$$I(z) = i\frac{Q_0 a\omega}{G}\left(1 - \frac{\cos(Gz/a)}{\cos G}\right), \tag{2.82}$$

where $Q_0 = ib\,d\,k\,H_0\sqrt{-\varepsilon_m}/\left[4\sqrt{2}\ln^{3/2}(d/b)\right]$. Using the Lorentz's gauge, we can rewrite the equation for the electric potential

$$\phi(r) = \int q(\mathbf{r}_1)\frac{\exp(ikR_1)}{R_1}d\mathbf{r}_1 - \int q(\mathbf{r}_2)\frac{\exp(ikR_2)}{R_2}d\mathbf{r}_2, \tag{2.83}$$

where $q(r_1)$ and $q(r_2)$ are electric charges distributed over the surface of the rods 1 and 2, $R_1 = |\mathbf{r} - \mathbf{r}_1|$, $R_2 = |\mathbf{r} - \mathbf{r}_2|$, and the integration is performed over the both rods (1 and 2). We then consider the electric field between the rods, i.e., in $\{z, x\}$ plane (see Fig. 2.13), and assume that $|x| \ll a$, $|z| < a$, while the distances to the rods $d_1 = |x - d/2| \gg b$ and $d_2 = |x + d/2| \gg b$. The subsequent integration of Eq. (2.83) over the wire's cross-section of the rod provides an one-dimension form

$$\phi(x, z) = \int_{-a}^{a} Q(z_1)\left[\frac{\exp(ikR_1)}{R_1} - \frac{\exp(ikR_2)}{R_2}\right]dz_1, \tag{2.84}$$

where the linear charge density $Q(z_1)$ is obtained from $q(r_1)$ by the integration over the rod circumference and is given by Eq. (2.81) [$Q(z_1) = -Q(z_2)$], while the distances R_1 and R_2 take the forms $R_1 = \sqrt{d_1^2 + (z - z_1)^2}$ and $R_2 = \sqrt{d_2^2 + (z - z_1)^2}$. Two terms in the square brackets in Eq. (2.84) cancel when $|z - z_1| > d_1, d_2$. Since we assume that $kd \ll 1$ and $d \ll a$, we can approximate the exponents $\exp(ikR_1) \simeq \exp(ikR_1) \simeq 1$ and extend the integration in Eq. (2.84) from $z_1 = -\infty$ to $z_1 = \infty$. A resulting integral is solved explicitly and we obtain the analytical equation for the electric potential in the system of two nanowires

$$\phi(x, z) = 2\ln\left|\frac{d/2 + x}{d/2 - x}\right| Q(z), \tag{2.85}$$

where $Q(z)$ is given by Eq. (2.81).

An extrapolation of this result to the surface of the wires gives the potential drop $U(z) = \phi(d/2 - b, z) - \phi(-d/2 + b, z) = 4\ln(2d/b)Q(z)$. Thus, we obtain that the inter-wire capacitance $C = Q(z)/U(z) = 4\ln$

(d/b) is indeed a constant and is independent of the coordinate z, which fully agrees with the estimate presented below in Sec. 2.5.3.

The vector potential $\mathbf{A} = \{0, 0, A\}$ is calculated in a similar way

$$A(x,z) \simeq \frac{1}{c}\int_{-a}^{a} I(z_1)\left[\frac{\exp(ikR_1)}{R_1} - \frac{\exp(ikR_2)}{R_2}\right] dz_1$$
$$\simeq \frac{1}{c}\int_{-\infty}^{\infty} I(z_1)\left[\frac{1}{R_1} - \frac{1}{R_2}\right] dz_1 \simeq \frac{2}{c}\ln\left|\frac{d/2+x}{d/2-x}\right| I(z), \quad (2.86)$$

where the electric current is given by Eq. (2.82). Extrapolating to the vector potential A to the surface of the first wire ($x = d/2 - b$) we obtain $2cA = LI$, where the inter-wire inductance $L \simeq 4\ln(d/b)$. Note the inductance L is also independent on the coordinate z. Since the inter-wire capacitance C and inductance L both remain constant along the wire direction, the Maxwell equation reduces to an ordinary differential equation (2.77).

The electric field $\mathbf{E} = -\boldsymbol{\nabla}\phi + ik\mathbf{A}$ is calculated from the potentials (2.85) and (2.86) as

$$E_x = -\frac{2Q_0\, d}{(d/2)^2 - x^2}\sin\left(\frac{G\,z}{a}\right)\sec(G), \quad (2.87)$$

$$E_z = -\frac{2Q_0}{a\,G}\ln\left|\frac{d/2+x}{d/2-x}\right|$$
$$\times\left[G^2\cos\left(\frac{G\,z}{a}\right)\sec(G) - a^2k^2\left(1 - \cos\left(\frac{G\,z}{a}\right)\sec(G)\right)\right], \quad (2.88)$$

where we still assume that $|x| \ll a$, $|z| < a$, $|x - d/2| \gg b$ and $|x + d/2| \gg b$. The transverse electric field E_x changes its sign with the coordinate z vanishing at $z = 0$. Yet, if the ratio $|E_x| / |E_z|$ is estimated near the resonance ($G \approx \pi/2$), the average $|E_x| / |E_z| \sim a/d \gg 1$, indicating that the transversal electric field is much larger than the longitudinal field at *MPR*. In the wire proximity, then transverse field E_x increases even more: $|E_x| \sim Q_0/b$. The potential drop ΔU between the points $x_1 = d/2 + l/2$ and $x_2 = d/2 - l/2$ ($2b < l \ll d$) equals to $-2lQ(z)/d$ as it follows from Eq. (2.88). The corresponding electric field $E_{out} \simeq -2Q(z)/d$. This field is the external field for the wire at the coordinate $y = d/2$. The internal transverse potential drop across the wire is then $|U_{in}| \simeq b|E_{out}| / |\varepsilon_m| \simeq 2|Q(z)|b/(d|\varepsilon_m|)$, where metal dielectric constant is assumed to be large $|\varepsilon_m| \gg 1$. The problem of calculating the internal transverse field closely resembles the classical problem of finding the field induced in a dielectric cylinder by another charged cylinder placed parallel, with an elegant solution to be found in Ref. [Landau *et al.*, 1984], Sec.7. In the discussed case, $|\varepsilon_m|$, $d/b \gg 1$ so that we get the above obtained estimate for U_{in}. The ratio of the potential drop U_{in} across a wire to the potential drop $U(z)$ between the wires is then equal to

$$\left|\frac{U_{in}(z)}{U(z)}\right| \simeq \frac{b}{d\,|\varepsilon_m|\ln(d/b)} \ll 1. \tag{2.89}$$

For any practical purpose we can neglect U_{in} in comparison with U, which allows us to reduce the problem of charge and current distribution in the two-wire system to the ordinary differential equation (2.77) for the electric current $I(z)$. Condition (2.89) is important for the developed analytical theory of *MPR* in the system of two thin rods. However, we can envision a system (for example, two closely packed metal nanowires or hemispheres), which reveals the *MPR* but condition (2.89) is not fulfilled.

To clarify the nature of the resonance, it is instructive to compute the ratio of the electric and magnetic resonance energies:

$$\frac{\mathcal{E}_E}{\mathcal{E}_M} \sim c^2 \frac{C^{-1}\int |Q(z)|^2\,dz}{L\int |I(z)|^2\,dz} \approx \frac{g^2}{k^2} \approx 1 - \frac{2}{\ln(d/b)k^2b^2\,|\varepsilon_m|}, \tag{2.90}$$

where we have assumed that the spatial frequency g, given by Eq. (2.78), is close to the resonance ($Ga \approx \pi/2$) and used the expressions for the capacitance $C \simeq [4\ln(d/b)]^{-1}$ and the inductance $L \simeq 4\ln(d/b)$ derived in Sec. 2.5.3. In the electrostatic limit $\left|8C\left[(kb)^2\varepsilon_m\right]^{-1}\right| \gg 1$, we obtain $U_E/U_M \gg 1$ justifying the name given to the regime. Because of the symmetry of the electric potential problem, it is clear that such polarization cannot be induced by an uniform electric field. Therefore, the discussed resonance can be classified as the dark mode [Stockman *et al.*, 2001].

We would like to emphasize that electric fields is not potential in that type of the nanoantenna. It is taken for granted that the magnetic field is a solenoidal field. What is more surprising the electric field is also solenoidal in the two stick resonator despite the fact that its size is much smaller than the wavelength. Indeed, it follows from Eqs. (2.87) and (2.88) the electric field E_x depends on the coordinate "z" and electric field E_z depends on the coordinate "x". The subwavelength solenoidal electric field is the essence of *MPR*.

The electric current $I(z)$ is found from Eq. (2.77) and is used to calculate the magnetic moment of the wire pair: $\mathbf{m} = (2c)^{-1}\int [\mathbf{r}\times\mathbf{j}(\mathbf{r})]\,d\mathbf{r}$, where $\mathbf{j}(\mathbf{r})$ is the current density. After the integration is performed over the both nanowires, we obtain

$$m = \frac{1}{2} H_0 a^3 \ln(d/b)\, (kd)^2\, \frac{\tan G - G}{G^3}. \tag{2.91}$$

The metal permittivity ε_m has a large negative value in the optical/near-IR range while its imaginary part is small; therefore, the magnetic moment m has a resonance at $G \approx \pi/2$ (see Eq. 2.78) when m attains large values. For a typical metal we can use the Drude formula (2.80) for ε_m, where the relaxation parameter is small $\omega_\tau/\omega \ll 1$. Then the normalized magnetic polarizability α_M near the *MPR* has the following form:

$$\alpha_M = \frac{4\pi m}{H_0 V} = \frac{16\, a\, d\, \omega_p}{\lambda^2\, \omega_r \sqrt{2\ln(d/b)}}\, \frac{1}{1 - \omega/\omega_r - i\omega_\tau/(2\omega_r)}, \tag{2.92}$$

where $V = 4abd$ is the volume, and $\omega_r = b\,\pi\,\omega_p\sqrt{2\ln(d/b)}/(4a)$ is *MPR* frequency. Note that the magnetic polarizability α_M contains a pre-factor $da/\lambda^2 \ll 1$ that is small for sub-wavelength nanostructures. However, near the resonance $G = \pi/2$, the enhancement factor can be very large for optical and infrared frequencies because of the high quality of the plasmon resonances for $\omega_r \ll \omega_p$. Therefore, the total pre-factor in Eq. (2.92) can be of the order of one, thereby enabling the excitation of a strong *MPR*.

Although the electric field near MPR is high, it is primarily concentrated in the direction perpendicular to the wires as it was discussed after Eqs. (2.87) and (2.88). If the wavevector of the propagating field is in the wire plane but is perpendicular to the wire length, then the *MPR* described above does not strongly affect the electric field component, which is parallel to the wires. The integral from the electric field, generated between the wires at the magnetic resonance condition, is an exact zero as it follows from Eqs. (2.87) and (2.88). Envisioning such a composite material that consists of similar wire pairs, one can expect that the magnetic plasmon resonance will not contribute to the dielectric permittivity in the direction parallel to the wires. Such a medium is therefore not bi-anisotropic as pointed out in Ref. [Marques *et al.*, 2002] and can be described by the two separate effective parameters: ε and μ.

Consider a metal nanoantenna of a horseshoe shape, obtained from a pair of nanowires by shorting their ends (see Fig. 2.14). When the quasi-static condition $\left|8C\left[(kb)^2\varepsilon_m\right]^{-1}\right| \gg 1$ holds in the horseshoe nanoantenna, the electric current $I(z)$ can be obtained from Eq. (2.77), where the boundary condition changes to $I_{z=a} = (dI/dz)_{z=0} = 0$ and, as above, $a \gg d \gg b$. It is easy to check that the magnetic polarizability α_M is still given by

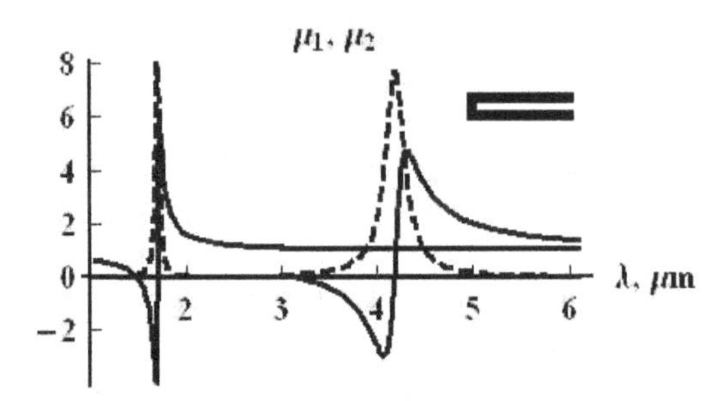

Fig. 2.14 Optical magnetic permeability $\mu = \mu_1 + \mu_2$ (μ_1– continuous line, μ_2–dashed line) estimated from Lorenz-Lorentz formula for the composite containing ⊏ - shaped silver nanoantennas; volume concentration $p = 0.3$; left curves: $a = 200$ nm, $d = 50$ nm, $b = 13$ nm; right curves: $a = 600$ nm, $d = 90$ nm, $b = 13$ nm. $d = 50$ nm, $b = 13$ nm. The dielectric susceptibility for silver is estimated from the Drude formula (3.1).

Eq. (2.92) in this case, where a is now to the total length of the horseshoe nanoantenna. Thus the horseshoe nanoantenna provides the same magnetic polarizability α_M at the twice shorter length. The magnetic permeability for a metamaterial, having silver horseshoe nanoantennas oriented in one direction ("z" direction in Fig. 2.13) and organized in the periodic square lattice, is shown in Fig. 2.17. To simulate it, we have taken the optical parameters of silver from Refs. [Johnson and Christy, 1972] and [Kreibig and Volmer, 1995]. As seen from the figure, the negative magnetism can be observed, in the near-infrared part of the spectrum, including the region of standard telecommunication wavelengths ($1.5\,\mu m$).

2.5.2 *Numerical simulations of two-dimensional nanowire structures*

To obtain a material, which is magnetically active in the optical region, it is convenient to employ (and much easier to model) a "two-dimensional" meta-material, composed of the nanoantennas bent in a horseshoe shape in $(x,\ y)$-plane and infinitely extended in "z"-direction. When the quasi-static condition $\left|k^2 bd\varepsilon_m\right| \ll 1$ is fulfilled (b and d are the thickness and the distance between the opposing walls, respectively), the *MPR* frequency ω_r is defined by the equation $G_2 = 2a\sqrt{-2/\left(\varepsilon_m bd\right)} = \pi/2$. The resonant magnetic field is shown in Fig.2.15, where the finite-element code FEMLAB-3 (COMSOL Inc.) was used to calculate the field distribution. Note that the magnetic field inside the horseshoe is large and has the *opposite* sign with

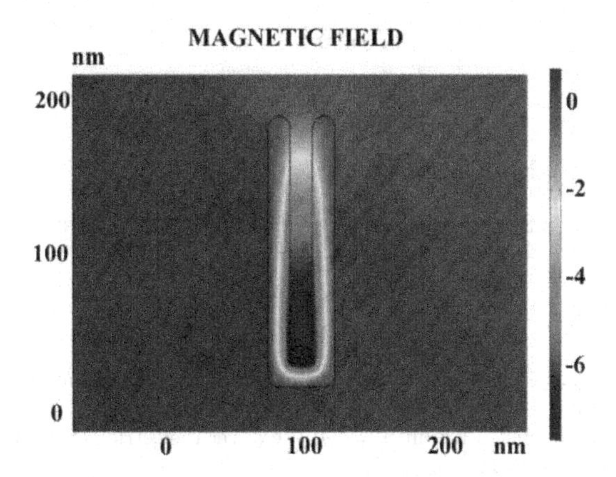

Fig. 2.15 A magnetic plasmon resonance in a silver horseshoe nanoantenna. The nanoantenna is placed in the maximal external field, directed perpendicular to the plane of paper. The incident wavelength is $\lambda = 1.5$ μm; the silver dielectric constant is estimated from the Drude formula (3.1).

respect to the external field H_0, resulting in a negative magnetic permeability in a close proximity to the magnetic plasmon resonance. To estimate the effective magnetic permeability we use the approach developed in Refs. [Sarychev *et al.*, 2000b], [Sarychev and Shalaev, 2000]. The latter gives $\mu_z = 1 + p\,(sH_0)^{-1} \int (H_{in} - H_0)\, ds = (32/\pi)\, a^2\, p\lambda^{-2}\, (\pi/2 - G_2)^{-1}$ for a plasmonic crystal composed by the horseshoe nano-antennas, where H_{in} is the magnetic field inside a horseshoe, p is the concentration of the nanoantennas organized in a square lattice, and the integration is performed over the area $s = da$. For a good,with optical standpoint, metal, such as gold or silver, μ_z becomes large and negative for $\omega > \omega_r$. For example, the magnetic permeability (both real and imaginary parts) of a metamaterial composed of silver horseshoes has a sharp resonance for $\omega = \omega_r$ shown in Fig. 2.16.

The real part μ_1 of the optical magnetic permeability turns negative for $\omega > \omega_r$. Figure 2.16 reveals the spectral range $\lambda \geq 1$ μm where $\mu_1 < -1$ while the relative loss is small: $\delta = \mu_2/\mu_1 \ll 1$. Obtaining the regime of minimal loss is crucial for successful application of LHM as the perfect lens material. The loss could, in principle, be further reduced by cooling the metal nanoantennas to cryogenic temperatures. Simple estimate shows that even at the liquid nitrogen temperatures, the electron mean free path is on the order of the horseshoe' characteristic size. The optical properties of metals are not well understood when the mean free path becomes larger

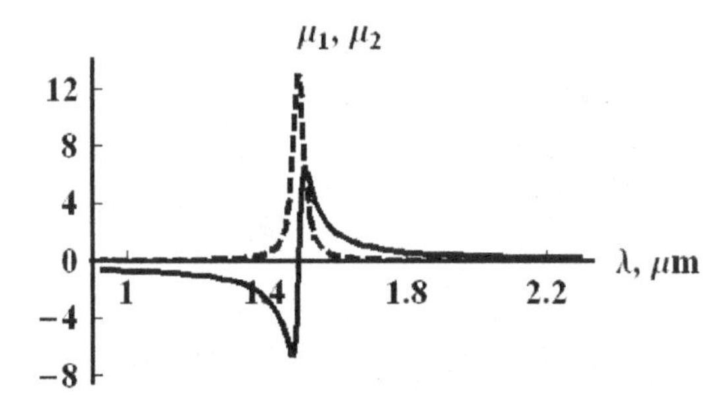

Fig. 2.16 Optical magnetic permeability $\mu = \mu_1 + \mu_2$ (μ_1- continuous line, μ_2–dashed line) of the composite containing silver nanoantennas shown in Fig. 2.15 organized in a square lattice; volume concentration $p = 0.4$; the silver dielectric constant is estimated from the Drude formula (3.1).

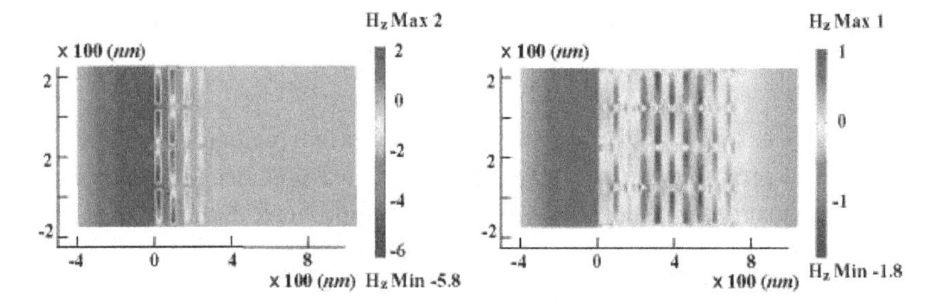

Fig. 2.17 Wave propagation through a two-dimensional (infinitely extended in the normal to the page direction) plasmonic crystal near the plasmon magnetic resonance. The crystal is composed of the horseshoe-shaped silver nano-antennas shown in Fig. 2.15, volume filling ratio $p = 0.4$, $\lambda = 1.4$ μm. Magnetic field $\mathbf{H}$ of the incident em wave is directed along $z-$axis perpendicular to plane of picture; $H_z = 1$ in the incident wave. (Left) Spaces between the horseshoes are filled with vacuum: no propagation. (Right) Spaces between the horseshoes are filled with a hypothetical $\varepsilon = -3$ material: wave propagates into the structure.

than the nanoantenna size. For these reasons, we further consider metamaterials only at the room temperatures.

To illustrate the effective magnetic properties of horseshoe metamaterials, we have simulated EM wave propagation in a plasmonic crystal made of silver nano-antennas, shown in Fig. 2.15. (The first three horseshoe columns are well pronounced in Fig. 2.17.) The EM wave is incident on the crystal from the left, the vacuum wavelength $\lambda = 1.4\,\mu$m. The wave is evanescent in the crystal, as can be clearly seen from Fig. 2.17 (left), with an overall transmittance of $T < 10^{-6}$. This is explained by the negative magnetic crystal permeability that could be rather large at the resonance (see Fig. 2.16.) One way of making the crystal transparent is to fill the space *between* the columns of the horseshoes by a material with the negative dielectric constant. In principle, such a modification can lead to a double-negative meta-material, which is confirmed by numerical simulations. The results are shown in Fig. 2.17 (right). Thus the addition of a negative-ε material turns otherwise negative-μ meta-material into transparent LHM. Note that the negative-ε material was added only between the adjacent columns of horseshoe-shaped nano-antennas. No additional material was placed in the exterior of the nanoantenna. This is done intentionally because the modification of the nanostructure region where most of the magnetic field is concentrated also affects the effective magnetic permittivity [Pokrovsky and Efros, 2002].

Interestingly enough, the horseshoe nanoantennas exhibit a double-negative behavior when they are closely packed. For these purposes, we have designed a two-dimensional dense periodic structure consisting of the nanoantennas, pointing up and down. The one-half of an elementary cell is shown in Fig. 2.18(a); the other half is obtained by 180° degree rotation in (x, y)-plane. The structure repeats itself in the "x" and "y"-directions; the separation between antenna centers is 80 nm, the horizontal periodicity is 160 nm, and the vertical periodicity is 400 nm; with the structure shown in Fig. 2.17). We then calculate the dispersion relation $\omega(k_x)$ for the electromagnetic wave propagating through the periodic structure in

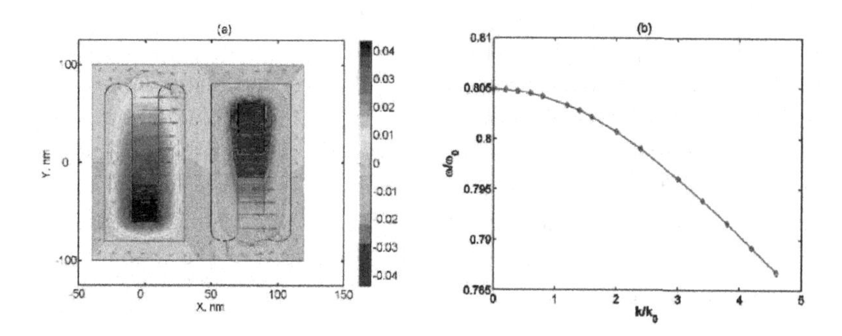

Fig. 2.18 A plasmonic crystal composed of the horseshoe-type metal nanoantennas. Separation between the antenna's centers is 80 nm. Magnetic (color and contours) and electric (arrows) fields inside a periodic array of horseshoe-shaped nanoantennas at the cutoff condition ($k_x = 0$). (b) Dispersion relation ω versus k-vector component k_x for a left-handed electromagnetic wave.

the "x"-direction by solving the Maxwell's equation for magnetic field H_z numerically. For computational simplicity, we have assumed a hypothetical lossless plasmonic material with the frequency-dependent dielectric permittivity $\varepsilon = 1 - \omega_p^2/\omega^2$, where $2\pi c/\omega_p = 225$ nm. The frequency ω and the wavevector k are normalized to $\omega_0 = 2\pi c/\lambda_0$ and $k_0 = 2\pi/\lambda_0$, respectively, where $\lambda_0 = 1.5\mu$m.

Remarkably, one of the propagating modes shown in Fig. 2.18(b) exhibits the properties typical for left-handed materials: Its group velocity $v_{gr} = \partial\omega/\partial k$ opposes the phase velocity. Fig. 2.18(a) shows the magnetic field profile and the electric field distribution inside the elementary cell for $k_x = 0$ (magnetic cutoff condition for $\mu = 0$). From the figure, we conclude that the magnetic field is concentrated in each horseshoe and for the adjacent horseshoes, the magnetic fields have opposing signs. The dominant field is E_x which does not contribute to the Poynting flux in the propagation direction. There is only a half of the elementary cell depicted in Fig. 2.18. In the other half, E_x field has opposite direction so that the average E_x is zero. (Arrows in Fig. 2.18 indicate the value the electric field). The figure clearly indicates that the *MPR* electric field is mainly confined inside the horseshoes and is negligible in the metal. The electric field is primarily potential (i.e. can be derived from an electrostatic potential), but has a non-vanishing solenoidal component that produces the magnetic field. The potential drop between the metal arms is much larger than the potential drop between external and internal interfaces of the arm. This behavior of the *MPR* electric field, obtained from direct computer simulations, resembles our analytical results for the two-wire system (see discussion at Eqs. (2.87) and (2.88)). The fact that the dominant electric field E_x does not change its sign in the cell along the direction of the propagation indicates that the investigated mode exhibits the negative dispersion not because of the band-folding effect, which is common in regular photonic crystals. The left-handed mode behavior occurs in the vicinity of $\lambda = 1.88$ μm, which is close to the *MPR* resonance. The condition of negative ε, which is necessary for the negative refractive index, is actually induced by the proximity of the dipole-type electric plasmon resonances.

2.5.3 *Capacitance and inductance of two parallel wires*

Below we derive equations for the capacitance C and inductance L between two parallel wires of the radius b and length a separated by the distance d. We assume that $b \ll d$ and $d \ll a$. We also assume that the dielectric constant of the wires is large in the absolute value $|\varepsilon_m| \gg 1$ whereas the skin depth $\delta \sim 1/\left(k\sqrt{|\varepsilon_m|}\right) \gg b$. To find the capacitance C we first calculate the electric potential Φ_a at the coordinate vector $\mathbf{r}_a$, representing the wire surface (point a in Fig. 2.13):

$$\Phi_a = \int q_1(\mathbf{r}_1) \frac{\exp(ikr_{a1})}{r_{a1}} d\mathbf{r}_1 + \int q_2(\mathbf{r}_2) \frac{\exp(ikr_{a2})}{r_{a2}} d\mathbf{r}_2, \tag{2.93}$$

where $r_{a1} = |\mathbf{r}_a - \mathbf{r}_1|$, $r_{a2} = |\mathbf{r}_a - \mathbf{r}_2|$, q_1 and q_2 are the electric charges distributed over the wire surface; and we integrate over the surface of the first (a, d) and second (b, c) rods in Fig. 2.13. For further consideration, we choose the coordinate system $\{x, y, z\}$ with z axis along the (a, d) rod, origin in the center of the system and the x axis connecting the axes of the rods so that y axis is perpendicular to the plane of two rods. We introduce the vector $\mathbf{d} = \{d, 0, 0\}$ between the wires and two dimensional unit vector $\boldsymbol{\rho}(\phi) = \{\cos\phi, \ \sin\phi\}$ in $\{x, y\}$ plane, where ϕ is the polar angle. Then the vectors in Eq. (2.93) can be written as $\mathbf{r}_a(\phi_a, z_a) = \{b\,\boldsymbol{\rho}(\phi_a) + \mathbf{d}/\mathbf{2}, \ z_a\}$, $\mathbf{r}_1(\phi_1, z_1) = \{b\,\boldsymbol{\rho}(\phi_1) + \mathbf{d}/\mathbf{2}, \ z_1\}$, and $\mathbf{r}_2(\phi_2, z_2) = \{b\,\boldsymbol{\rho}(\phi_2) - \mathbf{d}/\mathbf{2}, \ z_2\}$. As it follows from the symmetry, $q_1(\phi, \ z) = -q_2(\phi + \pi, \ z)$. (Recall that we consider the antisymmetric mode, at which the electric currents are equal in the wires but flows in the opposite directions.) We rewrite Eq. (2.93) by splitting on

$$\Phi_a \equiv \Phi_a^{(0)} + \Phi_a^{(1)}, \tag{2.94}$$

$$\Phi_a^{(0)}(z_a, \phi_a) = \int_{\phi=0}^{\pi} \int_{z=-a}^{a} q(\phi, z)$$

$$\left[\frac{1}{\sqrt{\Delta z^2 + b^2 \Delta\boldsymbol{\rho}_1^2}} - \frac{1}{\sqrt{\Delta z^2 + (b\Delta\boldsymbol{\rho}_2 - \mathbf{d})^2}}\right] dz\, bd\phi, \tag{2.95}$$

$$\Phi_a^{(1)} \equiv \int q(\mathbf{r}_1) \frac{\exp(ikr_{a1}) - 1}{r_{a1}} d\mathbf{r}_1 - \int q(\mathbf{r}_2) \frac{\exp(ikr_{a2}) - 1}{r_{a2}} d\mathbf{r}_2, \tag{2.96}$$

where $\Delta z = z_a - z$, $\Delta\boldsymbol{\rho}_1 = \boldsymbol{\rho}_a - \boldsymbol{\rho} = \{\cos\phi_a - \cos\phi, \ \sin\phi_a - \sin\phi\}$, $\Delta\boldsymbol{\rho}_2 = \boldsymbol{\rho}_a + \boldsymbol{\rho} = \{\cos\phi_a + \cos\phi, \ \sin\phi_a + \sin\phi\}$. Note that dimensionless vectors $\Delta\boldsymbol{\rho}_1$ and $\Delta\boldsymbol{\rho}_2$ satisfy the condition $|\Delta\boldsymbol{\rho}_1|$, $|\Delta\boldsymbol{\rho}_2| < 2$ so that the second terms in the radicals of Eq. (2.95) are much less than a.

The electric currents in the wires(rods) and the corresponding electric charges q change with the "z"-coordinate on the scales of an order of $\sim a$, which is much larger than the distance d between the rods. Therefore we can neglect the variation of the electric charge along the coordinate "z" for $|\Delta z| < d$. On the other hand, the term in the square brackets in Eq. (2.95) vanishes as $\sim d^2 / |\Delta z|^3$ for $|\Delta z| > d$. This allows us to replace the charge $q(z, \phi)$ by its value $q(z_a, \phi_a)$ in the observation point $\mathbf{r}_a$ in Eq. (2.95), obtaining

$$\Phi_a^{(0)}(z_a, \phi_a) = \int_{\phi=0}^{2\pi} q(z_a, \phi) \int_{z=-a}^{a}$$

$$\left[\frac{1}{\sqrt{\Delta z^2 + b^2 \Delta \boldsymbol{\rho}_1^2}} - \frac{1}{\sqrt{\Delta z^2 + (b \Delta \boldsymbol{\rho}_2 - \mathbf{d})^2}} \right] dz\, b d\phi. \tag{2.97}$$

The accuracy of such replacement is on the order of $(d/a)^2 \ll 1$. Since we consider the quasistatic limit when the distance between the rods $d \ll \lambda$ and the metal dielectric constant $|\varepsilon_m| \gg 1$, the potential lines in (x, y)-plane are close to the static case. Therefore we can safety assume that the angle distribution of the electric charge $q(z, \phi)$ is the same as it would be in the static case of two infinite metal cylinders, $q(z, \phi) = Q(z) \sqrt{(d/b)^2 - 4} / (2\pi\ (d + 2b \cos \phi))$, where $Q(z)$ is the electric charge per unit length of the rod so that $\int_{\phi=0}^{\phi=2\pi} q(z, \phi)\, b\, d\phi = Q(z)$. Then the integral in Eq. (2.97) inherits

$$\Phi_a^{(0)}(z) = Q(z) \operatorname{arccosh}\left(\frac{d^2}{2\, b^2} - 1 \right) + O\left((d/a)^2 \right), \tag{2.98}$$

where the second term includes all corrections to the integral (2.95) due to finite size of the system. For thin wires, $b \ll d$, considered here, the potential $\Phi_a^{(0)}$ approximates as

$$\Phi_a^{(0)}(z) \simeq Q(z)\, 2 \left[\ln (d/b) \right]. \tag{2.99}$$

The second term $\Phi_a^{(1)}$ in Eq. (2.94) is small in the limit of $a \ll \lambda$, i.e., $(kr_{a1}, kr_{a2}) \ll 1$. The real part of $\Phi_a^{(1)}$ gives the potential $\Phi_a^{(0)}$a small correction on the order of $(d/a)^2$, which can be neglected. The imaginary part is important regardless of its absolute value since it gives radiative loss. To estimate the loss we assume that $b/d \ll 1$ and neglect the angular dependence of the charge distribution. We expand Eq. (2.96) in series of $k's$

and linearly approximate $Q(z) \simeq q_1 z$ (recall that $Q(z)$ is an odd function of z) obtaining

$$\Phi_a^{(1)}(z) \simeq -iQ(z)(ak)^3(kd)^2/45, \tag{2.100}$$

where we neglect the terms the higher-order terms on k as well as all the terms up to the second order of $(bk)^2$.

Due to the symmetry of the system, the potential difference $U = \Phi_a - \Phi_b$ between points a and b (see Fig. 2.13 ; $z_a = z_b$) equals to $U = 2\Phi_a$. The capacitance C defined as $C = Q(z)/U(z)$ is then given by

$$\frac{1}{C} \simeq 2\,\text{arccosh}\left[(d/b)^2/2 - 1\right] - i\frac{2}{45}(ak)^3(kd)^2 \tag{2.101}$$

$$\simeq 4\ln(\frac{d}{b}) - i\frac{2}{45}(ak)^3(kd)^2, \tag{2.102}$$

where the first term represents the capacitance between two parallel infinite cylinders (see [Landau *et al.*, 1984] Ch. 3) while the second term gives the radiative loss due to the retardation effects.

To find the inductance L between the wires we first calculate the vector potential A_a at a space-vector $\mathbf{r}_a$, representing the wire points. In derivation, we neglect the edge effects and assume that the vector potential is parallel to the wire axis, obtaining

$$A_a = \frac{1}{c}\int j(\mathbf{r})\left(\frac{\exp(ikr_{a1})}{r_{a1}} - \frac{\exp(ikr_{a2})}{r_{a2}}\right)d\mathbf{r}, \tag{2.103}$$

where $r_{a1} = |\mathbf{r}_a - \mathbf{r}|$ and $r_{a2} = |\mathbf{r}_a - \mathbf{r} + \mathbf{d}|$, $j(\mathbf{r}_1)$ is the current density and the integration is completed over the volume of the first wire. Since we are considering the quasistatic case when the skin effect is small ($kb\sqrt{|\varepsilon_m|} \ll 1$), the electric current distributes uniformly over the cross-section of the wire and $j(\mathbf{r}) = I(z)/(\pi b^2)$. Following the procedure above for calculating the electric potential, the vector potential is expressed as $A_a = A_a^{(0)} + A_a^{(1)}$, where

$$A_a^{(0)} = \frac{1}{c}\int \frac{I(z)}{\pi b^2}\left(\frac{1}{r_{a1}} - \frac{1}{r_{a2}}\right)d\mathbf{r}, \tag{2.104}$$

$$A_a^{(1)} = \int \frac{I(z)}{\pi b^2}\left(\frac{\exp(ikr_{a1}) - 1}{r_{a1}} - \frac{\exp(ikr_{a2}) - 1}{r_{a2}}\right)d\mathbf{r}. \tag{2.105}$$

The term $A_a^{(0)}$ can be estimated in the similar way as $\Phi_a^{(0)}$. As the result, we obtain the vector potential $A_a^{(0)}$ averaged over the wire cross-section:

$$A_a^{(0)}(z) \simeq \frac{I(z)}{2c}[4\ \ln(\frac{d}{b}) + 1], \tag{2.106}$$

where $I(z)$ is electric current, and we have again neglected terms on the order of $(b/a)^2$ and $(d/a)^2$. Eq. (2.105) can be expanded in the series of k, obtaining that the linear term is zero, k^2-term ($\sim (kd)^2$) gives a small correction to $A_a^{(0)}$ and the third-order term gives the radiative loss, namely

$$A_a^{(1)} \simeq i(kd)^2 k \frac{1}{c} \int I(z)\, dz \sim 2i \frac{I(z)}{c}(kd)^2 ka. \tag{2.107}$$

While deriving Eq. (2.107), we have neglected the current variation over the wire length. The inductance L is then obtained form the equation $A_a - A_b = 2A_a = (L/c)I(z)$:

$$L = 4\ \ln(\frac{d}{b}) + 1 + 4i(kd)^2 ka. \tag{2.108}$$

Again, similar to the two-wire capacitance case, the first two terms represent a system of two parallel infinite wires: the self-inductance per unit length in this case ([Landau *et al.*, 1984], Ch. 34.) This estimate as well as Eq. (2.102) are certainly invalid near the rod ends but it does not affect the current distribution $I(z)$ and magnetic moment calculations.

We can now compare the radiation loss (given by imaginary parts of capacitance C and inductance L) and the ohmic loss in the metal wires. In near infrared spectral region, the dielectric constant ε_m for a "good" optical metal (*Ag, Au, etc.*) can be estimated from the Drude formula (2.80) as $\varepsilon_m(\omega) \sim (\omega_p/\omega)^2 (1 - i\omega_\tau/\omega)^{-1}$, where ω_p is plasma frequency and $\omega_\tau \ll \omega \ll \omega_p$ is the relaxation rate. The real part of the rod resistance $R_{ohm} \sim 8\left(\omega_\tau/\omega_p^2\right)\left(a/b^2\right)$ should be then compared with "radiation" resistance $R_{rad} \sim (kd)^2(ka)^2/c$. For the silver nanowires considered here, the ohmic loss either larger ($R_{ohm} > R_{rad}$) or much larger ($R_{ohm} \gg R_{rad}$) than the radiation loss. Therefore we can neglect the imaginary parts of the capacitance C and inductance L:

$$L \simeq \frac{1}{C} \simeq 4 \ln \frac{d}{b}. \tag{2.109}$$

This estimate has the logarithmic accuracy; with an error on the order of $(4 \ln d/b)^{-1}$. Note that the radiation loss depend on the parameter ka

in a crucial manner. The magnetic plasmon resonance becomes very broad when $ka > 1$, placing a rather sever constraint on the wire length $2a$.

Overall, we have described a new phenomenon of a magnetic plasmon resonance in the metallic horseshoe-shaped split rings. This resonance is distinctly different from the geometric LC-resonance described earlier for split rings because it is determined by the plasmonic properties of the metal. The magnetic plasmon resonance paves the way to designing metallic meta-materials that are magnetically active in the optical and near-infrared spectral ranges. Presented above three-dimensional analytic calculations and two-dimensional numerical simulations thus reveal that resonantly enhanced magnetic moments can be induced in very thin (thinner than the skin depth) horseshoe with typical dimensions much shorter than the wavelength (on the order of 100 nanometers). Periodic arrays of such horseshoe-shaped nanoantennas can be used to design left-handed meta-materials by exploiting the proximity of electric resonances in the dielectric permittivity ε and magnetic permeability μ.

2.6 Planar Nanowire Composites

In this section we consider a planar composite that is composed of a regular array of parallel nanowire pairs (see Fig. 2.19). The array is illuminated by a plane electromagnetic wave, perpendicular to the plane of the composite. First, we show that the field scattered by nanowire pairs can be approximated in a far optical field by the effective dipole and magnetic moments even when the size of the pair is comparable with the incident field wavelength λ. Then we consider the explicit optical properties of a

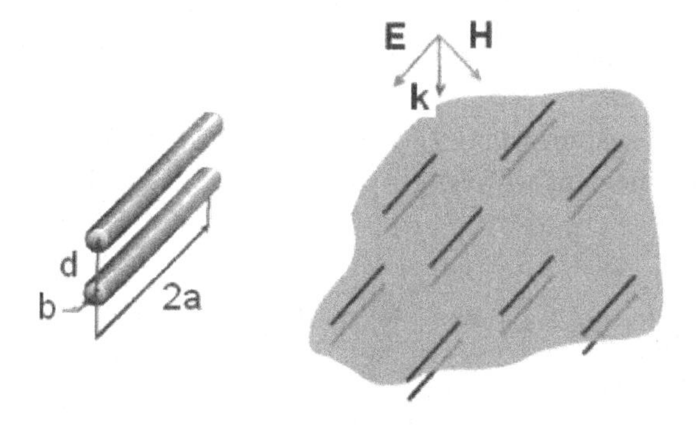

Fig. 2.19 A layer filled with pairs of parallel nanowires.

layer array for such nanowire pairs. Electric and magnetic fields from a nanowire pair with dimensions $2a \times d \times 2b$ (see Fig. 3.16) are derived from the vector potential $\mathbf{A}$. We chose the axes origin in the center of two wires. The coordinates of the observation point equal to $\mathbf{R} = \mathbf{n}R$. We suppose that the distance R is much larger than the stick length $2a$. Then the distance from the point $\mathbf{r}$ ($r < a$) to the observation point $\mathbf{R}$ approximates as $|\mathbf{R} - \mathbf{r}| \simeq R - (\mathbf{n} \cdot \mathbf{r})$. At large distances, $R \gg \lambda, a, b, d$, the vector potential takes the standard form

$$\mathbf{A} = \left(e^{ikR}/cR\right) \int e^{-ik(\mathbf{n}\cdot\mathbf{r})} \mathbf{j}(\mathbf{r})\ d\mathbf{r},$$

where $\mathbf{j}(\mathbf{r})$ is the current density inside the nanowires and $\mathbf{n}$ is the unit vector in the observation direction. We also introduce the vector $\mathbf{d}$, directed from one nanowire to another. Since the coordinate system has its origin in the center of the system the centers of the wires have the coordinates $\mathbf{d}/2$ and $-\mathbf{d}/2$, respectively. The electromagnetic wave is incident in the wire plane, perpendicular to the wires (see Fig. 2.19), i.e., the wavevector $\mathbf{k} \| \mathbf{d}$. Then, the vector potential $\mathbf{A}$ can be written as

$$\mathbf{A} = \frac{e^{ikR}}{cR} \left[e^{-\frac{ik}{2}(\mathbf{n}\cdot\mathbf{d})} \int_{-a}^{a} e^{-\frac{ik}{2}(\mathbf{n}\cdot\mathbf{z})} \mathbf{j}_1(z)\ dz + e^{\frac{ik}{2}(\mathbf{n}\cdot\mathbf{d})} \int_{-a}^{a} e^{-ik\mathbf{n}\cdot\mathbf{z}} \mathbf{j}_2(z)\ dz \right], \tag{2.110}$$

where $\mathbf{j}_1$ and $\mathbf{j}_2$ are the wire currents, and $\mathbf{z}$ is the coordinate along the wires ($\mathbf{z} \perp \mathbf{d}$). As it is known, the dipole components of multipole expansion dominate the scattering from a thin antenna even when the antenna size is comparable with the wavelength (see, e.g., [Jackson, 1998], [Balanis, 2005].) Therefore we can approximate the term $e^{-ik\mathbf{n}\cdot\mathbf{z}}$ in Eq. (2.110) by unity. Note that for the forward and backward scattering, which are responsible for the effective properties of the medium, this term is exactly equals to one.

We consider a system where the distance d between the wires is much smaller than the incident wavelength and expand Eq. (2.110) in series over d. This results in

$$\mathbf{A} = \frac{e^{ikR}}{cR} \left[\int_{-a}^{a} (\mathbf{j}_1 + \mathbf{j}_2)\ dz - \frac{ik}{2} (\mathbf{n} \cdot \mathbf{d}) \int_{-a}^{a} (\mathbf{j}_1 - \mathbf{j}_2)\ dz \right]. \tag{2.111}$$

The first term in the square brackets in Eq. (2.111) gives the effective dipole moment $\mathbf{P}$ for the two-nanowire system and its contribution to the scattering can be written as $\mathbf{A}_d = -ik\left(e^{ikR}/R\right)\mathbf{P}$, where

$$\mathbf{P} = \int \mathbf{p}(\mathbf{r})\, d\mathbf{r}, \tag{2.112}$$

and $\mathbf{p}(\mathbf{r})$ is the local polarization, and the integration is performed over the volume of both wires. The second term in Eq. (2.111) gives the magnetic dipole and quadrupole contributions to the vector potential:

$$\mathbf{A}_{mq} = \frac{ike^{ikR}}{R}\left[\left[\mathbf{n}\times\mathbf{M}\right] - \frac{\mathbf{d}}{2c}\int_{-a}^{a}\left(\mathbf{n}\cdot(\mathbf{j}_1 - \mathbf{j}_2)\right)\, dz\right], \tag{2.113}$$

where $\mathbf{M}$ is the magnetic moment of two wires,

$$\mathbf{M} = \frac{1}{2c}\int \left[\mathbf{r}\times\mathbf{j}(\mathbf{r})\right]\, d\mathbf{r}, \tag{2.114}$$

and the integration is performed over the volume of the both wires [as in Eq. (2.112)].

We show here the results of our numerical simulations for optical properties of gold nanowires (Fig. 2.20) (see [Podolskiy *et al.*, 2003]). According to the simulations, both the dielectric and magnetic moments excited in the nanowire system are opposite to the excited field when the wavelength of the incident electromagnetic wave falls below the resonance. Thus in this frequency range, a composite material made of parallel nanowire pairs have both negative the dielectric permittivity and magnetic permeability and acts as a left-handed material. These results are in good qualitative agreement with Eqs. (2.58) and (2.77) which were derived for metal sticks with high aspect ratio. Note that quality of the electric and magnetic resonances increase progressively with decreasing space between nanowires.

We consider now the transmittance and reflectance of a planar nanowire composite when em wave is impinged normal to its plane. We take into account the dipole $\mathbf{P}$ and magnetic $\mathbf{M}$ moments given by Eqs. (2.112) and (2.114), respectively, since they are responsible for the main contribution to forward and backward scattering. The second term in Eq. (2.113) describes a quadrupole contribution, which vanishes for the forward direction (see discussion in [Podolskiy *et al.*, 2003].)

The Maxwell equations for the composite can be written in the following form

$$\operatorname{curl}\mathbf{E} = ik\mathbf{H}, \qquad \operatorname{curl}\mathbf{H} = \frac{4\pi}{c}\mathbf{j} - ik\mathbf{E}, \tag{2.115}$$

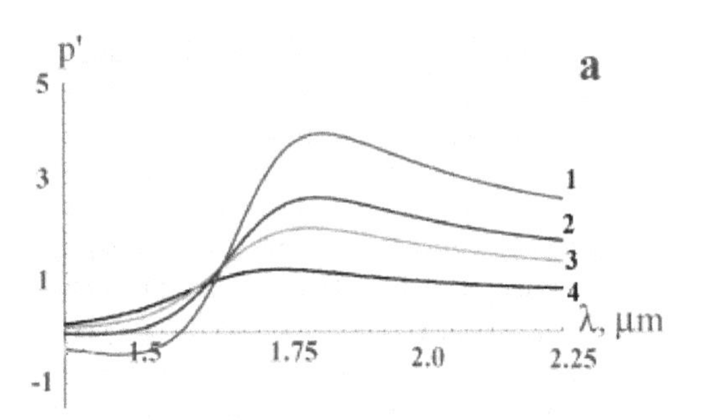

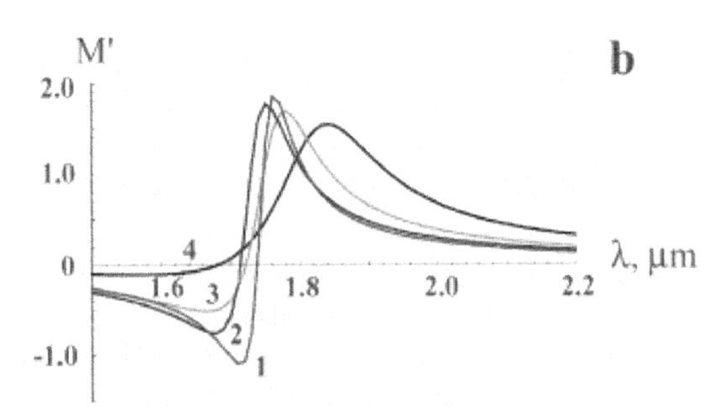

Fig. 2.20 Dielectric (a) and magnetic (b) moments in nanowire pairs as functions of wavelengths; distance between nanowires in the pairs is varied: $d = 0.15$ μm (1), $d = 0.23$ μm (2), $d = 0.3$ μm (3), and $d = 0.45$ μm (4); $a = 0.35$ μm and $b = 0.05$ m for all the plots. The moments are normalized to the unit volume.

where $\mathbf{j}$ is the current in the nanowires. We split the current in two parts $\mathbf{j} = \mathbf{j}_P + \mathbf{j}_M$. Here $\mathbf{j}_M$ is the circular current in the nanowire pair and can be presented as $\mathbf{j}_M = c \operatorname{curl} \mathbf{m}$ with the vector $\mathbf{m}$ vanishing outside the composite. Then Eq. (2.115) can be rewritten as

$$\operatorname{curl} \mathbf{E} = ik \left(\mathbf{H}' + 4\pi \mathbf{m} \right), \qquad \operatorname{curl} \mathbf{H}' = \frac{4\pi}{c} \mathbf{j}_P - ik\mathbf{E}, \tag{2.116}$$

where $\mathbf{H}' = \mathbf{H} - 4\pi\mathbf{m}$. Assuming that $z = 0$ is the principal plane of the composite and the EM wave is incident along z axis, we invoke Eq. (2.116) and perform averaging over the $\{x, y\}$ plane with further integration over the space between the two reference planes placed in front ($z = -h$) and behind ($z = h$) of the composite. The distance h is chosen so that $d \ll h \ll$

$1/k$. After the integration, Eq. (2.116) takes the form

$$\mathbf{E}_2 - \mathbf{E}_1 \simeq ik4\pi dp\mathbf{M}_1, \qquad \mathbf{H}_2 - \mathbf{H}_1 \simeq -ik4\pi dp\mathbf{P}_1, \tag{2.117}$$

where p is the filling factor, i.e., the ratio of the area covered by the nanowires and the total area of the film; $\mathbf{E}_1 = \mathbf{E}(-h)$, $\mathbf{E}_2 = \mathbf{E}(h)$, $\mathbf{H}_1 = \mathbf{H}(-h)$, $\mathbf{H}_2 = \mathbf{H}(h)$; $\mathbf{P}_1 = \mathbf{P}/(4abd)$ and $\mathbf{M}_1 = \mathbf{M}/(4abd)$ are the specific dipole and magnetic moments of the nanowire pairs, respectively. These moments are given by Eqs. (2.112) and (2.114) which are normalized to the volume $V = 4abd$ of the pairs.

The moments $\mathbf{P}_1$ and $\mathbf{M}_1$ are proportional to the effective electric and magnetic fields, respectively. For the dilute case ($p \ll 1$) considered here, we can write $\mathbf{P}_1$ and $\mathbf{M}_1$ as $4\pi p\mathbf{P}_1 = \varepsilon(\mathbf{E}_2 + \mathbf{E}_1)/2$ and $4\pi p\mathbf{M}_1 = \mu(\mathbf{H}_2 + \mathbf{H}_1)/2$, where the coefficients ε and μ have the meaning of the effective dielectric constant and magnetic permeability of the nanowire plane composite. Then Eq. (2.117) take the following form

$$\mathbf{E}_2 - \mathbf{E}_1 \cong ikd\mu(\mathbf{H}_1 + \mathbf{H}_2), \qquad \mathbf{H}_2 - \mathbf{H}_1 = -ikd\varepsilon(\mathbf{E}_1 + \mathbf{E}_2). \tag{2.118}$$

We match Eqs. (2.118) at $z = -h$ with a plane wave solution

$$E = E_0\left[\exp(ikz) + r\exp(-ikz)\right]$$

that holds behind the film ($z < -h$) and match Eqs. (2.118) at $z = h$ with the solution $E = E_0 t\exp(ikz)$ that holds in front of the film, where r and t are reflection and transmission coefficients, respectively, E_0 is the amplitude of the impinged wave. This matching results in two equations for r and t. Solutions to these equations allow us to find the reflection R and transmittance T coefficients of the nanowire composite in the following form

$$R = \left|\frac{2dk(\varepsilon - \mu)}{(-2 + i\,dk\varepsilon)(-2 + i\,dk\mu)}\right|^2, \tag{2.119}$$

$$T = \left|\frac{4 + d^2k^2\varepsilon\mu}{(-2 + i\,dk\,\varepsilon)(-2 + i\,dk\mu)}\right|^2. \tag{2.120}$$

When $\varepsilon = \mu$ the reflectance vanishes while the transmittance is given by $T = |(2 + i\,dk\varepsilon)/(2 - i\,dk\varepsilon)|^2$. If $\varepsilon = \mu$ and it is a real number, the reflectance $T = 1$. Still, the interaction of the EM wave with the composite results in the phase shift, equal to 2 $\arctan(dk\varepsilon/2)$, for the transmitted wave. The phase shift is positive if $\varepsilon = \mu > 0$ and the shift is negative when $\varepsilon = \mu < 0$. The latter case corresponds to the left-handed material. Thus, a negative phase of the transmitted EM wave pinpoints the left-handedness of the composite.

In this chapter, we have presented a detailed study of electrodynamics of the metal-dielectric composites consisting of elongated conducting inclusions — conducting sticks — embedded into a dielectric host. Conducting stick composites have new and unusual properties at large frequencies when surface waves are excited near the stick (in optics, they are called plasmon polaritons, in microwave — antenna resonances). The effective dielectric permittivity has strong resonances at some frequencies, where its real part vanishes at the resonance and acquires negative values for the frequencies above the resonance. The dispersion behavior does not depend on the stick conductivity and takes an universal form when the stick conductivity approaches the infinity. The surface plasmon polaritons can be localized in such composites.

We have shown that composites consisting of nonmagnetic inclusions may have a large magnetic response in the optical spectral range. The effective magnetic response is strong in a composite comprising pairs of parallel nanowires and results from the collective interaction of the nanowire pairs with the external magnetic field. The composite materials based on the plasmonic nanowires can have a negative refraction index and act as left-handed materials in the optical range of the spectrum.

Chapter 3

Semicontinuous Metal Films

3.1 Introduction

The optical properties of semicontinuous metal films have been intensively studied both experimentally and theoretically. Random metal-dielectric films, also known as semicontinuous metal films, are usually produced by thermal evaporation or metal spattering onto an insulating substrate as it is schematically shown in Fig. 3.1.

As the film grows, the metal filling factor increases and coalescences occur, so that irregularly shaped clusters are formed on the substrate resulting in occurrence of $2D$ fractal structures. The sizes of these structures diverge in a vicinity of the percolation threshold. A percolating cluster of metal particles is formed when a continuous conducting path appears between the ends of the sample. The metal-insulator transition (the percolation threshold) is very close to this point, even in the presence of quantum tunneling.

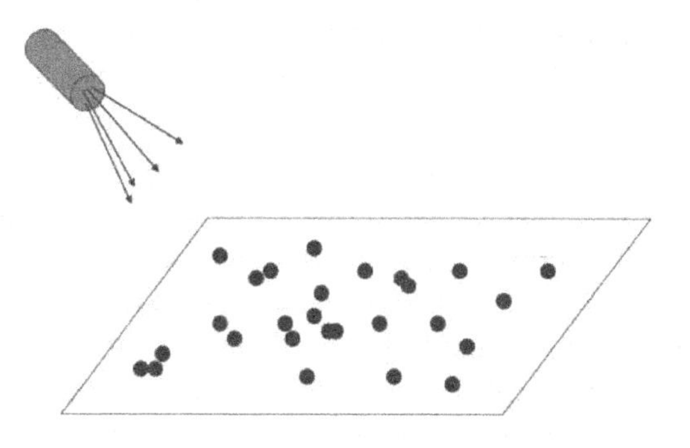

Fig. 3.1 Preparation of semicontinuous film.

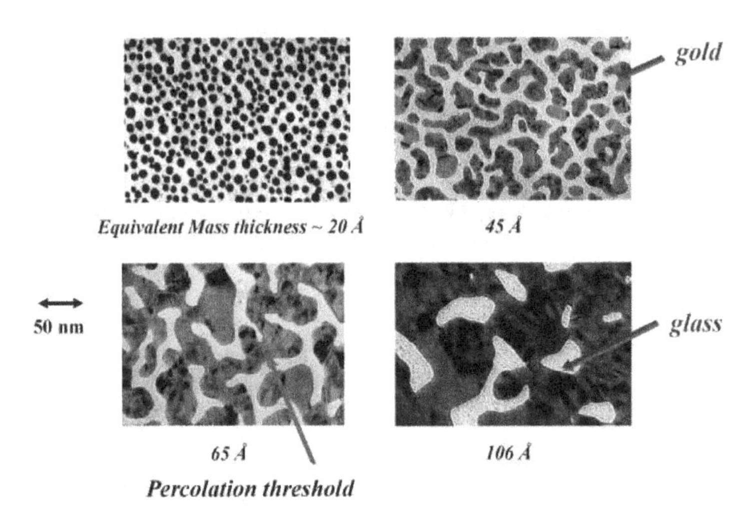

Fig. 3.2 Semicontinuous gold film on glass substrate for different metal concentration p; at percolation threshold $p = p_c$ continuous gold channel spans the system.

At higher surface coverage, the film is mostly metallic with few voids of irregular shapes being present. As further coverage increase, the film becomes uniform. The evolution of the semicontinuous film with increasing metal concentration is shown in Fig. 3.2.

Semicontinuous metal films are being investigated for many years since they have appeal for numerous important applications such as optical filters, solar energy concentrators, and others. However, the most intriguing and important optical discovery has come up recently [Gresillon *et al.*, 1999a]. It appears that the local electric fields fluctuate in these films on the giant scales as it is demonstrated in Fig. 3.3. The local intensities $I(x,y) = |\mathbf{E}(x,y)|^2$, where x and y are the coordinates in the film surface plane, may exceed the intensity I_0 of the incident filed by many orders on magnitude.

Giant electromagnetic field fluctuations and related enhancement of nonlinear optical phenomena in metal-dielectric composites like semicontinuous metal films have recently become an area of active studies, because of many fundamental problems involved and high potential for various applications. At zero frequency, strong nonlinearity may result in breaking down conducting elements when the electric current exceeds some critical value [de Arcangelis *et al.*, 1985; Duxbury *et al.*, 1987; Khang *et al.*, 1988; Vinogradov *et al.*, 1989a; Garanov *et al.*, 1991; Vinogradov *et al.*, 1992]. If the external electric field exceeds this value known as a critical field, a crack

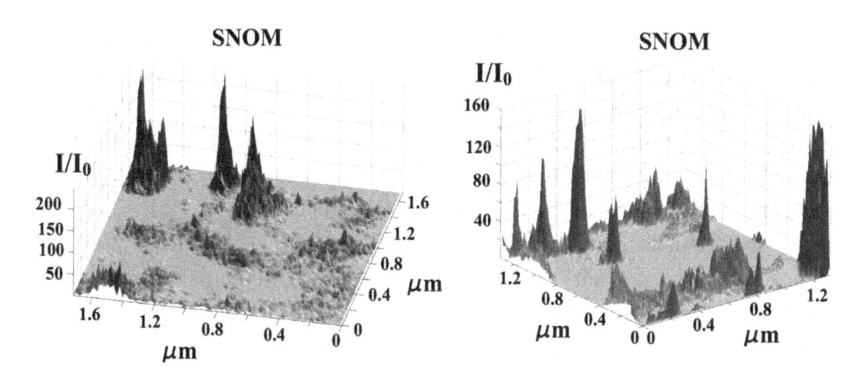

Fig. 3.3 Experimental images of the localized optical excitations in a percolation ($p = p_c$) gold-on-glass film; results are obtained by scanning near-field optical microscope; wavelengths $\lambda = 780$ nm.

within material spreads over the system. The critical field decreases to zero when the concentration of the conducting component approaches the percolation threshold. This indicates that percolation composites become progressively more responsive to the external field when the percolation threshold is approached. The simplest fuse model can be applied here, which is appropriate for description of fractures in disordered media and related problems of weak material tensility (The latter has to be compared to the strength of the atomic bonds.) [Herrmann and Roux, 1990; Meakin, 1991; Duxbury and Leath, 1994]. It states that the tension concentrates near these weak points of the material so that the material crack begins forming at these weak points and then spread over the entire material.

Another example of unusual nonlinear behavior have been observed for the ac and dc conductivities in a percolation mixture of carbon particles embedded in the wax matrix[Bardhan and Chakrabarty, 1994; Nandi and Bardhan, 1995; Bardhan, 1997]. In this case, neither carbon particles nor wax matrix have any intrinsic nonlinearity in the conductivity; nevertheless, the conductivity of a macroscopic composite sample doubles when the applied voltage is increased by just a few volts. Such a strong nonlinear response can be attributed to the quantum tunneling between conducting (carbon) particles, which is a distinguished feature of the electric transport phenomena in the composite materials near the percolation threshold. [Sarychev and Brouers, 1994].

The current and electric fields are concentrated in few "hot" spots. This allows to alter the local conductance significantly due to the large local fields while the external field can be relatively small. In general, percolation systems are very sensitive to the external electric field since their transport and optical properties are determined by a rather sparse network of conducting channels, and the field concentrates in the weak points of the conducting channels. This implies that composite materials have much larger nonlinear susceptibilities (with respect to the susceptibility of the constituent particles) at zero and finite frequencies.

The distinguishable feature of the percolation composites, to amplify nonlinearity of their components, has been recognized very early [Duxbury *et al.*, 1987; Aharony, 1987; Stroud and Hui, 1988; Shalaev and Stockman, 1987; Butenko *et al.*, 1988; Vinogradov *et al.*, 1989a; Bergman, 1989]. The behavior of nonlinear conductivity and susceptibility had been intensively studied in last two decades (see, for example, [Herrmann and Roux, 1990; Bergman and Stroud, 1992; Duxbury and Leath, 1994; Hui, 1996; Shalaev, 1996; Ma *et al.*, 1998; Stroud, 1998; Sarychev and Shalaev, 2000], [Hui *et al.*, 2004a]). In optics, even weak nonlinearity can lead to qualitatively new effects. For example, the generation of higher harmonics can be strongly enhanced in the percolation composites and bistable behavior of the effective conductivity occurs when the conductivity switches between two stable values ([Levy and Bergman, 1994; Bergman *et al.*, 1994; Levy *et al.*, 1995; Dykhne *et al.*, 2003]).

The local field fluctuations can be strongly enhanced in the optical and infrared spectral regions for a composite containing metal particles that are characterized by the dielectric constant with negative real and small imaginary parts. Indeed, the electric field E in an isolated metal particle can be estimated as $E \sim E_0/\left(\varepsilon_m + 2\right)$, where ε_m is metal permittivity and E_0 is external field [Landau *et al.*, 1984, Sec. 8], The field E is resonantly amplified when $\varepsilon_m\left(\omega\right) \approx -2$. This phenomenon, known as surface plasmon resonance, becomes responsible for the enhancement of the local optical field in metal semicontinuous films and originates in metal granules and the film clusters [Cohen *et al.*, 1973; Clerc *et al.*, 1990; Bergman and Stroud, 1992; Shalaev, 1996; Sarychev and Shalaev, 2000; Sarychev and Shalaev, 2002]. The enhancement of the local optical field in the periodic array of metal disks have been discussed in Ch.1.1.

The strong fluctuations of the local electric fields lead to enhancement of various nonlinear effects. Nonlinear percolation composites thus have great potential for practical applications [Flytzanis *et al.*, 1992]. They can be implemented as media with intensity-dependent dielectric functions and, in particular, as nonlinear filters and optical bistable elements. The optical response of nonlinear composites can be tuned, for example, by controlling the volume fraction and morphology of the constituting particles.

In spite of the large research efforts, the local field distributions and corresponding nonlinearities were poorly known for percolation metal-dielectric composites until the end of the 20th century. This was particularly true for the spectral regions where the plasmon resonances typically occur in metal granules. However, when the metal grains embedded in a linear host have small volume concentration $p \ll 1$, the effective nonlinear response of the whole composite can be calculated explicitly [Roussignol *et al.*, 1985; Olbright *et al.*, 1987]. As one can expect now, the composite's nonlinearities are enhanced at the frequency ω_r, which corresponds to the plasmon resonance of a single metal grain. Numerical calculations for a finite concentrations p also give a considerable enhancement in the narrow frequency range near ω_r [Stroud and Zhang, 1994; Zhang and Stroud, 1994]. The calculations also show that the system sizes typically used in the numerical models of the nineties (see for example, [Bergman *et al.*, 1990]) were not large enough to make quantitative conclusions about the nonlinear properties for the frequencies ω close to the resonance frequency ω_r. This happened because the local field fluctuations are typically separated on large spatial scales ξ_e when $\omega \leq \omega_r$ [Sarychev and Shalaev, 2000; Sarychev and Shalaev, 2001; Sarychev and Shalaev, 2002].

To avoid direct numerical calculations, the effective medium theory (EMT) that has the virtue of mathematical and conceptual simplicity [Bruggeman, 1935] was extended for the nonlinear response of percolation composites, [Zeng *et al.*, 1989; Bergman, 1991; Bergman and Stroud, 1992; Levy and Bergman, 1993; Wan *et al.*, 1996; Hui, 1996; Hui *et al.*, 1997], and fractal clusters [Hui and Stroud, 1994]. For linear cases, predictions of the effective medium theory can be easily understood. They offer quick insight into problems that are difficult to attack by other means. However, EMT has disadvantages typical for the all mean-field theories, namely, it diminishes the role of fluctuations in the system. One example: It assumes that the local electric fields are the same inside the volume occupied by each component of the composite. The electric fields in different components are determined self-consistency. Another disadvantage is the generalization to the nonlinear case, which is ambiguous.

For the static case, the results of the last modification of the nonlinear EMT [Wan *et al.*, 1996; Hui *et al.*, 1997] are in the best agreement with comprehensive computer simulations performed for the two-dimensional

(2D) percolation composites [Levy and Bergman, 1993; Wan *et al.*, 1996; Hui *et al.*, 1997]. The original approach that combines the effective medium theory and spectral representation [McPhedran and Milton, 1981; Milton, 1981; Bergman and Stroud, 1992], has been developed by [Ma *et al.*, 1998]. In spite of the success, the application of this approach to any kind of nonlinear EMT remains questionable for the frequency ranges corresponding to the plasmon resonances in metal grains. The computer simulations and experimental results show that the local field distributions contain sharp peaks, sparsed on the distances much larger than a characteristic metal grain size (computer simulations: [Blacher *et al.*, 1995; Brouers *et al.*, 1997b; Baskin *et al.*, 1997; Brouers *et al.*, 1997a; Shalaev *et al.*, 1997] [Brouers *et al.*, 1998; Shalaev and Sarychev, 1998; Shalaev *et al.*, 1998; Gadenne *et al.*, 1998; Gresillon *et al.*, 1999a] [Sarychev *et al.*, 1999a; Sarychev *et al.*, 1999b; Gresillon *et al.*, 1999b; Sarychev *et al.*, 1999c; Sarychev and Shalaev, 2000] [Shubin *et al.*, 2000; Sarychev *et al.*, 2000a; Breit *et al.*, 2001; Ducourtieux *et al.*, 2001] [Shalaev *et al.*, 2001; Podolskiy *et al.*, 2002b; Shalaev *et al.*, 2002; Genov *et al.*, 2003a; Sarychev and Shalaev, 2002] [Sarychev and Shalaev, 2003; Genov *et al.*, 2003b]; experimental results: [Lagarkov *et al.*, 1997c; Gresillon *et al.*, 1999a; Gresillon *et al.*, 1999b; Gadenne *et al.*, 2000; Ducourtieux *et al.*, 2001], [Breit *et al.*, 2001; Gadenne and Rivoal, 2002; Seal *et al.*, 2003; Seal *et al.*, 2005].) Such pattern agrees qualitatively with numerical calculations and experimental results in metal fractals [Stockman *et al.*, 1994; Shalaev, 1996], [Stockman, 1997a] [Stockman, 1997b; Shalaev, 2000]. This implies that the local electric field by no means can be the same in all the metal grains of the composite. Therefore, the main assumption of the effective medium theory fails for the frequency range corresponding to the plasmon resonance in the semicontinuous films.

A new theory of electromagnetic field distribution and nonlinear optical processes in metal-dielectric composites has been developed recently [Sarychev *et al.*, 1995; Brouers *et al.*, 1997b; Lagarkov *et al.*, 1997c; Brouers *et al.*, 1997a; Brouers *et al.*, 1998; Gresillon *et al.*, 1999a], [Sarychev *et al.*, 1999a; Sarychev *et al.*, 1999c; Sarychev and Shalaev, 2000; Shubin *et al.*, 2000; Genov *et al.*, 2003a], [Genov *et al.*, 2005], [Seal *et al.*, 2005], [Seal *et al.*, 2006]. The new approach is based on the percolation theory. It relies on the fact that the problem of optical excitations in percolation composites is similar to the Anderson localization in quantum mechanics. The theory predicts that surface plasmons (SP) are typically localized and gives relatively simple expressions for the enhancements of local fields and corresponding nonlinear optical responses. It is important to stress out that the characteristic SP localization length is much smaller than the incident optical wavelength; in that sense, the predicted sub-wavelength localization of the SP differs from the well known localization of light due to strong scattering in a random homogeneous medium [Sheng, 1995].

3.2 Giant Field Fluctuations

In metal-dielectric percolation composites, the effective *dc* conductivity σ_e decreases with decreasing the volume concentration of metal component p and vanishes when the concentration p approaches a specific concentration p_c known as the percolation threshold [Clerc *et al.*, 1990; Bergman and Stroud, 1992; Stauffer and Aharony, 1994]. In the vicinity of the percolation threshold p_c, the effective conductivity σ_e is determined by an infinite cluster of percolating (conducting) channels. For concentration p smaller than the percolation threshold p_c, the effective *dc* conductivity $\sigma_e = 0$, implying that the system is dielectric-like. Therefore, metal-insulator transition takes place at $p = p_c$. Since the metal-insulator transition associated with the percolation represents a geometrical phase transition, one can anticipate that both the electric current and the field fluctuations are scale-invariant and large.

In percolation composites, however, the fluctuation pattern appears to be quite different from that for a second-order transitions, where the fluctuations are characterized by the long-range correlations with relative magnitudes of an order of unity regardless of any spatial point of the system [Stanley, 1981; Chaikin and Lubensky, 1995]. Contrary to that, the local electric fields are concentrated at the edges of large metal clusters for the case of *dc* percolation so that the field maxima (due to large fluctuations) are separated by distances on the order of the percolation correlation length ξ_p, which diverges when the metal volume concentration p approaches the percolation threshold p_c [Aharony *et al.*, 1993; Stroud and Zhang, 1994; Stauffer and Aharony, 1994]. Recall that the structure of the metal-dielectric composites is discussed in Sec. 1.2.

We show below that the difference in fluctuations becomes even more striking in the optical spectral range, where the local fields have the resonance nature: The relative field enhancements can be as large as 10^5 for the linear cases and 10^{20} and more for nonlinear responses, with the distances between the peaks much larger than the percolation correlation length ξ_p. In optical and infrared spectral regions, the surface plasmon resonances in metal particles governs the properties of metal-dielectric composites. To gain the insight on the high-frequency behavior of metal composites, we will consider a simple Drude model, which semi-quantitatively reproduces the basic optical properties of the metals (see, e.g., [Ashcroft and Mermin, 1976]).

In this approach, the dielectric constant of metal grains can be approximated by the Drude formula

$$\varepsilon_m(\omega) = \varepsilon_b - (\omega_p/\omega)^2/\left[1 + i\omega_\tau/\omega\right], \tag{3.1}$$

where ε_b is contribution to ε_m due to the inter-band transitions, $\omega_p = 4\pi ne^2/m$ is the plasma frequency (n is electron density, e and m are the corresponding electron charge and the mass), and $\omega_\tau = 1/\tau \ll \omega_p$ is the characteristic relaxation rate. In the high-frequency region, considered here, metal loss is relatively small, $\omega_\tau \ll \omega$. Therefore, the real part ε_m' of the metal dielectric function ε_m is much larger by an absolute value than the imaginary part ε_m'' ($\left|\varepsilon_m'\right|/\varepsilon_m'' \cong \omega/\omega_\tau \gg 1$), and ε_m' is *negative* for all the frequencies ω smaller than the renormalized plasma frequency,

$$\tilde{\omega}_p = \omega_p/\sqrt{\varepsilon_b}. \tag{3.2}$$

Thus, the metal conductivity $\sigma_m = -i\omega\varepsilon_m/4\pi$ can be characterized by the dominant imaginary part in the frequency range $\tilde{\omega}_p > \omega \gg \omega_\tau$. The imaginary part of conductivity has positive sign and it is inverse proportional to the frequency meaning that the metal conductivity σ_m exhibits the inductive behavior in optical range of frequencies. Thus the metal grains can be thought of as inductances (further abbreviated as L), while the dielectric gaps can be represented by the capacitances (C). The percolation composite is then represented by a set of randomly distributed L and C elements while the collective surface plasmons excited by an external field can be thought of as resonances and their combination of different $L-C$ circuits. In such description the giant local field fluctuations can be simulated as the excited surface plasmon eigenstates.

Below we mainly consider silver and gold semicontinuous films since the most experiments are performed with noble materials. In the Drude formula 3.1 for the metal dielectric function, the optical constants are taken from [Johnson and Christy, 1972] and [Kreibig and Volmer, 1995] (see also

[Palik, 1985]):

	ε_b	ω_p, eV	ω_τ, eV
Silver	5	9.1	0.02
Gold	6.5	9.3	0.03

(3.3)

The films are to be deposited on a glass substrate with the dielectric constant $\varepsilon_d = 2.2$ and we will use this value for all the numerical simulations below. The optical properties of metal-dielectric films show anomalous behavior that is absent for the bulk metal and dielectric components. For example, the anomalous absorption in the near-infrared spectral range leads to an unusual behavior of the composite transmittance and reflectance. (Typically, the transmittance is much higher than that of continuous metal films, whereas the reflectance is much lower, see [Cohen *et al.*, 1973; Niklasson and Granquist, 1984; Botten and McPhedran, 1985; Gadenne *et al.*, 1988; Gadenne *et al.*, 1989; Yagil *et al.*, 1991; Yagil *et al.*, 1992; Noh *et al.*, 1992; Seal *et al.*, 2003]). Near and well below the conductivity threshold, the anomalous absorptance can be as large as 50%.

A number of the effective-medium theories were proposed for the calculation of the optical properties of semicontinuous random films, including the Maxwell-Garnet [Garnett, 1904] and Bruggeman [Bruggeman, 1935] approaches, and their various modifications [Bergman and Stroud, 1992; Yagil *et al.*, 1992]. The renormalization group method is also widely used to calculate effective dielectric response of semicontinuous metal films near the percolation threshold (see [Vinogradov *et al.*, 1988; Depardieu *et al.*, 1994] and references therein). However, none of these theories allows to calculate the field fluctuations and the effects resulting from these fluctuations.

Because semicontinuous metal films are of the great interest due to their fundamental physical properties and various applications, it is important to study the statistical properties of the electromagnetic fields in the near-field zone of these films. To simplify the theory, we assume that the electric field is homogeneous in the direction perpendicular to the film plane. This assumption means that the skin depth for the metal grains, $\delta \cong c/(\omega\sqrt{|\varepsilon_m|})$, is much larger than the grain size, a, so that the quasistatic approximation holds. Note that the role of the skin depth effect can be very important, resulting, in many cases, to the strong alterations of the electromagnetic response. These effects will be discussed in Sec. 3.7. Yet, the quasistatic approximation significantly simplifies theoretical consideration of the field fluctuations and describes well the optical properties of semicontinuous films providing qualitative (and in some cases, quantitative) agreement with an experimental data.

Below, we neglect the skin effect so that a semicontinuous film can be considered as a $2D$ object. In the optical frequency range, when the frequency, ω, is much larger than the relaxation rate of the metallic component, ω_τ, a semicontinuous metal film can be represented by a two-dimensional L–R–C lattice. As in the above, the capacitance C stands for the gaps between metal grains that are filled by the dielectric material (the substrate), characterized by the dielectric constant ε_d. The inductive (L-R) elements represent the metallic grains, which have the dielectric function $\varepsilon_m(\omega)$ given by Eq. (3.1) for the Drude model. In the high-frequency range considered here, the loss in metal grains are small, $\omega \gg \omega_\tau$. Therefore, the absolute value of the real part of the metal dielectric function is much larger than that of the imaginary part and it is negative for the frequencies ω, which are below the renormalized plasma frequency, $\tilde{\omega}_p = \omega_p/\sqrt{\varepsilon_b}$. As a result, the metal conductivity is almost purely imaginary and metal grains can be simulated as the L-R elements, with the active component much smaller than the reactive one.

If the skin-depth effect cannot be neglected, i.e. the skin depth δ is smaller than the metal grain size a, the simple quasistatic presentation of a semicontinuous film as a two-dimensional array of the L–R and C elements is not valid. Still, we can use the L–R–C model in other limiting case: When the skin effect is strong, $\delta \ll a$ [Sarychev *et al.*, 1995; Sarychev and Shalaev, 2000]. In this case, the loss in metal grains are small, regardless of the ratio ω/ω_τ, whereas the effective metal-grain inductance depends on the grain size and its shape rather than on the material constants of the metal. Properties of metal-dielectric composites beyond the quasistatic approximation will be discussed in details in Sec. 3.7.

To gain the initial insight on the problem it is instructive to consider the film properties at the percolation threshold, $p = p_c$, where the exact result for the effective dielectric constant $\varepsilon_e = \sqrt{\varepsilon_d \varepsilon_m}$ holds in the quasistatic case [Dykhne, 1971] (see discussion in Ch. 1.2). If we neglect the metal loss ($\omega_\tau = 0$), the metal dielectric function ε_m becomes negative for the frequencies, which are smaller than the renormalized plasma frequency, $\tilde{\omega}_p$. If we also neglect possible small loss in the dielectric substrate, assuming that ε_d is real and positive, then, ε_e is purely imaginary for $\omega < \tilde{\omega}_p$. Thus, a film consisting of loss-free metal and dielectric grains is absorptive for $\omega < \tilde{\omega}_p$. The effective absorption of a loss-free film implies that the electromagnetic

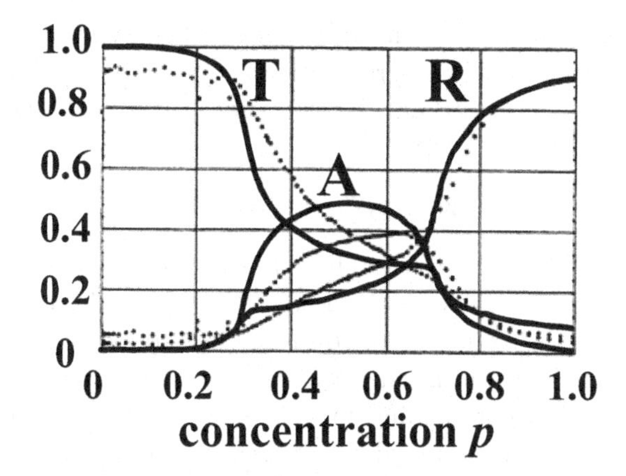

Fig. 3.4 Optical properties of semicontinuous gold film as functions of metal concentration p : A (absorptance), R (reflectance), and T (transmittance); wavelength $\lambda = 2.2\ \mu m$; dots show the experimental results [Yagil *et al.*, 1991; Yagil *et al.*, 1992], solid lines calculations [Sarychev *et al.*, 1995]; the thickness of the gold grains $d = 20$ nm.

energy is stored in the system and the local fields can increase unlimitedly. In reality, the local fields in the metal films are, of course, finite because of the loss. However, if the loss is small, one anticipates very strong field fluctuations. We, therefore, can anticipate that the giant field fluctuations exist at the percolation threshold p_c and at the metal concentrations p near p_c, where the effective absorptance achieves $40 - 50\%$ (see Fig. 3.4) so that the electromagnetic energy can accumulate in the film.

3.2.1 *Lattice model*

To calculate the local field in the film we will assume that a percolation composite is illuminated by the light and consider the local optical field distribution. We consider nanocomposites where a typical metal grain size a is much smaller than the wavelength of the light in the visible and infrared spectral ranges, λ. When it is the case, we can introduce the potential $\phi(\mathbf{r})$ for the local electric field. Then the local current density $\mathbf{j}$ can be written as $\mathbf{j}(\mathbf{r}) = \sigma(\mathbf{r})\,(-\nabla\phi(\mathbf{r}) + \mathbf{E}_0)$, where $\mathbf{E}_0$ is the applied field and $\sigma(\mathbf{r})$ is the local conductivity. In the considered here the quasistatic case the problem of finding the field distribution reduces to the solution of the Poisson equation, which is represented by the current conservation law $\operatorname{div}\mathbf{j} = 0$, so that

$$\nabla \cdot (\sigma(\mathbf{r})\,[-\nabla\phi(\mathbf{r}) + \mathbf{E}_0]) = 0. \tag{3.4}$$

In Eq. (3.4), the local conductivity $\sigma(\mathbf{r})$ takes either values for metal and dielectric components, σ_m or σ_d, respectively.

It is convenient to rewrite Eq. (3.4) in terms of the local dielectric constant $\varepsilon(\mathbf{r}) = 4\pi i\sigma(\mathbf{r})/\omega$ as follows

$$\nabla \cdot [\varepsilon(\mathbf{r})\nabla\phi(\mathbf{r})] = \mathcal{E}, \tag{3.5}$$

where $\mathcal{E} = \nabla \cdot [\varepsilon(\mathbf{r})\mathbf{E}_0]$. The external field $\mathbf{E}_0$ is chosen to be real, while the local potential $\phi(\mathbf{r})$ is complex-valued since the metal dielectric constant ε_m is complex $\varepsilon_m = \varepsilon'_m + i\varepsilon''_m$ in the optical and infrared spectral regions.

Because of difficulties in finding a closed-form solution to the Poisson Eq. (3.4) or (3.5), a tight binding model, in which metal and dielectric particles are represented by metal and dielectric bonds of a cubic lattice, has been extensively used. After such discretization, Eq. (3.5) acquires the form of Kirchhoff's equations defined on a cubic lattice [Bergman and Stroud, 1992]. We write the Kirchhoff's equations in terms of the local dielectric function and assume that the external electric field $\mathbf{E}_0$ is directed along the "z"-axis. Thus, we obtain the following set of the equations

$$\sum_j \varepsilon_{ij}\left(\phi_j - \phi_i\right) = \sum_j \varepsilon_{ij}E_{ij}, \tag{3.6}$$

where ϕ_i and ϕ_j are the electric potentials determined at the sites of the cubic lattice and the summation is over the nearest neighbors of the i-th site. The electromotive force (EMF) E_{ij} takes value E_0a_0, for the bond $\langle ij\rangle$ in the positive z direction (where a_0 is the spatial period of the cubic lattice) and $-E_0a_0$, for the bond $\langle ij\rangle$ in the $-z$ direction; $E_{kj} = 0$ for the other four bonds at the ith site. As a result, the composite is simulated by a resistor-capacitor-inductor network represented by Kirchhoff's equations (3.6) while the EMF forces E_{ij} represent the external electric field applied to the system.

Performing the transition from the continuous medium described by Eq. (3.4) to the random network described by Eq. (3.6), we have assumed, as usually [Bergman and Stroud, 1992; Stauffer and Aharony, 1994] that bond permittivities ε_{ij} are statistically independent and a_0 is equal to the metal grain size, $a_0 = a$. In the considered case of a two-component metal-dielectric random composite, the permittivities ε_{ij} take values ε_m and ε_d, with probabilities p and $1-p$, respectively. Assuming that the bond permittivities ε_{ij} are statistically independent in Eq. (3.6), we can simplify considerably the computer simulations as well as the analytical consideration of the local optical fields in the composite. We note that the critical material properties are universal, i.e. they are independent of the details of the model, e.g., of the possible correlations of permittivity ε_{ij} in different bonds.

3.2.2 *Numerical method*

Although there presently exist many efficient numerical methods for calculating the effective conductivity of the composite materials (see [Genov *et al.*, 2003a; Clerc *et al.*, 2000; Tortet *et al.*, 1998; Bergman and Stroud, 1992]), they typically do not allow calculations of the field distributions. Below we describe a computer approach, which is based on a real space renormalization group (RSRG) method [Blacher *et al.*, 1995; Brouers *et al.*, 1997a; Brouers *et al.*, 1997b], first suggested by Reynolds, Klein, Stanley, and Sarychev [Reynolds *et al.*, 1977; Sarychev, 1977] and then extended to study the conductivity [Bernasconi, 1978] and permeability of oil reservoirs by Aharony [Aharony, 1994].

This approach can be adopted for finding the field distributions in the following way. First, we generate a square lattice of the L–R (metal) and C (dielectric) bonds, using a random number generator. As seen in Fig. 3.5, such the lattice can be pictured as the set of the "rectangular" elements. One of such elements, labelled as ABCDEFGH, is shown in Fig. 3.5(a). In the first stage of the RSRG procedure, each of these elements is replaced by the two Wheatstone bridges, as demonstrated in Fig. 3.5(b). After the transformation, the initial square lattice is converted to another square lattice, where the distance between the sites are now twice larger. The each bond between the two nearest neighboring sites is the Wheatstone bridge. Note that there is a one-to-one mapping between the "x" (horizontally aligned) bonds in the initial lattice and the "x" bonds in the triangular-like

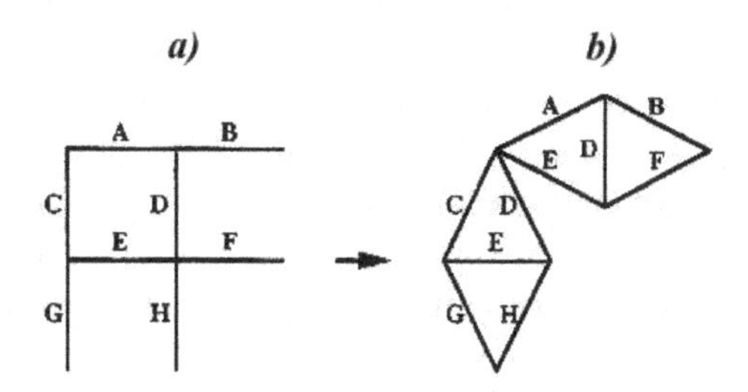

Fig. 3.5 The real space renormalization scheme.

bridges of the transformed lattice, as seen in Fig. 3.5. The same one-to-one mapping also exists between the "y" (vertical) bonds. The transformed lattice is also a square lattice, and we can use the RSRG transformation again to this new lattice. We continue RSRG procedure until desired size (l) of the system is reached. As a final result, we will receive two large Wheatstone bridges in the "x" and "y" directions instead of the original lattice. Each of the two bridges will have a hierarchical structure that consists of smaller bridges whose size is partitioned between 2 to $l/2$. Because the one-to-one mapping was preserved at each step of the transformation, the direct correspondence also exists between the elementary bonds of the transformed lattice in each generation and the bonds of the original lattice.

After the RSRG transformation, we solve the Kirchhoff's equations in the presence of an external field to determine the fields and the currents in all the bonds of the final lattice. Due to the hierarchical lattice structure, these equations can be solved exactly. Then, we trace back one-to-one correspondence between the elementary bonds of the final and original lattice to find the field distributions in the initial lattice as well as its effective conductivity. The number of operations to obtain the full local fields distribution is proportional to l^2, which is orders of magnitude smaller then l^7 operations needed in the transform-matrix method [Bergman and Stroud, 1992], and l^3 operations needed in the Frank-Lobb algorithm [Frank and Lobb, 1988]; none of which gives the local field distributions. Using the proposed method, it takes less than a minute of computational time of ordinary PC to obtain the effective conductivity and field distributions for the system of 1000×1000 elements.

Clearly, the RSRG procedure is not exact since the effective connectivity of the transformed system does not repeat exactly the connectivity of the initial square lattice. To check the accuracy of the RSRG, we solved the $2D$ percolation problem using this method. Namely, we calculated the effective parameters of a two-component composite with the real metallic conductivity being much larger than the real conductivity of the dielectric component, $\sigma_m \gg \sigma_d$. We have obtained the percolation threshold $p_c = 0.5$. The effective conductivity at the percolation threshold was very close to tabulated result, $\sigma(p_c) = \sqrt{\sigma_m \sigma_d}$. The results repeated the exact solution for $2D$ composites [Dykhne, 1971]. This is not surprising since the RSRG procedure preserves the self-duality of the initial system. The critical exponents obtained by the RSRG are also close to the known exponent values from the percolation theory discussed in Sec. 1.2. This assures us that the local fields obtained in the RSRG approach are close to the actual ones.

Recently, we have developed an *exact* numerical method [Genov *et al.*, 2003a; Genov *et al.*, 2003b], which is based on the same ideas of the real space renormalization. The new method gives exact values of the local fields and has efficiency of l^6-operations. Comparison of the RSRG results and exact numerical results shows that RSRG method gives very good approximation for the giant fluctuations of the local electric field. Numerical results for the local field distribution and various linear and nonlinear optical effects for semicontinuous metal films, presented in this chapter and in Ch. 3.7 are obtained by using both of these methods.

3.2.3 *Field distributions on semicontinuous metal films*

As mentioned earlier, we model a film by a square lattice consisting of metallic bonds with conductivity $\sigma_m = -i\varepsilon_m\omega/4\pi$ of concentration p (for L - R bonds) and the and dielectric bonds with the conductivity $\sigma_d = -i\varepsilon_d\omega/4\pi$ and concentration $1-p$ (C bonds). The applied field is set to unity $E_0 = 1$, whereas the local fields inside the system are complex-valued. The dielectric constant of silver is described by Drude expression (3.1) with the parameters given in Table 3.3. We still use $\varepsilon_d = 2.2$ for a glass substrate. In Fig. 3.6, we show the field distributions $|E(\mathbf{r})/E_0|^2$ for the plasmon resonance frequency $\omega = \omega_r$ that corresponds to the condition $\mathrm{Re}(\varepsilon_m(\omega_r)) = -\varepsilon_d$. The value of the frequency ω_r, which gives the resonance of an isolated metal particle, is slightly bellow the renormalized plasma frequency $\tilde{\omega}_p$ defined in Eq. (3.2). For silver particles, the resonance condition is fulfilled at $\lambda \approx 0.4$ μm. [For a two-dimensional, i.e., z -independent, problem, particles can be thought of as infinite cylinders in the z -direction. In the quasistatic approximation, these cylinders resonate at the frequency $\omega = \omega_r$, which corresponds to the condition $\varepsilon_m(\omega_r) = -\varepsilon_d$, for the field polarized in the (x, y) -plane]. The results presented in Fig. 3.6 are given for various metal fractions p. For $p = 0.001$, metal grains weakly interact so that all the local intensity peaks have similar height and their presence indicates the metal particle locations. Note that similar distribution is obtained for $p = 0.999$ when the role of metal particles is played by the dielectric voids. For $p = 0.01$, the metal particles already strongly interact in spite of their relatively small concentration so that the field distribution is rather inhomogeneous. The local maxima of the intensity greatly differ by the amplitudes from one location to another.

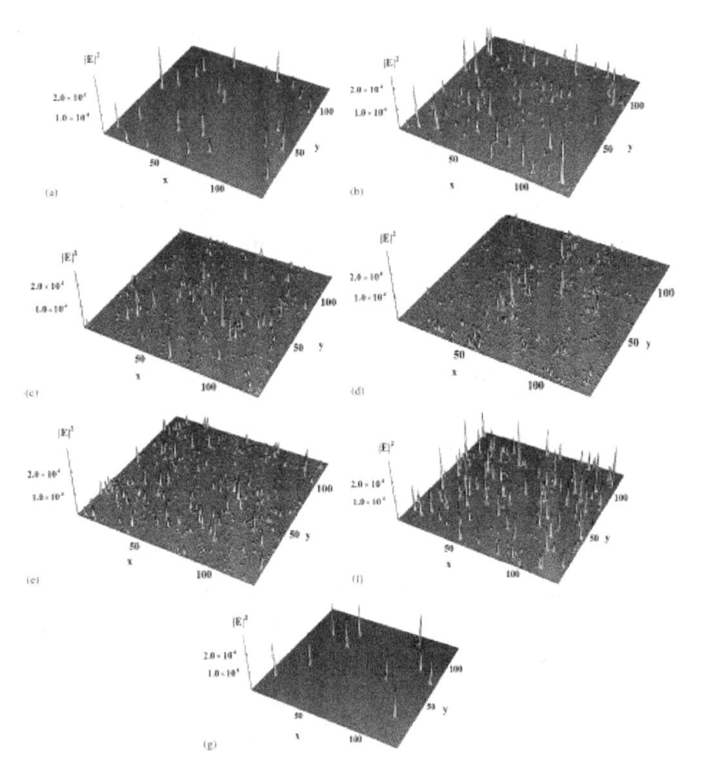

Fig. 3.6 Distribution of the local field intensities $|\mathbf{E}(x,y)|^2$ on a metal (silver) semicontinuous film for $\varepsilon_m' = -\varepsilon_d = -2.2$ ($\lambda \cong 365$ nm) for different metal concentrations p: (a) $p = 0.001$, (b) $p = 0.01$, (c) $p = 0.1$, (d) $p = 0.5$, (e) $p = 0.9$, (f) $p = .99$, and (g) $p = 0.999$.

At metal concentrations $p = 0.1$ and, especially, $p = 0.5$, the metal grains form clusters of strongly interacting particles. These clusters resonate on frequencies different than that one for an isolated particle, therefore, the local intensity peaks are smaller on average than in the case of an isolated particles. The peak height distribution is again very inhomogeneous. Note that spatial scale for the local field distribution is much larger then the metal grain size a (which is chosen to be unity in all the figures). Therefore the main assumption of the effective medium theory that the local fields are the same for all the metal grains within the composite fails for the plasmon resonance frequencies and non-vanishing concentrations p. Again, we emphasize a strong resemblance in the field distributions for p and $1-p$ (cf. Figs. 3.6(a) and 3.6(g), 3.6(b) and 3.6(f), 3.6(c) and 3.6(e)).

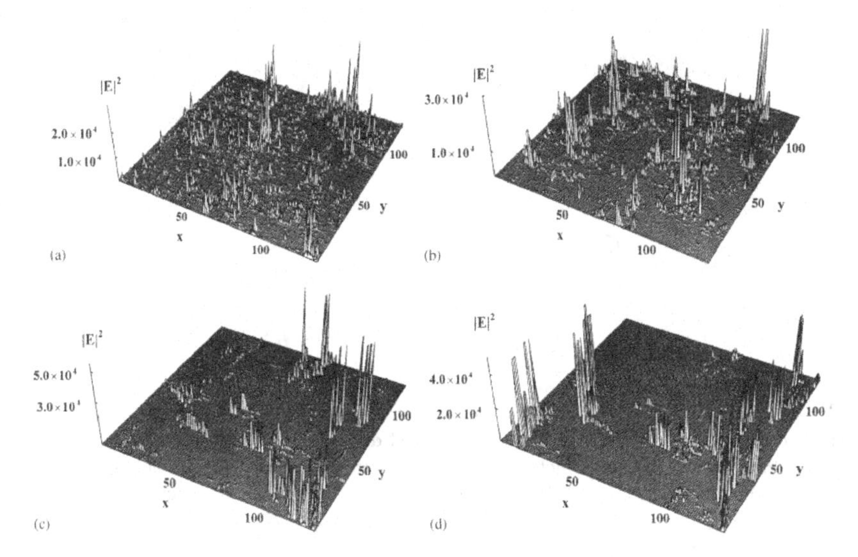

Fig. 3.7 Distribution of the local intensity $|\mathbf{E}(x,y)|^2/|\mathbf{E}_0|^2$ in a semicontinuous films at the percolation threshold for different wavelengths: (a):$\lambda = 0.5$ μm, (b): $\lambda = 1.5$ μm, (c): $\lambda = 10$ μm, and (d): $\lambda = 20$ μm.

For larger wavelengths, a single metal grain is far off the plasmon resonance. Nevertheless, the local field fluctuations are even larger than those at the plasmon resonance as clearly seen from Figs. 3.7(a)–(d). At these wavelengths, clusters of the conducting particles (rather than individual particles) resonate with the external field. Therefore, it is not surprising that the local field distributions are quite different from those in Fig. 3.6. In Fig. 3.7, we show the field distributions at percolation threshold $p = p_c = 0.5$ for $\lambda = 0.5$ μm (2a), $\lambda = 1.5$ μm (2b), $\lambda = 10$ μm (2c), and $\lambda = 20$ μm (2d). Note that the field peak intensities increase with a wavelength, reaching very high values, $\sim 10^5 I_0$. The spatial separations between the peaks also increase with a wavelength. These results contradict again with the effective medium theory that predicts strong field fluctuations [Stroud and Zhang, 1994] only in the vicinity of plasmon resonance frequency ω_r.

The origin of the giant field fluctuations in the visible and infrared spectral regions can be understood from the following consideration. Optical properties of metal are determined by the plasma frequency (see Eq. (3.1)). For the "good" optical metal like silver, gold, aluminum, copper, etc., the plasma frequency ω_p is much larger than the optical frequencies and

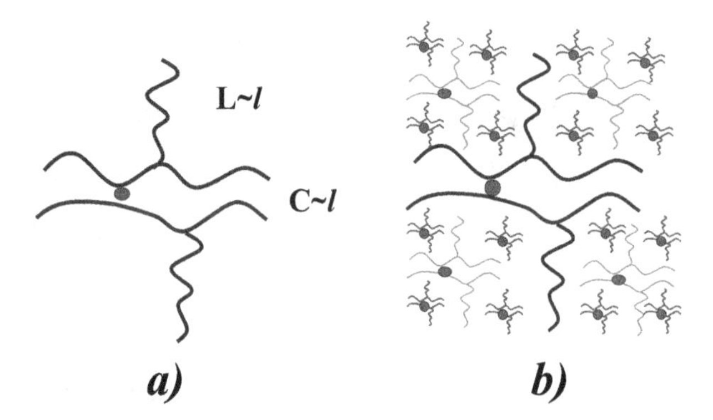

Fig. 3.8 (a) Two adjacent metal clusters of size l; external electric field is applied in vertical direction; it concentrates in hot spot (red dot) — point of closest proximity of two clusters. (b) Local field fluctuations in semicontinuous metal film; hot spots are shown as red dots; clusters of different sizes resonate at different frequency.

corresponds the energies of about 10 eV. The absolute value of the metal permittivity is much larger than unity $|\varepsilon_m| \gg 1$. Under these conditions the electric current (real one and the displacement one) mainly flows through the metal clusters and in the inter-cluster gaps of the semicontinuous film (cf. Fig. 3.2). The local electric field concentrates at the points of closest proximity between two adjacent clusters (hot spots) as it is shown in Fig. 3.8 by the red dots. Consider now the film with the metal concentration p just below the percolation threshold. The metal clusters of various sizes are present in the system, which is depicted in Fig. 3.8(b). It has been shown that the metal clusters near the percolation threshold have complicated self-similar droplet structure [Sarychev and Vinogradov, 1981; Sarychev and Vinogradov, 1983], however its resistance and inductance are determined by a certain critical pass, which spans over the entire cluster. It is shown in Fig. 3.8. The inductance of the cluster is proportional to its size l (recall that metal has inductive conductivity in the optical range of frequencies), and the capacitance of the intercluster gap is also proportional to the size l. Therefore the resonance frequency $\omega_r \propto 1/\sqrt{LC}$ for the clusters of size l is inversely proportional to its size

$$\omega_r \propto \frac{1}{\sqrt{LC}} \propto \frac{1}{l}. \tag{3.7}$$

Suppose that the clusters of size l_1, shown in blue in Fig. 3.8(b), resonate at frequency ω_{r1}. Then the green-colored clusters of size $l_2 = 2l_1$ resonate at

$\omega_{r2} \simeq \omega_{r1}/2$ and the brown clusters of size $l_3 = 4l_1$ resonate at $\omega_{r3} \simeq \omega_{r1}/4$. For each resonance, the local electric field concentrates in hot spots between the resonating clusters. Thus, the distance between the maxima of intensity of local electric field increases with decreasing frequency (increasing the wavelength), which indeed can be observed in Fig. 3.7. The resonance quality factor increases with increasing the local inductance, therefore, it is not surprising that the amplitude of the peaks in Fig. 3.7 increases with decreasing the frequency. To quantitative predictions for the giant fluctuations, we present a scaling theory for the local field distributions in the next section.

The first experiment on the field distribution in the semicontinuous metal films has been performed in a microwave region [Lagarkov *et al.*, 1997c] and will be discussed in Sec. 3.7. For the visible spectral range, the local field distribution have been measured by [Gresillon *et al.*, 1999a; Gresillon *et al.*, 1999b; Gadenne *et al.*, 2000; Gadenne *et al.*, 2000; Ducourtieux *et al.*, 2001] and by [Seal *et al.*, 2003; Seal *et al.*, 2002; Seal *et al.*, 2005] using scanning near-field optical microscopy (SNOM), which provides sub-wavelength resolution. The typical pattern of local intensity is shown in Fig. 3.3 for the metal concentrations p corresponding to the percolation threshold. It is qualitatively similar to the calculated local intensities, shown in Fig. 3.7. Since the SNOM was operated in the tapping mode the detected local signals were averaged over the tip-surface separations ranging between 0 and 100 nm. Because of it and because of the finite size of the tip, the detected field intensities were $\sim 10-100$ times less than the actual field measured directly on the film surface (which can be probed, for example, by using surface-adsorbed molecules). Aside of this, the detected field distribution shown in Fig. 3.3 is very similar to the one predicted by theory (Figs. 3.6 and 3.7). If, however, the SNOM averaging effect is taken into account, adjusted simulations agree very well with experimental data [Seal *et al.*, 2003].

The near-field spectroscopy of the percolation films was also performed by [Gadenne *et al.*, 2000; Gresillon *et al.*, 1999b; Gresillon *et al.*, 1999a], by parking the SNOM tip at different positions on the film surface and varying the wavelength. This local nano-spectroscopy allows one to determine the local resonances of nm-size areas right underneath the tip; the nanostructures at different spatial locations resonate at different wavelengths leading to different local near-field spectra. These spectra characterize wavelength-dependence of the field hot spots associated with the localized surface plasmon (SP) modes. The spectra consist of several peaks with spectral width

$\Delta\lambda \sim 10 \div 100$ μm, and they greatly depend on spatial location of the point where the near-field tip is parked. Even a small local position shift of $\sim$ 100 nm results in a completely different spectrum, which is by itself a strong manifestation of the surface plasmon localization. We note that for continuous metal (or dielectric) films neither sub-wavelength hot spots nor their local spectra can be observed, because, in that case, the optical excitations will be delocalized.

In conclusion, the near-field imaging and the spectroscopy of random metal-dielectric films near the percolation suggest the localization of optical excitations in small nm-sized, hot-field areas. The observed pattern of the localized modes and their spectral dependencies are in agreement with theoretical predictions and numerical simulations. The hot spots of a percolation film represent very large local fields (due to strong fluctuations); spatial positions of these spots strongly depend on the frequency of the incident light. Near-field spectra observed and calculated at various points of the surface consist of several spectral resonances whose spectral locations depend on the probing areas of the sample. All these features are only observable in the near-field optical zone. In the far zone, one observes images and spectra, in which the hot spots and the spectral resonances are averaged out.

3.3 Localization of Surface Plasmons

In this section, we consider the surface plasmons (SP), which are excited in metal-dielectric composites. We will investigate SP in two-dimensional ($2D$) systems (e.g., semicontinuous metal films), and also in three dimensional ($3D$) composites. We will again use the lattice model discussed in Sec. 3.2.1 to reduce the problem of the local field distribution to the solution of the Kirchhoff's equations on a cubic (square) lattice.

3.3.1 *Localization length and average intensity of local electric field*

For further consideration we assume that the cubic lattice, where the Kirchhoff's equations are defined, has a very large but finite number of sites N. We rewrite Eq. (3.6) in matrix form with the Hamiltonian $\widehat{\mathrm{H}}$ [Sarychev *et al.*, 1999b], [Sarychev *et al.*, 1999c], [Gresillon *et al.*, 1999a], [Genov *et al.*, 2003a], and [Genov *et al.*, 2005], defined in terms of the local dielectric constants,

$$\widehat{\mathrm{H}}\,\phi = \mathcal{E}, \tag{3.8}$$

where ϕ is a vector of the local potentials $\phi = \{\phi_1, \phi_2, \ldots, \phi_N\}$ determined in all N sites of the lattice, and the vector $\mathcal{E}$ is also determined in N sites of the lattice, namely, $\mathcal{E}_i = \sum_j \varepsilon_{ij} E_{ij}$, which is followed from Eq. (3.6). The Hamiltonian $\widehat{\mathrm{H}}$ is a $N \times N$ matrix that has off-diagonal elements $\mathrm{H}_{ij} = -\varepsilon_{ij}$ and diagonal elements defined as $\mathrm{H}_{ii} = \sum_j \varepsilon_{ij}$, where j now refers to the nearest neighbors of the i-th site. The off-diagonal elements H_{ij} take values $\varepsilon_d > 0$ and $\varepsilon_m = (-1 + i\kappa)\,|\varepsilon_m'|$ with probability p and $1 - p$ respectively. The loss factor $\kappa = \varepsilon_m'' / |\varepsilon_m'|$ is small so that $\kappa \ll 1$. The diagonal elements H_{ii} are distributed between $2D\varepsilon_m$ and $2D\varepsilon_d$ as $2D$ is the number of the nearest neighbors in D dimensional cubic lattice; D is the dimensionality of the space.

It is convenient to break the Hamiltonian $\widehat{\mathrm{H}}$ onto a sum of two Hermitian Hamiltonians $\widehat{\mathrm{H}} = \widehat{\mathrm{H}}' + i\kappa\widehat{\mathrm{H}}''$, where the term $i\kappa\widehat{\mathrm{H}}''$ $(\kappa \ll 1)$ represents the loss in the system. Hereafter, we refer to operator $\widehat{\mathrm{H}}'$ as to Kirchhoff's Hamiltonian and denote it KH. Thus, the problem of the field distribution in the system, i.e. the problem of finding solution to Kirchhoff's Eqs. (3.6) or (3.8), becomes the eigenfunction problem for the Kirchhoff's Hamiltonian, where the loss is treated as the perturbation. The matrix $\widehat{\mathrm{H}}'$ of KH is symmetric and all its elements are real so that the all eigenvalues Λ_n are real for the eigenvalue problem $\widehat{\mathrm{H}}'\Psi_n = \Lambda_n \Psi_n$.

Since the real part ε_m' of metal dielectric function ε_m is negative, $\varepsilon_m' < 0$, and the permittivity of dielectric host is positive, $\varepsilon_d > 0$, the manifold of the KH eigenvalues Λ_n contains eigenvalues which are equal (or close) to zero. Then eigenstates Ψ_n, which correspond to eigenvalues $|\Lambda_n| \ll |\varepsilon_m|, |\varepsilon_d|$, are strongly excited by the external field and eventually pronounced as the giant field fluctuations, representing the resonant SP modes. If we assume that the eigenstates excited by the external field are also localized, we receive the familiar picture of strongly localized field intensities.

Now we consider the behavior of the eigenfunctions Ψ_n of the KH $\widehat{\mathrm{H}}'$ in more details: for the resonance case $\varepsilon_m' = -\varepsilon_d$. Since a solution to Eq. (3.6) does not change when multiplying ε_m and ε_d by the same factor, we can normalize the system and set $\varepsilon_d = -\varepsilon_m = 1$. We also put for simplicity $p = 0.5$ for the metal concentration. (The local field corresponding to this case is shown in Figs. 3.6(d) and 3.7(a).)

Suppose we found all the eigenvalues Λ_n and the eigenfunctions Ψ_n of $\widehat{\mathrm{H}}'$. Then we can express the potential ϕ in terms of the eigenfunctions $\phi = \sum_n A_n \Psi_n$ and substitute it in Eq. (3.8). Doing so, we obtain the following equation for coefficients A_n:

$$(i\kappa b_n + \Lambda_n) A_n + i\kappa \sum_{m \neq n} \left(\Psi_n \left| \widehat{\mathrm{H}}'' \right| \Psi_m \right) A_m = \mathcal{E}_n, \tag{3.9}$$

where $b_n = \left(\Psi_n \left| \widehat{\mathrm{H}}'' \right| \Psi_n \right)$, and $\mathcal{E}_n = (\Psi_n | \mathcal{E})$ is a projection of the external field on the eigenstate Ψ_n. (The product of two vectors, e.g., Ψ_n and $\mathcal{E}$ is defined here in a usual way, as $\mathcal{E}_n = (\Psi_n|\mathcal{E}) \equiv \sum_i \Psi^*_{n,i} \mathcal{E}_i$, where the sum is now taken over all the lattice sites.) Since all the parameters in the Hamiltonian $\widehat{\mathrm{H}}''$ are on the order of unity, the matrix elements $b_n \sim 1$ and are approximated by some constant b. We then assume that eigenstates Ψ_n are localized within spatial domains $\xi_A(\Lambda)$, where $\xi_A(\Lambda)$ is the characteristic localization length, which depends on the eigenvalue Λ. Then, the sum in Eq. (3.9) converges and can be treated as a small perturbation. In the zeroth-order approximation of perturbation theory,

$$A_n^{(0)} = \mathcal{E}_n / (\Lambda_n + i\kappa b). \tag{3.10}$$

The first-order correction to the n-th coefficient A_n is equal to

$$A_n^{(1)} = -i\kappa \sum_{m \neq n} \left(\Psi_n \left| \widehat{\mathrm{H}}'' \right| \Psi_m \right) \frac{A_m^{(0)}}{(\Lambda_m + i\kappa b)}. \tag{3.11}$$

Due to the resonance behavior of the denominator in Eq. (2.28), the most important eigenstates in this sum for $\kappa \to 0$ are those, which have $|\Lambda_m| \leq b\kappa$. Since the eigenstates Λ_n are distributed in the unit-length interval, the spatial density of the eigenmodes with $|\Lambda_m| \leq b\kappa$ vanishes as $a^{-D}\kappa \to 0$ at $\kappa \to 0$ and, therefore a spatial "distance" between the modes $\sim a\kappa^{-1/D}$ goes to the infinity. Therefore the ratio $\left| A_n^{(1)} / A_n^{(0)} \right|$ is exponentially small $\left| A_n^{(1)} / A_n^{(0)} \right| \sim \left| \kappa^{-1} \sum_{m \neq n,\, |\Lambda_m| < \kappa} \left(\Psi_n \left| \widehat{\mathrm{H}}'' \right| \Psi_m \right) \right| \propto \exp\left\{ -\left[a/\xi_A(\kappa)\right] \kappa^{-1/D} \right\}$ and can be neglected when $\kappa \ll \left[a/\xi_A(\kappa)\right]^D$. Under these conditions, the local potential $\phi(\mathbf{r}) = \sum_n A_n^{(0)} \Psi_n = \sum_n \mathcal{E}_n \Psi_n(r) / (\Lambda_n + i\kappa b)$ [see Eq. (3.10)] and the fluctuating part of the local field $\mathbf{E}_f = -\nabla\phi(\mathbf{r})$ is given by

$$\mathbf{E}_f(\mathbf{r}) = -\sum_n \mathcal{E}_n \nabla \Psi_n(\mathbf{r}) / (\Lambda_n + i\kappa b), \tag{3.12}$$

where the operator ∇ is acting within the lattice. The average field intensity is as follows

$$\left\langle |E|^2 \right\rangle = \left\langle |\mathbf{E}_f + \mathbf{E}_0|^2 \right\rangle = |E_0|^2 + \left\langle \sum_{n,m} \frac{\mathcal{E}_n \mathcal{E}_m^* \left(\nabla \Psi_n(\mathbf{r}) \cdot \nabla \Psi_m^*(\mathbf{r}) \right)}{\left(\Lambda_n + i\kappa b \right) \left(\Lambda_m - i\kappa b \right)} \right\rangle, \tag{3.13}$$

where $\mathbf{E}_0$ is external field and we have accounted for $\langle \mathbf{E}_f \rangle = \left\langle \mathbf{E}_f^* \right\rangle = 0$.

The matrix of Kirchhoff Hamiltonian has dimensions $N \times N$ and, correspondingly N eigenstates. We can order the eigenstates by their respective eigenvalue: $\Lambda_1 < \Lambda_2 < \cdots < \Lambda_n < \cdots < \Lambda_N$. The properties of a particular eigenstate, e.g., the localization length ξ_A, are determined by its eigenvalue Λ. However, the eigenfunction probability function $|\Psi_n(r)|^2$ describes each time a new, different, position in space for each realization of a macroscopically homogeneous random system. Therefore, we can independently average the numerator in the second term of Eq. (3.13) over all the positions of eigenstates Ψ_n and Ψ_m. Taking into account $\langle \nabla \Psi_n(\mathbf{r}) \rangle = 0$, we obtain

$$\mathcal{E}_n \mathcal{E}_m^* \left(\nabla \Psi_n(\mathbf{r}) \cdot \nabla \Psi_m^*(\mathbf{r}) \right) \simeq \left\langle |\mathcal{E}_n|^2 \left| \nabla \Psi_n(\mathbf{r}) \right|^2 \right\rangle \delta_{nm}, \tag{3.14}$$

and accordingly

$$\left\langle |E|^2 \right\rangle = |E_0|^2 + \sum_n \frac{\left\langle |\mathcal{E}_n|^2 \left| \nabla \Psi_n \right|^2 \right\rangle}{\Lambda_n^2 + (b\kappa)^2}. \tag{3.15}$$

To simplify further estimates we assume that the "integral" characteristics of the eigenstates Ψ_n, similar to the projection of Ψ_n on the vector coordinate $\mathcal{E}_n = (\Psi_n | \mathcal{E})$, depend mainly on the eigenvalue Λ_n: $\mathcal{E}_n \simeq \mathcal{E}(\Lambda_n)$. Then we replace the sum in Eq. (3.15) by the integration over the entire spectrum, obtaining

$$\left\langle |E|^2 \right\rangle = |E_0|^2 + \int \frac{|\mathcal{E}(\Lambda)|^2 \left\langle |\nabla \Psi(\Lambda)|^2 \right\rangle}{\Lambda^2 + (b\kappa)^2} \rho(\Lambda)\, d\Lambda, \tag{3.16}$$

where $\rho(\Lambda)$ is the density of the eigenstates

$$\rho(\Lambda) = \left\langle \sum_n \delta(\Lambda - \Lambda_n) \right\rangle, \tag{3.17}$$

Such replacement is possible since the system is macroscopically homogeneous.

To estimate the numerator of Eq. (3.16) we return to the continuous model given by Eq. (3.5), obtaining

$$|\mathcal{E}|^2 \sim \left| \int \Psi_n \left(\mathbf{E}_0 \cdot \nabla \varepsilon \right) d\mathbf{r} \right|^2 \sim \left| \int \varepsilon \left(\mathbf{E}_0 \cdot \nabla \Psi_n \right) d\mathbf{r} \right|^2 , \tag{3.18}$$

where in order to get the last expression we have the integral by parts and taken into account that the eigenstates Ψ_n are localized within the localization length $\xi_A(\Lambda)$. Since the local dielectric constants are on the order of unity, $|\varepsilon| \sim 1$, and the spatial derivative $|\nabla \Psi_n|$ is estimated as $|\Psi_n| / \xi_A(\Lambda)$, we find

$$|\mathcal{E}|^2 \sim \frac{|E_0|^2}{\xi_A^2(\Lambda)} \left| \int |\Psi_n(\mathbf{r})| \, d\mathbf{r} \right|^2 . \tag{3.19}$$

Because the eigenfunctions Ψ_n are normalized, i.e. $\int |\Psi_n(\mathbf{r})|^2 d\mathbf{r} = 1$, and are localized within $\xi_A(\Lambda)$ we estimate $\Psi_n \sim [\xi_A(\Lambda)]^{-D/2}$ in the localization domain. Substituting this estimate in Eq. (3.19) gives

$$|\mathcal{E}(\Lambda)|^2 \sim |E_0|^2 [\xi_A(\Lambda)]^{D-2} . \tag{3.20}$$

The term $\left\langle |\nabla \Psi(\Lambda)|^2 \right\rangle$ in the numerator of Eq. (3.16) can be estimated in the similar manner:

$$\left\langle |\nabla \Psi_n(\mathbf{r})|^2 \right\rangle \sim \xi_A(\Lambda)^{-2} \int |\Psi_n(\mathbf{r})|^2 d\mathbf{r} \sim \xi_A(\Lambda)^{-2} . \tag{3.21}$$

Substitution of Eqs. (3.20) and (3.21) into Eq. (3.15) gives the following estimate for the field intensity

$$\left\langle |E|^2 \right\rangle \sim |E_0|^2 + |E_0|^2 \int \frac{\rho(\Lambda) [\xi_A(\Lambda)/a]^{D-4}}{\Lambda^2 + (b\kappa)^2} d\Lambda, \tag{3.22}$$

where we have expressed the localization length is in terms of the lattice constant a.

It will be shown in the next subsection that the localization length $\xi_A(\Lambda)$ and the density of states $\rho(\Lambda)$ could have weak singularities near the point $\Lambda = 0$. For the rough estimate, we can neglect these singularities but take into account that the denominator in Eq. (3.22) has the essential singularity at $\Lambda = \pm ib\kappa$. Then the second moment of the local electric field

$$M_2 \equiv M_{2,0} = \left\langle |E|^2 \right\rangle / |E_0|^2 \tag{3.23}$$

is estimated as

$$M_2 \sim 1 + \rho\,(a/\xi_A)^{4-D} \int \frac{1}{\Lambda^2 + (b\kappa)^2} d\Lambda \sim \rho\,(a/\xi_A)^{4-D}\,\kappa^{-1} \gg 1, \quad (3.24)$$

provided that $\kappa \ll \rho\,(a/\xi_A)^{4-D}$ (we have set $\xi_A(\Lambda = 0) \equiv \xi_A, \rho\,(\Lambda = 0) \equiv \rho$ and $b \sim 1$). For semicontinuous metal films $(D = 2)$, we obtain

$$M_2 \sim \kappa^{-1}. \quad (3.25)$$

This equation gives the enhancement of the local fields in the film when $\varepsilon_m' = -\varepsilon_d$. For the silver semicontinuous film, we the corresponding wavelength is $\lambda_r \simeq 0.37$ μm and the loss factor $\kappa = \varepsilon_m''(\lambda_r) / |\varepsilon_m'(\lambda_r)| \simeq 0.02$ (see Eq. (3.1)). Assuming for simplicity that $\rho \sim 1$ and $\xi_A \sim a$, the enhancement estimates as ~ 50. Thus the average integral intensity of the local electric fields in the film is almost two orders on magnitude larger than the intensity of the incident light. The intensity becomes even larger for the smaller frequencies as it can be seen, for example, from Fig. 2.7.

We discussed above only the average (integral) intensity. Yet, the local electric field distribution is rather inhomogeneous on the film surface (see Figs. 2.6 and 2.7) and maximum values of the local intensity are much larger than the integral value. To estimate the peak intensity we note that the eigenstates with eigenvalue $|\Lambda| \lesssim \kappa$ give the main impact in the average intensity as it follows from Eq. (3.22). Therefore, only modes with $|\Lambda| \lesssim \kappa$ are effectively excited by the external field. The number of these modes estimates as $N_\kappa = N\rho\,\kappa$, where $\rho \equiv \rho\,(\Lambda = 0)$ is the density of states. The total number of the eigenmodes N coincides with the number of cites in the system so that the spatial distance between the two closest ones is approximately equal to a. Then the average distance in-between N_κ eigenmodes is about

$$\xi_e \sim a\,(N_\kappa /\!/ N)^{-1/D} \sim a\,(\rho\kappa)^{-1/D}. \quad (3.26)$$

Thus the local electric field can be described as a set of the peaks (each peak is localized within ξ_A) separated by the distance $\xi_e \gg \xi_A$. The peak amplitude E_m is

$$|E_m|^2 \sim \left\langle |E|^2 \right\rangle \xi_e^D / \xi_A^D \sim |E_0|^2\,\kappa^{-2}\,(a/\xi_A)^4\,. \quad (3.27)$$

For small loss $\kappa \to 0$, the amplitude grows infinitely $E_m \to \infty$.

It can be seen from Figs. 2.6(d) and 2.7(a) that the eigenstates are rather strongly localized in the semicontinuous metal films. Therefore we can estimate the intensity in a hot spot of the film as $|E_m|^2 \sim |E_0|^2 \kappa^{-2}$ assuming $\xi_A \sim a$. Substituting here the loss factor $\kappa \simeq 0.02$ for the silver we obtain $|E_m^{\star}|^2 \sim 10^3 |E_0|$, which is in a qualitative agreement with that field distribution in Figs. 2.6(d) and 2.7(a). Note that all the aforementioned speculations adopted in deriving Eqs. (3.24)–(3.27) hold when the field correlation length ξ_e is much larger than the Anderson localization length, i.e., $\xi_e \gg \xi_A$. This condition is fulfilled in the limit of small loss.

3.3.2 *High-order moments of local electric fields*

Now we consider arbitrary high-order field moments, defined as

$$M_{n,m} = \frac{1}{V E_0^m |E_0|^n} \int |E(\mathbf{r})|^n E^m(\mathbf{r})\, d\mathbf{r} \tag{3.28}$$

where, as above, $E_0 \equiv E^{(0)}$ (both notations are used interchangeably) is the amplitude of the external field, $|E|^n \equiv (\mathbf{E}\cdot\mathbf{E}^*)^{n/2}$, and $E^m \equiv (\mathbf{E}\cdot\mathbf{E})^{m/2}$, and the integration is performed over the total system volume V. The moments $M_{n,0}$ will be denoted for simplicity as

$$M_n \equiv M_{n,0} = \frac{1}{V |E_0|^n} \int |E(\mathbf{r})|^n\, d\mathbf{r}. \tag{3.29}$$

It is assumed that the volume average in Eqs. (3.28) and (3.29) is equivalent to the ensemble average, i.e., $M_{n,m} = \langle |E|^n E^m \rangle / |E_0|^n E_0^m$.

The high-order field moments $M_{2k,m} \propto \langle E^{k+m} E^{*k} \rangle$ represent enhancement of various nonlinear optical processes as it is discussed in Secs. 3.5 and 3.6; k here is the number of photons that are subtracted (annihilated) in one elementary act of the nonlinear optical process. This is because the complex conjugated field in the general expression for the nonlinear polarization implies photon subtraction, so that the corresponding frequency enters the nonlinear susceptibility with the sign minus [Boyd, 1992], [Landau *et al.*, 1984].

The enhancement of the Kerr optical nonlinearity G_K is proportional to $M_{2,2}$, third-harmonic generation (THG) enhancement is given by $|M_{0,3}|^2$, and surface-enhanced Raman scattering (SERS) is represented by $M_{4,0}$ (see Sec. 3.5). The integrals in Eq. (3.28) for $M_{0,3}$ and $M_{2,2}$, i.e. the local nonlinear field $g_3 = E^3(\mathbf{r})/E_0^3$ (THG) and $g_K = |E(\mathbf{r})|^2 E^2(\mathbf{r}) / \left(E_0^2 |E_0|^2\right)$ (Kerr optical effect), are shown in Figs. 3.9 and 3.10.

We are interested in a case when the fluctuating part of the local electric field $\mathbf{E}_f$ is much larger than the applied field $\mathbf{E}_0$. For simplicity, the applied

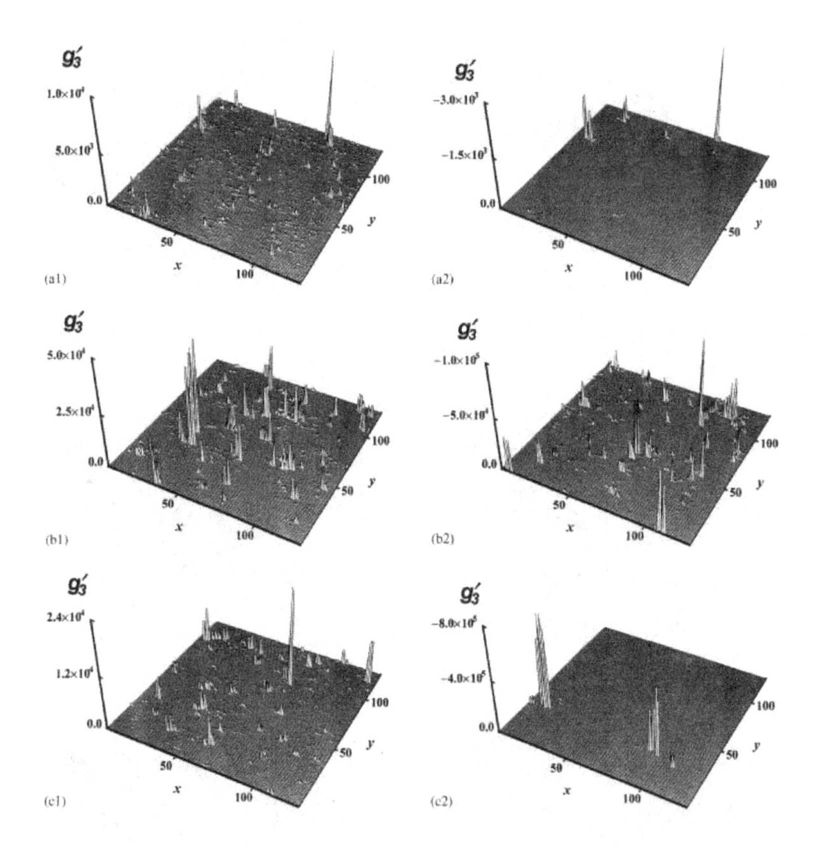

Fig. 3.9 Distribution of local "third harmonic field" (real part) $g_3' = \mathrm{Re}[E^3(\mathbf{r})]$ in semicontinuous silver films at wavelength $\lambda = 1.5$ μm, for different metal concentration p: $p = 0.3$ (a1 and a2), $p = p_c = 0.5$ (b1 and b2), $p = 0.7$ (c1 and c2); positive (a1, b1, c1) and negative (a2, b2, c2) values of the local nonlinear fields are shown in different figures. The applied field $E_0 = 1$.

field is taken real and its absolute value $|E_0| = 1$. We then substitute the expression for $\mathbf{E}_f$ given by Eq. (3.12) in Eq. (3.28) and obtain the following equation for the generalized moment $M_{2p,2q}$ (p and q are integers)

$$M_{n,m} = \begin{array}{l} \left\langle \sum_{h,i}^{N} \left[\frac{\mathcal{E}_h \mathcal{E}_i^* (\nabla \Psi_h \cdot \nabla \Psi_i^*)}{(\Lambda_h + ibk)(\Lambda_i - ibk)} \right]^{n/2} \right. \\ \left. \times \sum_{j,k}^{N} \left[\frac{\mathcal{E}_j \mathcal{E}_k (\nabla \Psi_j \cdot \nabla \Psi_k)}{(\Lambda_j + ibk)(\Lambda_k + ibk)} \right]^{m/2} \right\rangle \end{array} \tag{3.30}$$

where $\langle \cdots \rangle$ denotes the ensemble average, which is equivalent to the volume average and sums over all the eigenstates of KH $\widehat{\mathrm{H}}'$.

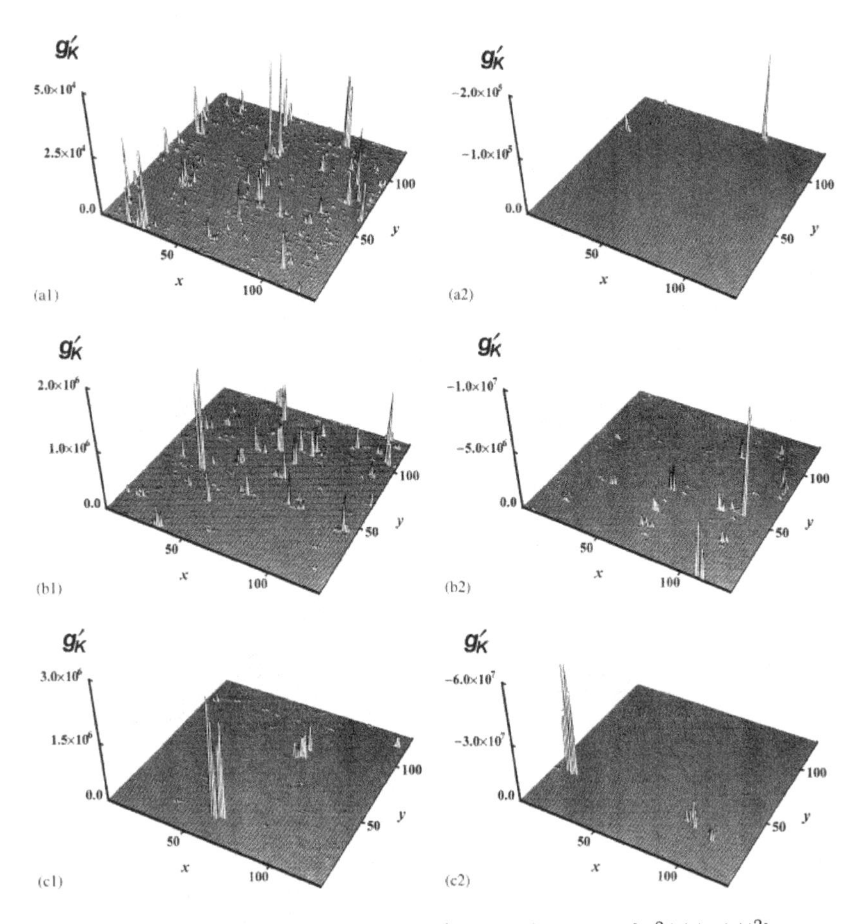

Fig. 3.10 Distribution of local "Kerr field" (real part) $g_K = \mathrm{Re}[E^2(r)\,|E(r)|^2]$ in semicontinuous silver films at wavelength $\lambda = 1.5\ \mu$m, for different metal concentration p: $p = 0.3$ (a1 and a2), $p = p_c = 0.5$ (b1 and b2), $p = 0.7$ (c1 and c2); positive (a1, b1, c1) and negative (a2, b2, c2) values of local field are shown in different figures. Applied field $E_0 = 1$.

Following the same approach that we have used to estimate the average intensity Eq. (3.22), we obtain

$$M_{n,m} \sim \int \frac{\rho(\Lambda)\,[a/\xi_A(\Lambda)]^{2(n+m)-D}}{\left[\Lambda^2 + (b\kappa)^2\right]^{n/2} (\Lambda + ib\kappa)^m} d\Lambda\,, \tag{3.31}$$

where we have neglected all the cross-terms in the eigenstate product while averaging Eq. (3.30) over the all spatial positions of the eigenfunctions $\Psi_n =$

$\Psi(\Lambda, \mathbf{r} - \mathbf{r}_n)$, with $\mathbf{r}_n$ being the "center" of the localized state Ψ_n. It can be shown that after integrating over all Λ, these cross-terms result in negligible [in comparison with the leading term given by Eq. (3.31)] contribution to $M_{n,m}$, for $\kappa \to 0$.

Assuming that the density of states $\rho(\Lambda)$ and the localization length $\xi_A(\Lambda)$ are both smooth functions of Λ in the vicinity of zero and taking into account that all parameters of the KH $\widehat{\mathrm{H}}'$ for studied $\varepsilon_d = -\varepsilon'_m = 1$ case are on the order of unity, we obtain the following estimate for the local field moments

$$M_{n,m} \sim \rho\,(a/\xi_A)^{2(n+m)-D}\,\kappa^{-n-m+1}; n+m > 1, n > 0, \tag{3.32}$$

where we have set $\rho = \rho(\Lambda = 0)$, $\xi_A = \xi_A(\Lambda = 0)$, and $b = 1$ for simplicity. The same estimate can also be obtained from the simple consideration: the local fields is a set of peaks (stretched over the distance ξ_A), with the magnitude E_m and the average distance between the peaks ξ_e, given by Eqs. (3.27) and (3.26).

It is important to stress out that the moment magnitudes in Eq. (3.32) do not depend on the number of "subtracted" (annihilated) photons in the elementary act of the nonlinear scattering. If there is at least one such a photon, then the poles of Eq. (3.31) are located in different complex semi-planes and the result of the integration is estimated by Eq. (3.32).

When all photons are added (in other words, all the frequencies contribute into the nonlinear susceptibility with the positive sign), i.e., when $n = 0$, we cannot estimate the moments $M_{0,m} \equiv E_0^{-m} V^{-1} \int E^m(\mathbf{r})\, d\mathbf{r}$ by Eq. (3.32) since the integral in Eq. (3.31) cannot be determined by the presence of the poles $\Lambda = \pm ib\kappa$. Yet all the functions in the integral are about unity, therefore the moments $M_{0,m}$ must be on the order of unity $M_{0,m} \sim O(1)$ for all $m > 1$. Note that the moment $M_{0,m}$ describes, in particular, enhancement G_{nHG} of n-th order harmonic generation through the relation $G_{nHG} = |M_{0,m}|^2$ (see, for example, [Sarychev and Shalaev, 2000], [Breit *et al.*, 2001]).

3.3.3 *Properties of the localized eigenmodes*

The electron localization, which occurs in a random media and was first predicted by Anderson [Anderson, 1958], is one of the most important concepts in the contemporary theory of disordered systems. Development of the scaling theory [Abrahams *et al.*, 1979] has improved and made possible more intuitive understanding of the phenomena governing the motion of the elementary particles in such media. It is now well established that in all $D = 1$ and $D = 2$ systems, which are described by non-correlated random potential distributions, all the electron states are exponentially localized [Lifshits *et al.*, 1988], [Kramer and MacKinnon, 1993], and [Kawarabayashi *et al.*, 1998]. The same statement is true for various levels of disorder in the limit of large systems. The localization of the electron wave function implies that each electron is bound to a particular region of space, and therefore its transport through the media is impeded. It is also believed that in the $D = 3$ case and for a certain strength of the disorder, Bloch states exist and a metal-dielectric transition takes place [Hofstetter and Schreiber, 1993]. Similar to the quantum-mechanics phenomenon, the localization can also occur for a classical wave propagation in the disordered media [Sheng, 1995]. In both cases, disorder prevents the establishment of extended solutions (Bloch states) due to the absence of the system translation symmetry.

It has been shown by [Boliver, 1992], [Hilke, 1994] that short-range correlations between the elements of the quantum-mechanical Hamiltonians result in the delocalization of the electron wave functions in an one-dimensional case. In these studies, the correlations are enforced separately for diagonal and off-diagonal matrix elements, while mixed cross-correlations between the elements are not considered. In this section, we investigate the excitation of collective electronic states, surface plasmons in random and periodic metal-dielectric films. Electromagnetic response of these systems is described by Kirchhoff Hamiltonian $\widehat{\mathrm{H}}$ (see Sec. 3.3.1, Eq. (3.8)) whose matrix representation is characterized by cross-correlated diagonal and off-diagonal elements: $\mathrm{H}_{ii} = -\sum_j \mathrm{H}_{ij}$. This property of KH originates from the charge conservation flow $\mathrm{div}\left[\varepsilon\left(\mathbf{r}\right)\mathbf{E}\left(\mathbf{r}\right)\right] = 0$, which includes material property of the system in contrast to the Anderson's problem of quantum mechanical description, where the current $j \sim (\Psi\nabla\Psi^* - \Psi^*\nabla\Psi)$ is defined by the wave function alone. The hidden symmetry of KH results in the localization of SP, which is quite different from the typical Anderson localization.

In order to simplify the treatment of the SP-excitation in metal-dielectric films, we still work in the regime of a single particle resonance, $\varepsilon_m' = -\varepsilon_d$, where ε_m' is a real part of the metal dielectric constant. We again normalize Eq. (3.5) by σ_d and use a new set of non-dimensional permittivities $\varepsilon_d = 1$ and $\varepsilon_m = -1 + i\kappa$, where for noble metals and the light in visible and infrared spectral regions, the loss is small $\kappa = \varepsilon_m''/|\varepsilon_m'| \ll 1$. Following the approach developed in Sec. 3.3.1, we seek the general solution of Eq. (3.8) as an expansion over the eigenstates Ψ_n of the SP eigenproblem,

$$\widehat{\mathrm{H}}'\Psi_n = \Lambda_n \Psi_n, \tag{3.33}$$

where $\widehat{\mathrm{H}}'$ is a real part of KH as it is defined after Eq. (3.8). We can solve the SP eigenvalue problem numerically by applying Neumann boundary conditions, thus assuring the conservation of the local currents at the film boundaries. For example, we use $\varepsilon(\mathbf{r})[\mathbf{n} \cdot \nabla \Psi_n(\mathbf{r})]|_{x=0} = \varepsilon(\mathbf{r})[\mathbf{n} \cdot \nabla \Psi_n(\mathbf{r})]|_{x=L}$ at the left and right boundaries of the film, where $\mathbf{n}$ is the unit vector, normal to the boundary.

To begin our analysis of the SP eigenproblem let us first examine some specific eigenmodes. For metal-dielectric films near the percolation threshold, we distinguish two limiting cases. In the first case, the SP eigenmode situated at the band edge is strongly localized and is shown in Fig. 3.11. However, the nature of the eigenstates at the band center (see Fig. 3.12) is different. These states are extended and they exhibit multifractal properties. It is constructive to compare behavior of SP in random and periodic metal-dielectric films. For example, Fig. 3.13 demonstrates a particular SP mode that is manifested in the periodic case. The periodic structure was simulated as the square lattice of round metal particles with metal coverage equal to 2/3. The important feature to recognize here is the presence of two length scales: one is corresponding to the macroscopically extended Bloch states and the second is to the local oscillations on the characteristic size of a single particle. We believe that the microscopic SP-eigenmode fluctuations correspond to a strong inhomogeneity of the electromagnetic

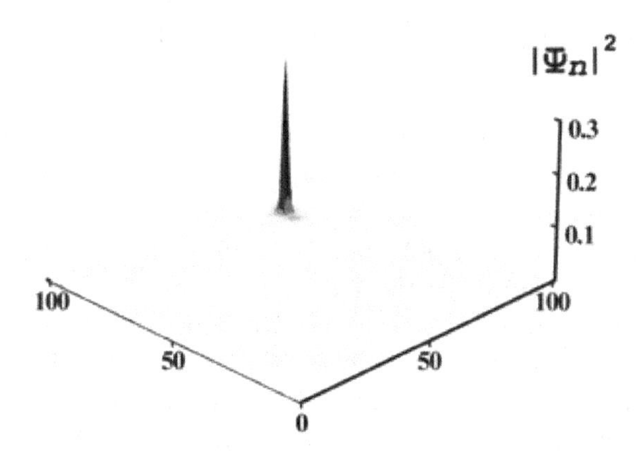

Fig. 3.11 Surface plasmon eigenmode in semicontinuous metal film; eigenvalues $\Lambda = -5.6945$ (localized).

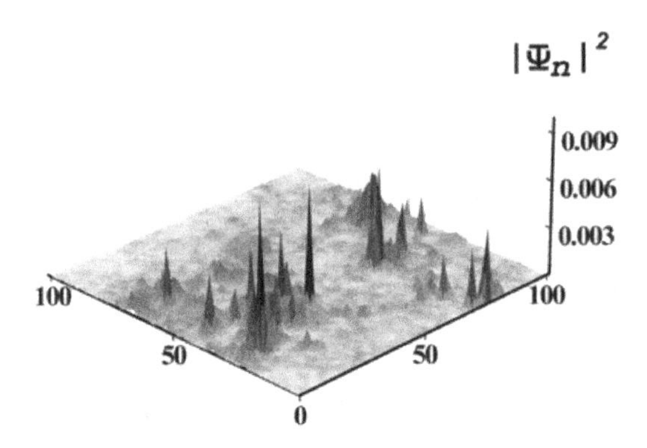

Fig. 3.12 Surface plasmon eigenmode in semicontinuous metal film; eigenvalue $\Lambda = 0.0044$ (delocalized).

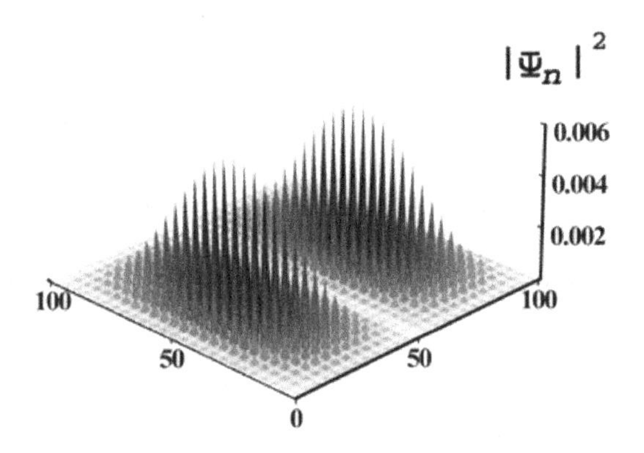

Fig. 3.13 Surface plasmon eigenmodes in periodic metal-dielectric films $\Lambda = -5.9974$ (periodic).

fields observed even for a perfectly ordered metal-dielectric films discussed in Sec.1.1.

The statistical properties of the SP eigenproblem can be investigated in terms of density of the states $\rho(\Lambda)$ and SP-localization length $\xi_A(\Lambda)$. Both characteristics are studied for the KH and for non-correlated Anderson Hamiltonian (AH), which has same elements as KH but the elements in AH are not correlated with each other. To simulate the AH, we rely on the fact that for each metal concentration p, the elements of KH matrix $\hat{\mathrm{H}}'$ take discrete values with a specific probability. These probability distributions are then used to build up the AH without enforcing correlations between any elements of AH matrix. In Figs. 3.14 and 3.15 we show that both correlated

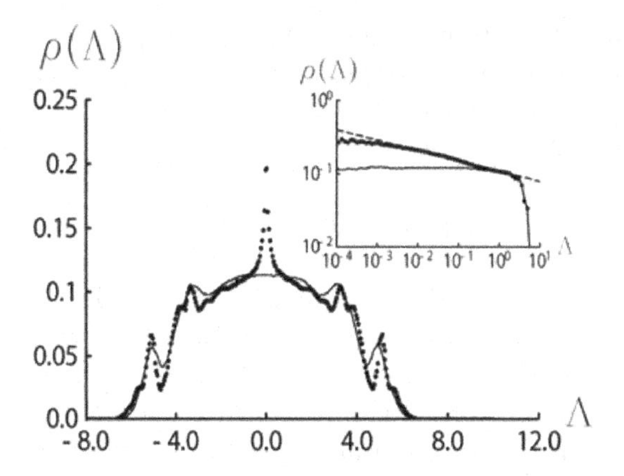

Fig. 3.14 Density of states $\rho(\Lambda)$ for KH (dots) and for corresponding Anderson problem (solid line), calculated at resonant condition $\varepsilon_d^* = -\varepsilon_m^* = 1$; band-center singularity is shown in log-log insets where power law fit (dashed line) with exponent $\gamma = 0.14$ is applied. Data are averaged over 100 different realizations of percolation samples each with size $L = 120$.

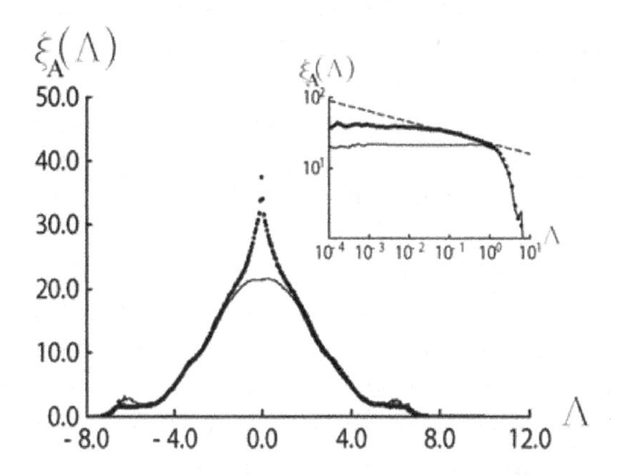

Fig. 3.15 Localization lengths $\xi(\Lambda)$ for the KH (dots) and for the corresponding Anderson problem (solid line), calculated at the resonant condition $\varepsilon_d^* = -\varepsilon_m^* = 1$; the band-center singularity is shown in the log-log insets where power law fit (dashed line) with exponent $\alpha = 0.14$ is applied. The data is averaged over 100 different realizations of percolation samples each with size $L = 120$.

and non-correlated eigenproblems have similar densities of states for most duration of the spectrum. However, a singularity at the band center is observed for the case of the KH. To understand better such peculiar behavior, we plot (see the insert in Fig. 3.14) the region of very small eigenvalues

($\Lambda \ll 1$) on the double logarithmic scale. In the first approximation, the density of states seems to diverge as a power law $\rho(\Lambda) \simeq A|\Lambda|^{-\gamma}$, where A is a normalization constant and $\gamma = 0.14 \pm 0.01$ is a critical exponent. However, a function $\rho(\Lambda) \simeq A[1 + \ln(|\Lambda|^{-\gamma})]$ with the logarithmic singularity also asymptotically fits the result. The two functions are virtually identical in the broad range of the arguments $e^{1/\gamma} \gg |\Lambda| \gg e^{-1/\gamma}$, and therefore in the scaling theory that follows, we use the power law relationship for the sake of simplicity in description. The non-correlated AH case, where the density of states is relatively uniform throughout the spectra and does not show any singularities, is also included in Fig. 3.14 (see the insert).

The role of the cross-correlations presented in the KH eigenproblem is studied in terms of the SP localization lengths $\xi_A(\Lambda)$. The localization length for each eigenmode is calculated using the gyration radius $\xi_A^2(\Lambda_n) = \int (\mathbf{r} - \langle \mathbf{r} \rangle_n)^2 |\Psi_n(\mathbf{r})|^2 d\mathbf{r}$, where $\langle \mathbf{r} \rangle_n = \int \mathbf{r} |\Psi_n(\mathbf{r})|^2 d\mathbf{r}$ is the "mass center" of the n-th mode and the integration is performed over the film surface. Our results for $\xi_A(\Lambda)$ are presented in Fig. 3.15. Similar to the density of states singularity (see Fig. 3.14), there is a singularity at the band center for the KH. The localization length diverges logarithmically when $\Lambda \to 0$, and also can be fitted with a power law $\xi_A(\Lambda) \sim |\Lambda|^{-\alpha}$, where $\alpha = 0.15 \pm 0.02$. The size effect is clearly visible for the extended (Bloch) states with $\xi_A(\Lambda) \geq L$.

The existence of delocalized states has been predicted for somewhat similar (but not the same) SP eigenproblem in [Stockman *et al.*, 2001], where it has been concluded that the modes play a dominant role in the material-light interaction. We, however, show that the role of the delocalized modes ($\xi_A(\Lambda) \geq L$) is more subtle. In the limit of large systems $L \to \infty$, the measure of the delocalized states rapidly falls as $\mu \sim e^{-L/\gamma} \ln(L)$ in the spectrum. (While deriving, we have adopted the same singularities for $\xi_A(\Lambda)$ and $\rho(\Lambda)$ so that the exponents are equal: $\alpha = \gamma$.) However, despite the zero measure of the delocalized states for $L \gg 1$, these states sill affect the optical properties of the composite. Indeed, substituting $\rho(\Lambda) \sim |\Lambda|^{-\gamma}$ and $\xi_A(\Lambda) \sim a|\Lambda|^{-\gamma}$ in Eq. (3.31) gives the following estimate for the moments

$$M_{n,m} \sim \kappa^{-\varkappa_{n,m}}, \tag{3.34}$$

where $\varkappa_{n,m} = (n+m-1)(1-2\gamma) + \gamma(D-1)$ is a positive scaling exponent ($n + m \geq 1$). Equation (3.34) gives the moments of the local electric fields $M_{n,m}$ for a composite at "resonance" condition $\varepsilon'_m/\varepsilon_d = -1$ whose loss factor is small, e.g. $\kappa = \varepsilon''_m/|\varepsilon_m| \ll 1$. (In the derivation of Eq. (3.34), we have used $\alpha = \gamma$, which agrees with our numerical calculations.) Note that the singularity's critical index γ affects the high-order field moments through its contribution to the index $\varkappa_{n,m}$. The estimate (3.32), which was obtained based on the assumption that all modes are localized, can be retrieved from Eq. (3.34) by setting $\gamma = 0$. Therefore Eq. (3.34) corrects the formula (3.32), by taking into account small amount of the states delocalized at $\Lambda = 0$.

We also estimate the peak intensity E_m for $D = 2$ case noticing that the average intensity of the local electric field is

$$\langle |E(\mathbf{r})|^2 \rangle = |E_0|^2 M_{2,0}^{\star} \sim |E_0|^2 \kappa^{-1+\gamma} \tag{3.35}$$

and the distance between the peaks can be estimated from the equation $(\xi_e/a)^{-2} \sim \int_0^{\kappa} \rho(\Lambda)\, d\Lambda$ as

$$\xi_e \sim \kappa^{-(1-\gamma)/2}. \tag{3.36}$$

Then the peak intensity in the semicontinuous films then follows:

$$|E_m|^2 \sim \left\langle |E|^2 \right\rangle \xi_e^2/\xi_A^2 \sim |E_0|^2 \kappa^{-2+4\gamma}, \tag{3.37}$$

where we account ξ_A for the average size of the peak. For small loss $\kappa \to 0$, the local field amplitude diverges $E_m \sim |E_0|\, \kappa^{-1+2\gamma} \to \infty$, representing the infinite range of fluctuations.

3.3.4 *Scaling theory of giant field fluctuations*

Above we have assumed $\varepsilon_m \simeq -\varepsilon_d$, which corresponds to the plasmon resonance condition in metal grains. The plasmon resonance for the metal nano-particles occurs in the violet or near-ultraviolet part of the spectrum. Peculiarly enough, the giant field fluctuations become more pronounced with decreasing frequency to the visible and infrared spectral bands, far away from the plasmon resonance frequency (see Fig. 3.7.) The qualitative explanation of this exciting phenomenon was given in Sec. 3.2.3. Below we present the scaling theory for the local electric field fluctuations in the percolation composites for visible and infrared part of the spectrum where the metal dielectric constant is large $|\varepsilon_m| \gg 1$. To estimate the local field fluctuations, we follow the approach developed by [Brouers *et al.*, 1997a], [Shalaev and Sarychev, 1998], [Brouers *et al.*, 1998], [Sarychev *et al.*, 1999c], which is based on Real Space Renormalization Group (RSRG) [Sarychev, 1977], [Reynolds *et al.*, 1977], [Bernasconi, 1978]. The RSRG approach has been already used in Sec. 3.2.2. We briefly recapture its main ideas below. Consider first a percolating metal-dielectric composite with the metal concentration p. We divide a system into cubes of size l and consider each cube as a newly renormalized element. All such cubes can be classified onto two types. A cube, which contains a continuous path of metallic particles throughout the whole cube, is being considered as a "conducting" element. A cube without such an "infinite" cluster is considered as a non-conducting, "dielectric," element. When the metal concentration p is larger than the percolation threshold p_c, the conducting cube concentration $p^*(l)$ in the renormalized system increases with its size l: $p^*(l) > p$. The concentration asymptotically approaches unity $p^*(l) \to 1$ for $l \to \infty$ as the system renormalizes to the conducting state. In the opposite case $p < p_c$ the renormalized metal concentration $p^*(l) < p$ and $p^*(l) \to 0$ for $l \to \infty$ so that the system renormalizes to the dielectric state. The condition $p^*(l) = p$ corresponds to the percolation threshold $p = p_c$.

We consider now the metal-dielectric composite at the percolation threshold. The effective dielectric constant of the "conducting" cube $\varepsilon_m(l)$ decreases with increasing its size l as $\varepsilon_m(l) \simeq l^{-t/\nu}\varepsilon_m$, whereas the effective dielectric constant of the "dielectric" cube $\varepsilon_d(l)$ increases with l as $\varepsilon_d(l) \simeq l^{s/\nu}\varepsilon_d$. We then define the percolation critical exponents t, s and ν, for the static conductivity, dielectric constant, and percolation correlation length, respectively to estimate $\varepsilon_m(l)$ and $\varepsilon_d(l)$. In $2D$-case, the exponents equal to $t \cong s \cong \nu \cong 4/3$. In $3D-$ systems, the exponents are equal to $t \simeq 2.0$, $s \simeq 0.7$, and $\nu \simeq 0.88$ (see discussion in 1.2). We have set the cube size l to

$$l = l_r = \left(|\varepsilon_m|/\varepsilon_d\right)^{\nu/(t+s)}, \tag{3.38}$$

where we have assumed that all the distances are expressed in terms of the particle size a. (In a lattice model, a is the lattice constant.) In a renormalized system, where each cube of size l_r is considered as a single element, the dielectric constant of new elements takes either value: $\varepsilon_m(l_r) = \varepsilon_d^{t/(t+s)} |\varepsilon_m|^{s/(t+s)} (\varepsilon_m/|\varepsilon_m|)$ for the element renormalized from the conducting cube, or $\varepsilon_d(l_r) = \varepsilon_d^{t/(t+s)} |\varepsilon_m|^{s/(t+s)}$ for the element renormalized from the dielectric cube. The ratio of the dielectric constants for these new elements is $\varepsilon_m(l_r)/\varepsilon_d(l_r) = \varepsilon_m/|\varepsilon_m| \cong -1 + i\kappa$, where the loss-factor $\kappa = \varepsilon_m''/|\varepsilon_m| \ll 1$ is the same as in the original system. Thus the field distribution in a two-component system depends on the ratio of the dielectric permittivities of the components.

After the renormalization, the problem becomes equivalent to the considered in Secs. 3.3.1 and 3.3.3 field distribution for the resonance case $\varepsilon_d = -\varepsilon'_m = 1$. In the resonance system, electric field is a set of peaks with the amplitudes on the order of $E^\star_m$, separated by the average distances $\xi^\star_e$ (see Eqs. (3.36) and (3.37)). Taking into account that the electric field renormalizes as $E^\star = El_r$ we obtain that the peaks in the renormalized system are

$$E_m \sim l_r E^\star_m, \tag{3.39}$$

where l_r is given by Eq. (3.38).

In the original system, each peak splits into a group of $n(l_r)$ peaks with the amplitude E_m located along a dielectric gap between the two metal clusters as shown in Fig. 3.8. The dielectric constant of a "dielectric" cube scales as $\varepsilon_d(l) \sim l^{s/\nu}\varepsilon_d$, therefore its capacitance scales as $\sim l^{s/\nu}\varepsilon_d l^{D-2}$. The gap "area" scales as the capacitance of the dielectric cube, so does the number of peaks

$$n(l_r) \sim {l_r}^{D-2+s/\nu}. \tag{3.40}$$

The average distance between the field maxima in the renormalized system is $\xi^\star_e$. In the original system, the average distance between the groups of the peaks is

$$\xi_e \sim l_r \xi^\star_e \sim \left(\frac{|\varepsilon_m|}{\varepsilon_d}\right)^{\nu/(t+s)} \left(\frac{|\varepsilon_m|}{\varepsilon''_m}\right)^{(1-\gamma)/2}. \tag{3.41}$$

This equation simplifies for semicontinuous metal films where the adopted earlier exponents are $t \cong s \cong \nu \cong 4/3$ so that

$$\xi_e \sim \left(\frac{|\varepsilon_m|}{\varepsilon_d}\right)^{1/2} \left(\frac{|\varepsilon_m|}{\varepsilon''_m}\right)^{(1-\gamma)/2}. \tag{3.42}$$

Note, that $\xi^\star_e \gg 1$ in resonance systems so that the field maxima are well isolated in the renormalized as well as in the original system.

By multiplying the amplitude of the field peaks E_m raised to the proper power on the number of the peaks in one group $n(l_r)$ and then normalizing them to the distance between the groups ξ_e, we obtain the following estimate for the local-field moments:

$$
\begin{aligned}
|M_{n,m}| &\sim \frac{n(l_r)}{\xi_e{}^D}\left(\frac{E_m}{E_0}\right)^{n+m} \\
&\sim \frac{l_r^{n+m} l_r{}^{D-2+s/\nu}}{l_r{}^D}\left[\left(\frac{E_m^\star}{E_0}\right)^{n+m}\frac{1}{(\xi_e^\star)^D}\right] \sim l_r{}^{n+m-2+s/\nu} M_{n,m}^\star \\
&\sim \left(\frac{|\varepsilon_m|}{\varepsilon_d}\right)^{\frac{(n+m-2)\nu+s}{t+s}} M_{n,m}^\star \\
&\sim \left(\frac{|\varepsilon_m|}{\varepsilon_d}\right)^{\frac{(n+m-2)\nu+s}{t+s}} \left(\frac{|\varepsilon_m|}{\varepsilon_m''}\right)^{(n+m-1)(1-2\gamma)+\gamma(D-1)}
\end{aligned}
\tag{3.43}
$$

where $M_{n,m}^\star$ is the field moment in the resonance system $\varepsilon_m = -\varepsilon_d$ [see Eq. (3.34)] and we have used Eq. (3.38) for l_r. The metal dielectric constant $|\varepsilon_m| \gg \varepsilon_d$ in the visible spectral range and progressively increases with decreasing frequency to infrared range. The corresponding enhancement factor

$$
\Xi = \left(\frac{|\varepsilon_m|}{\varepsilon_d}\right)^{\frac{(n+m-2)\nu+s}{t+s}} \tag{3.44}
$$

is also large. $\Xi \gg 1$ in the visible range and becomes progressively more significant with the frequency decreasing. This result is in the best agreement with computer simulations, shown in Fig. 3.7. Note that the silver permittivity achieves extremely large values $|\varepsilon_m| \cong 10^3$ at $\lambda = 4.3$ μm.

For the semicontinuous metal films $(D = 2)$, the critical exponents are $t \cong s \cong \nu \cong 4/3$. The enhancement factor equals to $\Xi_2 = (|\varepsilon_m|/\varepsilon_d)^{(n+m-1)/2}$. The field moments $M_{n,m}^\star$ in the resonance system can be estimated from Eq. (3.34). Substituting $M_{n,m}^\star$ and Ξ_2 in Eq. (3.43), we obtain the equation for the field moments in the semicontinuous metal films

$$
M_{n,m} \sim \left(\frac{|\varepsilon_m|}{\varepsilon_d}\right)^{\frac{n+m-1}{2}} \kappa^{-(n+m-1)(1-2\gamma)-\gamma}, \tag{3.45}
$$

where $|\varepsilon_m| \gg \varepsilon_d$, the loss factor $\kappa = \varepsilon_m''/\,|\varepsilon_m| \ll 1$. Therefore, we obtain that all the moments of the optical electric field are large: $|M_{n,m}| \gg 1$. Substituting expressions $|\varepsilon_m| \simeq (\omega_p/\omega)^2$, $\kappa \simeq \omega_\tau/\omega$, which follow from the Dude model (3.1) for $\omega \ll \omega_p$ and $\omega \gg \omega_\tau$, we obtain the following equation for the field moments in the semicontinuous metal films

$$
M_{n,m} \sim \left(\frac{\omega_p}{\omega_\tau}\right)^{n+m-1} \left(\frac{\omega_\tau}{\omega}\right)^{\gamma(2n+2m-3)}, \tag{3.46}
$$

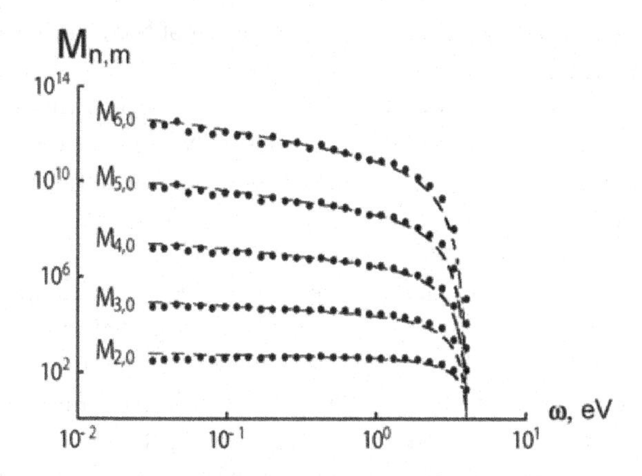

Fig. 3.16 Local field moments $M_{n,0}$ (dots) are calculated with numerical method and compared to analytical results with exponent $\gamma = 0.14$ (dashed lines). The numerical data is averaged over 20 realizations of percolation systems with size $L = 250$.

where the critical exponent $\gamma \cong 0.15$ defined in Sec. 3.3.3. It is interesting to note that delocalization results in the frequency dependence of the moments. If all the eigenstates of the KH Hamiltonian were localized, i.e., the critical exponent γ were equal to zero, the moments $M_{n,m} \sim (\omega_p/\omega_\tau)^{n+m-1}$ would not depend on the frequency.

To verify Eq. (3.45) and examine the frequency dependence of the moments $M_{n,m}$ one can apply the exact block elimination method [Genov *et al.*, 2003a] to calculate local electric fields in silver semicontinuous films in the frequency range $\omega_p > \omega > \omega_\tau$. The results are shown in Fig. 3.16. Clearly, there is an excellent correlation between the numerical simulations and theoretical results. In both cases the field moments converge toward unity for $\omega = \tilde{\omega}_p = \omega_p/\sqrt{\varepsilon_b}$, which corresponds to the condition $\varepsilon_m' \cong 0$ [see Eq. (3.1).] For larger frequencies, the silver permittivity becomes positive and the local fields do not enhance as much. The field moments $M_{n,m}$ gradually increase with decreasing the frequency below energies of 0.5 eV. The slope of $M_{n,0}(\omega)$ dependence increases with the order n of the moment, in agreement with Eq. (3.46).

It is important to note that the enhancement of the local electric fields, manifested through the high-order field moments, is directly measured in various linear and non-linear optical processes. For example, surface-enhanced Raman scattering is proportional to $M_{4,0}$, while the enhancement of Kerr optical nonlinearity is given by $M_{2,2}$ (see Secs. 3.5 and 3.6). The detailed experimental investigation of these processes and their spectral dependencies would result in further understanding of the localization properties of the SP eigenmodes.

Above, we have assumed for the sake of simplicity that the metal concentration p equals to the percolation threshold. Now we can estimate the concentration range $\Delta p = p - p_c$, where the adopted estimates for the local field moments are valid [Brouers *et al.*, 1997a], [Gadenne *et al.*, 1998]. The consideration was based on the RSRG argumentation, which holds provided that the scale of the renormalized cubes l_r is smaller than the percolation correlation length $\xi_p \cong a(|p - p_c|/p_c)^{-\nu}$. At the percolation threshold, where the correlation length ξ_p diverges, our estimates are valid in the wide frequency range that includes the visible, infrared, and far-infrared spectral ranges for typical metals. For any particular frequency from these intervals, we can estimate the concentration range Δp (where the giant field fluctuations occur) by substituting values for l_r and ξ_p, which results in an inequality

$$|\Delta p| \leq (\varepsilon_d / |\varepsilon_m|)^{1/(t+s)}, \tag{3.47}$$

where critical exponents s and t are defined as the above. Therefore, the local electric field fluctuates strongly in this concentration range and its moments $M_{n,m}$ are gigantically enhanced.

3.4 Anomalous Light Scattering from Semicontinuous Metal Films

In this section, we consider quantitatively the spatial distribution of the local field fluctuations and light scattering induced by these fluctuations. It is clear that the field fluctuations, shown in Figs. 3.6, and 3.7 can lead to the enhanced light scattering from the film. It is worth mentioning that the nature of fluctuations considered here and the corresponding light scattering are not linked to the fractal nature of metal clusters but to the distribution of local resonances in disordered metal-dielectric structures, which are homogeneous on a macroscopic scale. It appears that the local intensity of the electric field strongly correlates in space and its distribution is dominated by the field correlation length ξ_e introduced by Eqs. (3.36) and (3.41).

3.4.1 *Rayleigh scattering*

We consider now Rayleigh scattering induced by the giant field fluctuations [Brouers *et al.*, 1998]. A standard procedure leading to the characterization of the critical opalescence can be implemented [Stanley, 1981] to find the anomalous scattering from the metal-dielectric films. A semicontinuous film is illuminated by the light, which is normal to the film surface plane. The gaps between the metal grains are filled with a dielectric material of the substrate. The film can be considered as a two-dimensional array of metal and dielectric grains that are distributed over the surface. The incident electromagnetic field excites the surface currents $\mathbf{j}$. The local electromagnetic field is induced by these currents. The coordinate origin is chosen somewhere in the film. Then a contribution to the vector potential of the scattered fields $\mathbf{A}(\mathbf{R})$, (magnetic field $\mathbf{H}(\mathbf{R}) = [\nabla \times \mathbf{A}(\mathbf{R})]$) arising from the surface current $\mathbf{j}$ at film coordinate-vector $\mathbf{r}$ equals to

$$\mathbf{A}(\mathbf{R}, \mathbf{r})\, d\mathbf{r} = \frac{\mathbf{j}(\mathbf{r})}{c} \frac{\exp\left(ik|\mathbf{R} - \mathbf{r}|\right)}{|\mathbf{R} - \mathbf{r}|}\, d\mathbf{r}, \tag{3.48}$$

where $k = \omega/c$ is a wavevector, $\mathbf{R}$ is a radius-vector in space. The vector potential for the total scattered field equals to $\mathbf{A}(\mathbf{R}) = \int \mathbf{A}(\mathbf{R}, \mathbf{r})\, d\mathbf{r}$, where the integration is performed over the entire film's area. The physical film dimensions are typically small ($r \ll R$), therefore the exponent in Eq. (3.48) can be expanded in series as its argument $ik|\mathbf{R} - \mathbf{r}| \cong ikR - ik(\mathbf{n} \cdot \mathbf{r})$, where $\mathbf{n} = \mathbf{R}/R$ is the unit vector in the direction of a space vector $\mathbf{R}$. The distance r is also neglected in comparison with R in the denominator of Eq. (3.48). Thus the equations for the magnetic $\mathbf{H}$ and electric $\mathbf{E}$ fields at a distant point $\mathbf{R}$ take the following forms

$$\begin{aligned}
\mathbf{H}(\mathbf{R}) &= [\nabla \times \mathbf{A}(\mathbf{R})] \cong \frac{ik \exp\left(ikR\right)}{cR} \int [\mathbf{n} \times \mathbf{j}(\mathbf{r})] \exp\left[-ik(\mathbf{n} \cdot \mathbf{r})\right] d\mathbf{r}, \\
\mathbf{E}(\mathbf{R}) &= \frac{i}{k}[\nabla \times \mathbf{H}(\mathbf{R})] \\
&\cong \frac{-ik \exp\left(ikR\right)}{cR} \int [\mathbf{n} \times [\mathbf{n} \times \mathbf{j}(\mathbf{r})]] \exp\left[-ik(\mathbf{n} \cdot \mathbf{r})\right] d\mathbf{r},
\end{aligned} \tag{3.49}$$

where the integrals are taken over the film area. The magnetic field $\mathbf{H}(\mathbf{R})$ is perpendicular to the electric field $\mathbf{E}(\mathbf{R})$ as it follows from Eqs. (3.49). Their absolute values are equal $|\mathbf{E}(\mathbf{R})| = |\mathbf{H}(\mathbf{R})|$, which means that the scattered field can be considered locally as a plane wave, when the distances from the film is large. The total intensity S_t for the light scattered in the direction

$\mathbf{n} = \mathbf{R}/R$ equals to

$$\begin{aligned} S_t(\mathbf{n}) &= \frac{c}{4\pi} R^2 \frac{1}{2} Re\langle[\mathbf{E}(\mathbf{R}) \times \mathbf{H}^*(\mathbf{R})]\rangle \\ &= \frac{c}{8\pi} R^2 \langle \mathbf{E}(\mathbf{R}) \cdot \mathbf{E}^*(\mathbf{R})\rangle = \frac{c}{8\pi} R^2 \langle \mathbf{H}(\mathbf{R}) \cdot \mathbf{H}^*(\mathbf{R})\rangle \\ &= \frac{c}{8\pi} \frac{k^2}{c^2} \int \langle [\mathbf{n} \times \mathbf{j}(\mathbf{r}_1)] \cdot [\mathbf{n} \times \mathbf{j}^*(\mathbf{r}_2)]\rangle \exp\left[ik\mathbf{n} \cdot (\mathbf{r}_1 - \mathbf{r}_2)\right] d\mathbf{r}_1\, d\mathbf{r}_2, \end{aligned} \tag{3.50}$$

where sign "$*$" denotes a complex conjugation and the angular brackets stand for the ensemble averaging. Note that a typical semicontinuous metal film size is much larger than any intrinsic spatial scale, such as, for example, the field correlation length ξ_e. Therefore, the ensemble average can be included in the integrations over the film area in Eq. (3.50) without changing the result. It is assumed for simplicity that the incident light is natural (unpolarized) and that it is directed perpendicular to the film's surface. Then the product $\langle [\mathbf{n} \times \mathbf{j}(\mathbf{r}_1)] \cdot [\mathbf{n} \times \mathbf{j}^*(\mathbf{r}_2)]\rangle$ in Eq. (3.50) can be averaged over the all polarizations, which gives $\langle \mathbf{j}(\mathbf{r}_1) \cdot \mathbf{j}^*(\mathbf{r}_2)\rangle (1 - \sin^2\theta/2)$, where θ is the angle between the direction $\mathbf{n}$ and the normal to the film surface.

Replacing the local currents $\mathbf{j}(\mathbf{r})$ by their averaged values $\langle \mathbf{j}(\mathbf{r})\rangle$ in Eq. (3.50) gives the specular scattering S_s. The scattering in all other directions is obtained as

$$S(\theta) = S_t - S_s = \frac{c}{8\pi}\frac{k^2}{c^2}\left(1 - \frac{\sin^2\theta}{2}\right) \tag{3.51}$$
$$\int \left[\langle \mathbf{j}(\mathbf{r}_1) \cdot \mathbf{j}^*(\mathbf{r}_2)\rangle - |\langle \mathbf{j}\rangle|^2\right] \exp\left[ik\mathbf{n} \cdot (\mathbf{r}_1 - \mathbf{r}_2)\right] d\mathbf{r}_1\, d\mathbf{r}_2.$$

There exists a natural correlation length ξ_e [see Eqs. (3.36) and (3.41)] for the local field fluctuations and, therefore, for the current-current correlations. If this correlation length is much smaller than the wavelength of the incident light $\xi_e \ll \lambda = 2\pi/k$, Eq. (3.51) is simplified by replacing the exponent by unity ($\exp(ik\mathbf{n} \cdot (\mathbf{r}_1 - \mathbf{r}_2)) \approx 1$). This gives

$$S(\theta) = \frac{c}{8\pi}\frac{k^2}{c^2}\left(1 - \frac{\sin^2\theta}{2}\right)|\langle \mathbf{j}\rangle|^2 \int \left[\frac{\langle \mathbf{j}(\mathbf{r}_1) \cdot \mathbf{j}^*(\mathbf{r}_2)\rangle}{|\langle \mathbf{j}\rangle|^2} - 1\right] d\mathbf{r}_1\, d\mathbf{r}_2. \tag{3.52}$$

Since we consider the macroscopically homogeneous and isotropic films, the current-current correlations $\langle \mathbf{j}(\mathbf{r}_1) \cdot \mathbf{j}^*(\mathbf{r}_2)\rangle$ depend only on the inter-current distance $r = |\mathbf{r}_2 - \mathbf{r}_1|$. It is then convenient to introduce a correlation function, defined as

$$G(r) = \frac{\langle \mathbf{j}(\mathbf{r}_1) \cdot \mathbf{j}^*(\mathbf{r}_2) \rangle}{|\langle \mathbf{j} \rangle|^2} - 1 = \frac{Re\langle \mathbf{j}(0) \cdot \mathbf{j}^*(\mathbf{r}) \rangle}{|\langle \mathbf{j} \rangle|^2} - 1. \tag{3.53}$$

Substituting definition (3.53) into Eq. (3.52) and replacing the integrations over coordinates r_1 and r_2 by integrations over $r = r_2 - r_1$ and $r' = (r_2 + r_1)/2$, respectively, the following equation is obtained for the intensity of the scattered light

$$S(\theta) = A \frac{c}{8\pi} \frac{k^2}{c^2} (1 - \frac{\sin^2\theta}{2}) \; |\langle \mathbf{j} \rangle|^2 \, 2\pi \int_0^\infty G(r) r \, dr, \tag{3.54}$$

where A is the film area.

The intensity of the scattered light can be compared to the integral intensity (power) of the incident light $I_0 = A(c/8\pi)|\mathbf{E}_0|^2$, where $\mathbf{E}_0$ is the amplitude of the incident wave. For the normal incidence, the average electric field in the film equals to $\langle \mathbf{E} \rangle = t\mathbf{E}_0$, where t is the amplitude transmission coefficient of the film (see discussion in Refs. [Brouers *et al.*, 1997a] and [Brouers *et al.*, 1998]). Note that for semicontinuous metallic films at $p = p_c$ the transmittance $T = |t|^2 \approx 0.25$ in a wide spectral range, from the visible to the far-infrared spectral range [Yagil *et al.*, 1992] (see Fig. 3.3). The average surface current $\langle \mathbf{j} \rangle$ relates to the average electric field $\langle \mathbf{E} \rangle$ through the Ohm's law: $\langle \mathbf{j} \rangle = a\sigma_e \langle \mathbf{E} \rangle = a\sigma_e t \mathbf{E}_0$ valid for thin films, where $\sigma_e = -i\varepsilon_e\omega/(4\pi)$ is the effective conductivity, and thickness of the film is approximated by the size of a metal grain a.

By substituting $\langle \mathbf{j} \rangle = \sigma_e a t \mathbf{E}_0$ into Eq. (3.54), the ratio of the scattering intensity $S(\theta)$ to the total intensity of the incident light I_0 is obtained. The ratio is independent of the film geometry:

$$\tilde{S}(\theta) = \frac{S(\theta)}{I_0} = \frac{2\pi (ka)^2}{c^2} (1 - \frac{\sin^2\theta}{2}) T \, |\sigma_e|^2 \int_0^\infty G(r) r \, dr, \tag{3.55}$$

which can be rewritten as

$$\tilde{S}(\theta) = \frac{(ka)^4}{8\pi} (1 - \frac{\sin^2\theta}{2}) T \, |\varepsilon_e|^2 \, \frac{1}{a^2} \int_0^\infty G(r) r \, dr. \tag{3.56}$$

It follows from the equation that the portion of the incident light, which is not reflected, transmitted or adsorbed, is scattered from the film according to

$$S_{tot} = 2\pi \int \tilde{S}(\theta) \sin\theta \, d\theta = \frac{(ka)^4}{3} T \, |\varepsilon_e|^2 \, \frac{1}{a^2} \int_0^\infty G(r) r \, dr. \tag{3.57}$$

Equations (3.56) and (3.57) have a transparent physical meaning. The anomalous scattering like Rayleigh scattering is inversely proportional to the fourth power of the wavelength $\tilde{S} \propto S_{tot} \propto (ak)^4 \propto (a/\lambda)^4$, and it is much enhanced in semicontinuous films due to spatial current-current (field-field) correlations described by the correlation function $G(r)$ in Eqs. (3.54)–(3.57). The function $G(r)$ can have different behaviors for different frequencies. The factor $T\left|\varepsilon_e\right|^2$ in Eq. (3.57) also depends on the frequency and achieves large values in the infrared spectral range.

Scattering described by Eq. (3.57) can be compared with the scattering from an ensemble of metal grains, which interact with the electromagnetic field independently. The Rayleigh scattering cross section σ_R from a single metal grain estimates as $\sigma_R = (8\pi/3)\,(ka)^4\,a^2$ for $|\varepsilon_m| \gg 1$ [Landau *et al.*, 1984]. The portion of the scattered light, if all the grains were independent, would then equal to $S_{tot}^R \simeq p(8/3)\,(ka)^4$. Assuming $p = 1/2$, we obtain the following estimate for enhancement g of the scattering light (the latter occurs due to the field fluctuations):

$$g = \frac{S_{tot}}{S_{tot}^R} \sim \frac{T\left|\varepsilon_e\right|^2}{4a^2} \int_0^\infty G(r) r\, dr. \tag{3.58}$$

If the integral of (3.58) is determined by the largest distances where field correlations are essential, i.e. $r \sim \xi_e$, the scattering is enhanced up to infinity, when the loss vanish and $\xi_e \to \infty$. It is the case for $2D$ metal-dielectric films as it will be shown below. Certainly, the formalism above holds only if the incident wave is scattered insignificantly, $S_{tot} \ll 1$. Otherwise, it is necessary to take into account the feedback effects, i.e., the interaction of the scattered light with the film.

The function $G(r)$ for 1024×1024 $L - C$ system is calculated using the RSRG method, discussed in Sec. 3.2.2. The scattering function $S_{tot}(\omega)$ is obtained for gold semicontinuous metal film at the percolation threshold $p = p_c = 1/2$. The results are shown in Figs. 3.17 and 3.18. One can see in Fig. 3.17 that the scattering portion of the light intensity dramatically increases for frequencies below the frequency, at which the real part of the metal dielectric function ε_m' becomes negative (cf. Fig. 3.16). The scattering has a broad double-peak maximum and finally drops in the infrared spectral range. To understand this result, let us investigate a behavior of the correlation function $G(r)$ in more details below, using numerical calculations as well as the scaling arguments discussed in Sec. 3.3.4.

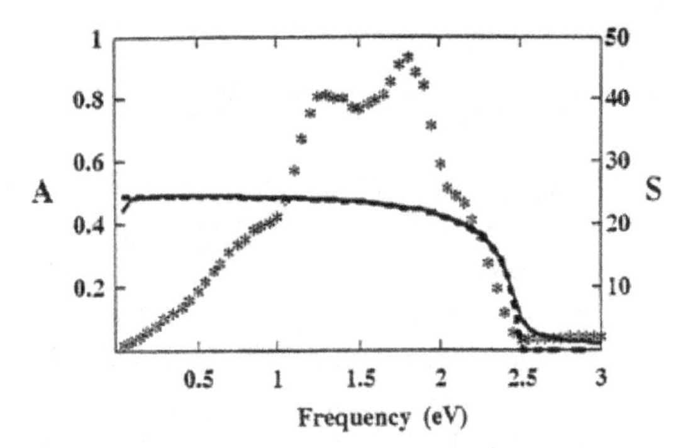

Fig. 3.17 Numerical results for absorptance of gold semicontinuous film at the percolation threshold ($p = p_c$) — continuous line; the same but film is loss free — dash line; average scattering $S = S_{tot}/4\pi$ from the film into solid angle — stars.

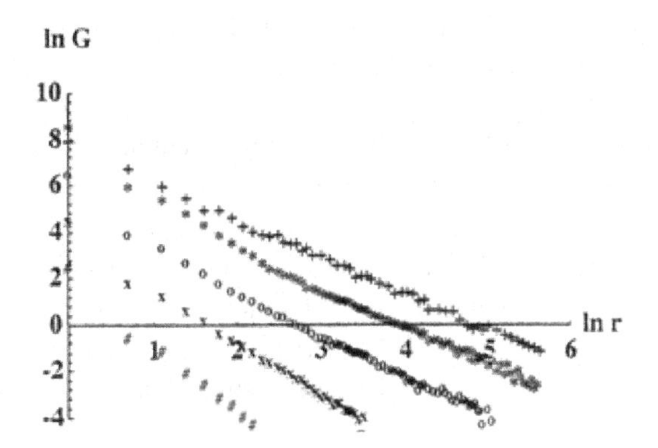

Fig. 3.18 Correlation function $G(r)$ for the resonance system $\varepsilon_m' = -\varepsilon_d$ at the percolation threshold $p = p_c$ calculated for different loss parameters $\kappa = \varepsilon_m''/|\varepsilon_m|$: # – $\kappa = 10^{-1}$, $\times$ – $\kappa = 10^{-2}$, $\bigcirc$ – $\kappa = 10^{-3}$, $*$ – $\kappa = 10^{-4}$; Correlation function $G(r)$ for $\varepsilon_m' = -100\ \varepsilon_d$ and $\kappa = 10^{-3}$ – +.

3.4.2 *Scaling properties of correlation function*

The correlation function $G(r)$ was calculated for the resonance film $\varepsilon_m' = -\varepsilon_d$, and for different values of loss parameters $\kappa = \varepsilon_m''/|\varepsilon_m'|$ [Brouers *et al.*, 1998]. The metal concentration was chosen to be $p = p_c = 1/2$. The system size was 1024×1024 elements and results were averaged over 100 different sample realizations for each value of κ. The function $G(r)$ is shown in Fig. 3.18, where the distance r is in units of the metal grain size a. It follows from the figure that for the scales $a < r < \xi_e$, the correlation function decays as

$$G(r) \sim M_2 \, (r/a)^{-(1+\eta)} \sim \kappa^{-1+\gamma} \, (r/a)^{-(1+\eta)} \,, \tag{3.59}$$

where $M_2 \sim \kappa^{-1+\gamma}$ is the second moment for the local electric fields given by Eq. (3.45) and is calculated for the system with $\varepsilon'_m = -\varepsilon_d$; the critical exponent η in Eq. (3.59) equals to $\eta = 0.8 \pm 0.1$, which determines the spatial correlation of the local fields. Substitution of the correlation function (3.59) in the expressions (3.56) and (3.57) for the scattering leads to the observation that the integrals diverge in the upper limit. Therefore, the scattering is determined at large distances by the function $G(r)$, where Eq. (3.59) still holds, i.e. at $r \sim \xi_e$. This suggests that the field fluctuations with spatial distances on the order of the field correlation length ξ_e but larger than the metal grain size a are responsible for the anomalous scattering from semicontinuous films.

Consider now the correlation function $G\,(r)$ for the case when $|\varepsilon_m| \gg \varepsilon_d$. We use the standard RSRG procedure for dividing the system into squares of size l and considering each square as a new element, as it has been discussed in Sec. 3.3.4. The square size l is set to be equal to the size l_r given by Eq. (3.38). Then the correlation function $G^\star$ in the newly renormalized system has the form of Eq. (3.59), while in the original system it has the form $G(r) \cong \; (l_r/a)^{\,1+\eta} G^\star(r)$ for $r \gg l_r$. The function follows the usual behavior of the current-current correlation function in a percolation system $G(r) \propto r^{-t/\nu}$ [Lagarkov *et al.*, 1992], for the distance $r \ll l_r$. By matching its asymptotes at $r = l_r$, the following approximations for the correlation function are obtained:

$$\begin{aligned} G(r) &\sim \; \kappa^{-1+\gamma} \left(\frac{l_r}{r}\right)^{t/\nu}, && a < r < l_r \,, \\ G(r) &\sim \kappa^{-1+\gamma} \left(\frac{l_r}{r}\right)^{1+\eta}, && l_r \; < r < \xi_e, \end{aligned} \tag{3.60}$$

where l_r and ξ_e are given by Eqs. (3.38) and (3.41) respectively. The correlation function calculated for the loss parameter $\kappa = 10^{-3}$ and the ratio $|\varepsilon_m|/\varepsilon_d = 10^2$, which corresponds $l_r = 10a$, is shown in Fig. 3.18. It demonstrates the excellent agreement with the scaling behavior of Eq. (3.60).

We now estimate the enhancement g for the Rayleigh scattering due to the field fluctuations in the semicontinuous metal films. Combining Eqs. (3.60) and (3.58), we obtain $g \sim |T\varepsilon_e|^2 \kappa^{-1+\gamma} l_r^{1+\eta} \xi_e^{1-\eta}/4$, where l_r and ξ_e are given by Eqs. (3.38) and (3.42), respectively. At the percolation

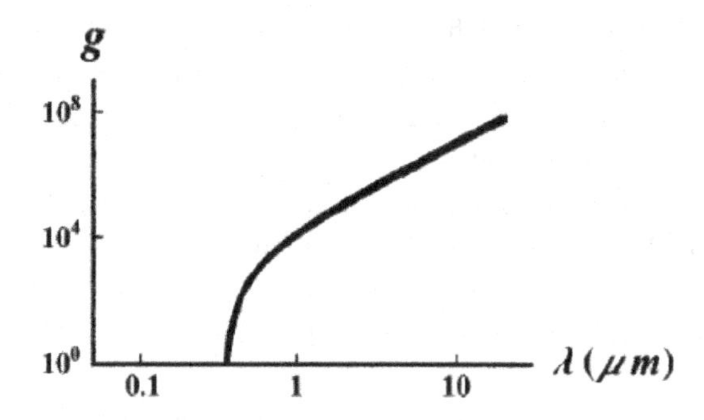

Fig. 3.19 Enhancement of Rayleigh scattering from semicontinuous gold film as function of wavelength; $p = p_c$.

threshold, the effective dielectric function of a semicontinuous metal film estimates as $\varepsilon_e = \sqrt{\varepsilon_d \varepsilon_m}$ [Dykhne, 1971] (see Secs. 1.2 and 4.4). Its substitution into the enhancement expression results in

$$g \sim \frac{T}{4} |\varepsilon_m|^2 \kappa^{(-1+\gamma)(1+(1-\eta)/2)} . \tag{3.61}$$

For the frequencies $\omega \ll \omega_p$, the metal dielectric permittivity for the Drude model [see Eq. (3.1)] is approximated by $\varepsilon_m \simeq -(\omega_p/\omega)^2 (1 - i\omega_\tau/\omega)$ and is further simplified: $|\varepsilon_m| \simeq (\omega_p/\omega)^2$. Using this and $\kappa = \omega_\tau/\omega$ in Eq. (3.61), we obtain

$$g \sim \frac{T}{4} \left(\frac{\omega_p}{\omega}\right)^4 \left(\frac{\omega}{\omega_\tau}\right)^{(1-\gamma)(1+(1-\eta)/2)} . \tag{3.62}$$

The scattering enhancement factor from a semicontinuous gold film is shown in Fig. 3.19, where we assume a typical value $T = 1/4$. The enhancement becomes large ($g \sim 5 \cdot 10^4$) at wavelength $\lambda = 1.5$ μm and continues to increase towards the far infrared spectral range. At first glance this continuous increase of the scattering enhancement g seems to contradict the anomalous scattering behavior shown in Fig. 3.17. This can be explained by noticing that Fig. 3.17 represents the anomalous scattering itself, whereas Fig. 3.19 represents the enhancement of the anomalous scattering with respect to Rayleigh scattering within assumption that metallic grains are independent. Rayleigh scattering is proportional to ω^4, whereas the anomalous scattering varies as $S \sim \omega^{(1-\gamma)(1+(1-\eta)/2)} \approx \omega$. The overall enhancement therefore increases as $g \sim \lambda^3$ for the infrared part of the spectrum.

In the formalism above, we have assumed that the wavelength λ was much larger than the grain size a and that λ was much larger than the spatial scale of the giant field fluctuations, $\lambda \gg \xi_e$. Consequently, the calculated Rayleigh scattering contains only a small portion of the incident light. The size of metal grains in semicontinuous metal films is usually on the order of few nanometers but it can be increased significantly by using a proper preparation method [Semin *et al.*, 1996] so that the field correlation length ξ_e [Eq. (3.42)] becomes larger than the incident wavelength λ. In this case, the scattering can be a dominating process in the light interaction with semicontinuous metal films. Then a critical opalescence can be observed in these films.

Although the possible analogy between the anomalous scattering and the critical opalescence was indicated, the important difference should be noted between these phenomena. The critical opalescence originates from the long-range fluctuations of material property, for example, the density fluctuations near a liquid-vapor critical point or the fractal structure of metal clusters (see [Stanley, 1981], [Stockman *et al.*, 1994], and [Shalaev *et al.*, 1991]). In contrast to this, the spatial distributions of metal grains in the semicontinuous metal films are random or correlated on the scales of a grain size $a \ll \lambda$. The anomalous light scattering originates from the long-range *field* fluctuations. Therefore, the incident electromagnetic wave plays a two-fold role: It first generates the giant fluctuations of the local electric fields, which then induce the anomalous scattering. The anomalous scattering is then, in a sense, a new kind of critical opalescence — the field opalescence.

3.5 Surface Enhanced Raman Scattering (SERS)

This section describes surface-enhanced Raman scattering (SERS) – one of the most intriguing optical effects discovered in the past 30 years (see, for example, [Shalaev, 2002], [Chang and Furtak, 1982], [Moskovits, 1985], [Kneipp *et al.*, 1997], [Poliakov *et al.*, 1996], [Kneipp *et al.*, 1999], [Nie and Emory, 1997], [Krug *et al.*, 1999], [Lyon *et al.*, 1998]). We present below a theory for Raman scattering, which is enhanced by strong fluctuations of the local fields [Brouers *et al.*, 1997a], [Sarychev and Shalaev, 2000]. Raman-active molecules are assumed to be uniformly distributed on the top of the film and serve only as a source for the initial Raman signal. To simplify the consideration we assume here that Raman molecules are centrosymmetric, so there is no specific directions in the system, and therefore, no polarization dependence. It is the semicontinuous metal films, which consist of metal grains randomly distributed on a dielectric substrate, where we have to pay main attention to the local electric field distribution. The gaps between the metal grains of the film are filled with the dielectric material of the substrate. Then the local conductivity of the film $\sigma(\mathbf{r})$ takes either the "metallic" values, $\sigma(\mathbf{r}) = \sigma_m$, in the metal grains or the "dielectric" values, $\sigma(\mathbf{r}) = -i\omega\varepsilon_d/4\pi$, outside of the metal, ω is the frequency of the external field. We again assume that the wavelength λ is much larger than a typical scale of inhomogeneity in the composite including the grain size a, the gaps between the grains, the percolation correlation length ξ_p, and the local field correlation length ξ_e [see Eq. (3.41)]. In this case, the local field $\mathbf{E}(\mathbf{r})$ is given by Eq. (3.4). This equation is solved to find the fluctuating potential $\phi(\mathbf{r})$ and the local field $\mathbf{E}(\mathbf{r})$ induced in the film by the external field $\mathbf{E}_0(\mathbf{r})$. The main results presented below hold for any metal-dielectric composite providing the local field strongly fluctuates.

It is instructive to assume first that the external field $\mathbf{E}_0(\mathbf{r})$ has a sharp pin-like distribution, $\mathbf{E}_0(\mathbf{r}) = \mathbf{E}_1\delta(\mathbf{r}-\mathbf{r}_1)$, where the $\delta(\mathbf{r})$ is the Dirac delta-function. The current density at arbitrary point $\mathbf{r}_2$ is given by the following linear relation

$$\mathbf{j}(\mathbf{r}_2) = \widehat{\Sigma}(\mathbf{r}_2, \mathbf{r}_1)\mathbf{E}_1, \tag{3.63}$$

which defines the non-local conductivity matrix $\widehat{\Sigma}(\mathbf{r}_2, \mathbf{r}_1)$. The matrix represents the system's response at coordinate point $\mathbf{r}_2$ for the field source located at the point $\mathbf{r}_1$. If an inhomogeneous external field $\mathbf{E}_0(\mathbf{r})$ is applied to the system, the local current at the point $\mathbf{r}_2$ equals to

$$\mathbf{j}(\mathbf{r}_2) = \int \widehat{\Sigma}(\mathbf{r}_2, \mathbf{r}_1)\mathbf{E}_0(\mathbf{r}_1)\, d\mathbf{r}_1, \tag{3.64}$$

where integration is performed over the total area of the system.

The non-local conductivity $\widehat{\Sigma}$ can be expressed in terms of the Green function G of Eq. (3.4):

$$\nabla \cdot \{\sigma(\mathbf{r}_2)\, [\nabla G(\mathbf{r}_2, \mathbf{r}_1)]\} = \delta(\mathbf{r}_2 - \mathbf{r}_1), \tag{3.65}$$

where the differentiation with respect the coordinate $\mathbf{r}_2$ is assumed. Comparing Eqs. (3.4) and (3.65) and using the definition of the non-local conductivity given in Eq. (3.63), the following equation is obtained

$$\Sigma_{\alpha\beta}(\mathbf{r}_2, \mathbf{r}_1) = \sigma(\mathbf{r}_2)\sigma(\mathbf{r}_1)\frac{\partial^2 G(\mathbf{r}_2, \mathbf{r}_1)}{\partial r_{2,\alpha}\, \partial r_{1,\beta}}, \tag{3.66}$$

where the Greek indices represent film surface coordinates (x and y).

As it follows from the symmetry of Eq. (3.65), the Green function is symmetric with respect to the interchange of its arguments: $G(\mathbf{r}_1, \mathbf{r}_2) = G(\mathbf{r}_2, \mathbf{r}_1)$. Then, Eq. (3.66) implies that the non-local conductivity is also symmetric:

$$\Sigma_{\alpha\beta}(\mathbf{r}_1, \mathbf{r}_2) = \Sigma_{\beta\alpha}(\mathbf{r}_2, \mathbf{r}_1). \tag{3.67}$$

The introduction of the nonlocal conductivity, $\widehat{\Sigma}$, considerably simplifies further calculations for the local field distributions. The symmetry of $\widehat{\Sigma}$ given by Eq. (3.67) is also important for the following analysis.

Since the incident EM wavelength is much larger than any spatial scale of the semicontinuous metal film, the external field $\mathbf{E}_0$ takes a constant value on the film surface. The local fields $\mathbf{E}(\mathbf{r}_2)$ induced by the external field $\mathbf{E}_0$ can be obtained by using Eq. (3.64) for the non-local conductivity $\widehat{\Sigma}$:

$$\mathbf{E}(\mathbf{r}_2) = \frac{1}{\sigma(\mathbf{r}_2)} \int \widehat{\Sigma}(\mathbf{r}_2, \mathbf{r}_1)\mathbf{E}_0 \, d\mathbf{r}_1. \tag{3.68}$$

The local field $\mathbf{E}(\mathbf{r}_2)$ excites Raman-active molecules distributed on the film surface. The Raman-active molecules, in turn, generate the Stokes fields $\mathbf{E}_s(\mathbf{r}_2) = \alpha_s(\mathbf{r}_2)\mathbf{E}(\mathbf{r}_2)$, which oscillate at the shifted frequency $\omega_s = \omega - \Omega$, where Ω is the Stokes frequency shift corresponding to the frequency of molecule oscillations (electronic or vibrational). Parameter $\alpha_s(\mathbf{r}_2)$ is the ratio of the Raman and the linear polarizabilities for the Raman-active molecule, located at the coordinate point $\mathbf{r}_2$. The Stokes fields $\mathbf{E}_s(\mathbf{r}_2)$ induce the local currents $\mathbf{j}_s(\mathbf{r}_3)$ in the composite, which are given by the equation similar to Eq. (3.68):

$$\mathbf{j}_s(\mathbf{r}_3) = \int \widehat{\Sigma}(\mathbf{r}_3, \mathbf{r}_2)\mathbf{E}_s(\mathbf{r}_2) \, d\mathbf{r}_2. \tag{3.69}$$

Since the Stokes-shifted frequency ω_s is typically close to the frequency of the external field $|\omega - \omega_s|/\omega \ll 1$ the non-local conductivities $\widehat{\Sigma}$, appearing in Eqs. (3.68) and (3.69), are considered to be the same.

The intensity of the electromagnetic wave I scattered from any inhomogeneous system is proportional to the current fluctuations in the system, as it has been discussed in Sec. 3.4:

$$I \propto \left\langle \left| \int \left[\mathbf{j}(\mathbf{r}) - \langle \mathbf{j} \rangle\right] \, d\mathbf{r} \right|^2 \right\rangle, \tag{3.70}$$

where the integration is taken over the entire system and the angular brackets $\langle \ldots \rangle$ denote the ensemble average. For Raman scattering, the averaging procedure means the averaging over the fluctuating phases of the incoherent Stokes fields generated by different Raman-active molecules. The averaged current densities oscillating at ω_s give zero for the currents $\langle \mathbf{j}_s \rangle = 0$. Then the intensity of Raman scattering I_s from a semicontinuous metal film acquires the following form

$$\begin{aligned} I_s &\propto \left\langle \left| \int \mathbf{j}_s(\mathbf{r})\, d\mathbf{r} \right|^2 \right\rangle \\ &= \int \langle \Sigma_{\alpha\beta}(\mathbf{r}_3, \mathbf{r}_2) \alpha_s(\mathbf{r}_2) E_\beta(\mathbf{r}_2) \\ &\quad \times \Sigma^*_{\alpha\gamma}(\mathbf{r}_5, \mathbf{r}_4) \alpha^*_s(\mathbf{r}_4) E^*_\gamma(\mathbf{r}_4) \rangle \; d\mathbf{r}_2\, d\mathbf{r}_3\, d\mathbf{r}_4\, d\mathbf{r}_5, \end{aligned} \tag{3.71}$$

where the summation over repeating Greek indices is implied. All the integration in Eq. (3.71) is performed over the entire film surface. Eq. (3.71) is averaged over the fluctuating phases of the Raman polarizabilities α_s. Since the Raman field sources are incoherent, the average results in:

$$\langle \alpha_s(\mathbf{r}_2) \alpha^*_s(\mathbf{r}_4) \rangle = |\alpha_s|^2 \delta(\mathbf{r}_2 - \mathbf{r}_4). \tag{3.72}$$

After that, Eq. (3.71) takes the form of

$$I_s \propto \int \left\langle \Sigma_{\alpha\beta}(\mathbf{r}_3, \mathbf{r}_2) \Sigma^*_{\mu\gamma}(\mathbf{r}_5, \mathbf{r}_2) \delta_{\alpha\mu} |\alpha_s|^2 E_\beta(\mathbf{r}_2) E^*_\gamma(\mathbf{r}_2) \right\rangle \, d\mathbf{r}_2\, d\mathbf{r}_3\, d\mathbf{r}_5\,, \tag{3.73}$$

where the Kronecker symbol $\delta_{\alpha\mu}$ is introduced to simplify further considerations. Since a semicontinuous film is macroscopically homogeneous, Raman scattering is independent of the orientation of the external field $\mathbf{E}_0$ assuming that the Raman molecules are purely symmetric (which is usually the case). Therefore Eq. (3.73) can be averaged over all the incident polarizations $\mathbf{E}_0$ without changing the result.

The averaging of the products $E_\beta(\mathbf{r}_2) E^*_\gamma(\mathbf{r}_2)$ and $E_{0,\alpha} E^*_{0,\mu}$ results in

$$\langle E_\beta(r_2) E^*_\gamma(r_2) \rangle_0 = \frac{1}{2} \, \langle |E(r_2)|^2 \rangle_0 \delta_{\beta\gamma} \tag{3.74}$$

$$\delta_{\alpha\mu} = 2 \frac{\langle E_{0,\alpha} E^*_{0,\mu} \rangle_0}{|E_0|^2}, \tag{3.75}$$

where the sign $\langle \ldots \rangle_0$ denotes averaging over all the orientations of the external field $\mathbf{E}_0$. Substituting Eqs. (3.74) and (3.75) in Eq. (3.73) and noting that the non-local conductivity $\widehat{\Sigma}$ is independent of the field polarizations, the following result is obtained for the intensity of the Raman signal

$$\langle I_s \rangle \propto \int \Sigma_{\alpha\beta}(r_3, r_2)\Sigma^*_{\mu\beta}(r_5, r_2)\frac{\langle E_{0,\alpha}E^*_{0,\mu}\rangle_0}{|E_0|^2} \times |\alpha_s|^2 \langle\, |E(r_2)|^2\rangle_0 \, d\mathbf{r}_2 \, d\mathbf{r}_3 \, d\mathbf{r}_5. \qquad (3.76)$$

For simplicity, the sign for ensemble averaging is omitted here. Now the symmetry of the non-local conductivity, given by Eq. (3.67), is used to rewrite Eq. (3.76) as

$$\langle I_s \rangle \propto \int \langle \Sigma_{\beta\alpha}(\mathbf{r}_2, \mathbf{r}_3)E_{0,\alpha}\Sigma_{\beta\mu}{}^*(\mathbf{r}_2, \mathbf{r}_5)E^*_{0,\mu}\rangle_0 \times \frac{|\alpha_s|^2}{|E_0|^2}\langle |E(\mathbf{r}_2)|^2\rangle_0 \; d\mathbf{r}_2 \, d\mathbf{r}_3 \, d\mathbf{r}_5 \, . \qquad (3.77)$$

Integration over the coordinates and the implementation of Eq. (3.64) both give

$$\langle I_s \rangle \propto \frac{|\alpha_s|^2}{|E_0|^2} \int |\sigma(\mathbf{r}_2)|^2 \; \langle |E(\mathbf{r}_2)|^2\rangle_0 \; \langle |E(\mathbf{r}_2)|^2\rangle_0 \; d\mathbf{r}_2. \qquad (3.78)$$

It is easy to show that this equation can be rewritten for macroscopically isotropic system in the following form

$$\langle I_s \rangle \propto \frac{|\alpha_s|^2}{|E_0|^2} \int |\sigma(\mathbf{r}_2)|^2 |E(\mathbf{r}_2)|^4 \; d\mathbf{r}_2. \qquad (3.79)$$

If there were no metal grains on the film, the local fields would not fluctuate and one would obtain the following expression for the Raman scattering

$$I_s^0 \propto \int |\sigma_d(\mathbf{r}_2)|^2 |\alpha_s|^2 |E_0(\mathbf{r}_2)|^2 \; d\mathbf{r}_2. \qquad (3.80)$$

Therefore, the enhancement of Raman scattering G_R due to presence of metal grains on a dielectric substrate is given by

$$G_R = \frac{\langle I_s \rangle}{I_s^0} = \frac{\langle |\sigma(\mathbf{r})|^2 |E(\mathbf{r})|^4\rangle}{|\sigma_d|^2 |E_0|^4} = \frac{\langle |\varepsilon(\mathbf{r})|^2 |E(\mathbf{r})|^4\rangle}{\varepsilon_d^2 |E_0|^4}. \qquad (3.81)$$

Note that the derivation of Eq. (3.81) is essentially independent of the dimensionality and morphology of the system.

Therefore, the enhancement of Raman scattering given by Eq. (3.81) holds for any inhomogeneous system provided the field fluctuations take place in the system. In particular, Eq. (3.81) gives the enhancement for the

Raman scattering from a rough metallic surface, provided that the wavelength is much larger than the roughness features. Eq. (3.81) can be also used to calculate enhancements for Raman scattering in three-dimensional percolation composites. In presented theory, the main sources for the Raman-active signal are the currents excited by the Raman molecules attached to or embedded into the metal grains. This explains why a significant enhancement for Raman scattering was observed even for relatively flat metal surfaces [Moskovits, 1985; Chang and Furtak, 1982; Tadayoni and Dand, 1991].

For the metal-dielectric percolation composites near the percolation threshold, the local electric field concentrates mainly in the dielectric gaps between the large metal clusters. Then can approximate $\varepsilon(\mathbf{r}) = \varepsilon_d$ in Eq. (3.81) and estimate SERS as $G_R \sim M_{4,0} = \langle |E(\mathbf{r})/E_0|^4 \rangle$. Equation (3.43) for the fourth moment gives the SERS

$$G_R \sim \left(\frac{|\varepsilon_m|}{\varepsilon_d} \right)^{\frac{2\nu+s}{t+s}} \left(\frac{|\varepsilon_m|}{\varepsilon_m''} \right)^{3-\gamma(7-D)}, \tag{3.82}$$

for the metal concentrations p in the vicinity of the percolation threshold p_c.

For $2D$ metal-dielectric composites (semicontinuous metal films), the critical exponents $s \approx t \approx \nu \approx 4/3$ and $\gamma = 0.14$. The Drude model (3.1) is invoked again for the frequencies $\omega \ll \omega_p$, and Eq. (3.82) results in

$$G_R \sim \varepsilon_d^{-3/2} \left(\omega_p/\omega_\tau \right)^3 \left(\omega_\tau/\omega \right)^{0.7}. \tag{3.83}$$

The enhancement weakly depends on the frequency: $G_R \sim \omega^{-0.7}$ (see Fig. 3.16). For typical good optical metals (*Ag, Au, Al, Cu*), the plasma frequency ω_p corresponds to about 10 eV while the relaxation frequency ω_τ is about $0.01 \div 0.1$ eV. Therefore, the enhancement G_R of the Raman signal has a large factor $(\omega_p/\omega_\tau)^3 \sim 10^6 \div 10^9$. For silver-on-glass semicontinuous films ($\omega_p = 9.1$ eV, $\omega_\tau = 0.021$ eV, $\varepsilon_d = 2.2$), the SERS factor $G_R \sim 10^6 \omega^{0.7}$, where the frequency is measured in eV.

The hyper-Raman scattering is the optical process when n photons of frequency ω are converted in one hyper-Stokes photon of the frequency $\omega_{hR} = n\omega - \Omega$. Following the general approach, described earlier in this section (also see [Brouers *et al.*, 1997a]), the surface enhancement of hyper-Raman scattering G_{hR} can be derived to equal to

$$G_{hR} = \frac{\langle |\varepsilon_{hR}(\mathbf{r})|^2 |\mathbf{E}_{hR}(\mathbf{r})|^2 |E(\mathbf{r})|^{2n} \rangle}{|\varepsilon_d|^2 \left| E_{0,hR} \right|^2 |E_0|^{2n}}, \tag{3.84}$$

where $\mathbf{E}_{hR}(\mathbf{r})$ is the local field excited in the system by the uniform probe field $\mathbf{E}_{0,hR}$ oscillating with ω_{hR}; $\varepsilon_{hR}(\mathbf{r})$ is dielectric constant at the frequency ω_{hR}. At $n = 1$ formula (3.84) describes the conventional SERS. The enhancement G_{hR} of the hyper-Raman scattering involves the high moments of the local fields. Therefore, it is enhanced even more than the regular SERS. The numerical estimates for G_{hR} could be found in [Sarychev and Shalaev, 2000].

3.6 Giant Enhancements of Optical Nonlinearities

In this Section, we consider enhancements for different optical nonlinearities, such as Kerr optical effect and generation of high optical harmonics. The Kerr optical nonlinearity results in nonlinear corrections (proportional to the light intensity) for the refractive index and the absorption coefficient (see, for example, [Boyd, 1992]). The Kerr-type nonlinearity is the third-order optical nonlinearity, which results in additional term for the electric displacement vector $\mathbf{D}$ as

$$D_{\alpha}^{(3)}(\omega) = \varepsilon_{\alpha\beta\gamma\delta}^{(3)}(-\omega;\omega,\omega,-\omega)E_{\beta}E_{\gamma}E_{\delta}^{*}, \tag{3.85}$$

where

$$\varepsilon_{\alpha\beta\gamma\delta}^{(3)}(-\omega;\omega,\omega,-\omega), \tag{3.86}$$

is the third-order nonlinear dielectric tensor [Boyd, 1992] [Landau *et al.*, 1984], $\mathbf{E}$ is an electric field at frequency ω, and the summation over repeated Greek indices is implied.

We consider a macroscopically homogeneous and isotropic composite. Then the third-order terms in the average electric displacement have the following generic form:

$$\left\langle \mathbf{D}^{(3)}(\mathbf{r})\right\rangle = \alpha|\mathbf{E}_0|^2\mathbf{E}_0 + \beta\, E_0^2\mathbf{E}_0^{*}, \tag{3.87}$$

where $\mathbf{E}_0$ is the amplitude of the external (macroscopic) electric field at frequency ω, $E_0{}^2 \equiv (\mathbf{E}_0 \cdot \mathbf{E}_0)$, α and β are some constants [not to be confused with the tensor components in Eq. (3.85)]. Note that the second term for the nonlinear displacement vector in Eq. (3.87) can result in change of the polarization of the incident light in an isotropic film. Equation (3.87) simplifies for the case of incident linear and circular polarizations. [Boyd, 1992]. For the linear polarization, the complex vector $\mathbf{E}_0$ reduces to the real vector. Then the expressions $|\mathbf{E}_0|^2\mathbf{E}_0$ and $E_0^2\ \mathbf{E}_0^*$ in Eq. (3.87) become the same and the equation can be rewritten as

$$\left\langle \mathbf{D}^{(3)}(\mathbf{r})\right\rangle = \varepsilon_e^{(3)}\left|E_0\right|^2\mathbf{E}_0, \tag{3.88}$$

where the nonlinear dielectric tensor $\varepsilon_e^{(3)}$ reduces to a scalar $\varepsilon_e^{(3)} = \alpha + \beta$. On the other hand $\langle \mathbf{D}^{(3)}(\mathbf{r})\rangle = \alpha|\mathbf{E}_0|^2\mathbf{E}_0$ for the circular polarization since $E_0^2 = 0$ in this case. For the sake of simplicity, we consider only a linearly polarized wave. Equation (3.88), being rewritten in terms of the nonlinear average current $\langle \mathbf{j}^{(3)}(\mathbf{r})\rangle$ and the effective Kerr conductivity $\sigma_e^{(3)} = -i\omega\varepsilon_e^{(3)}/4\pi$, takes the following form

$$\left\langle \mathbf{j}^{(3)}(\mathbf{r})\right\rangle = \sigma_e^{(3)}\left|E_0\right|^2\mathbf{E}_0. \tag{3.89}$$

This form of the Kerr nonlinearity is used in the discussion below.

First, we consider the case when the nonlinearities in metal grains $\sigma_m^{(3)}$ and dielectric $\sigma_d^{(3)}$ are close to each other $\sigma_m^{(3)} \approx \sigma_d^{(3)} \approx \sigma^{(3)}$. For example, the nonlinear response can originate from the dye molecules or quantum dots, that uniformly cover the semicontinuous film. When $\sigma_m^{(3)} \approx \sigma_d^{(3)} \approx \sigma^{(3)}$, the current in the composite is given by

$$\mathbf{j}(\mathbf{r}) = \sigma^{(1)}(\mathbf{r})\mathbf{E}'(\mathbf{r}) + \sigma^{(3)}|\mathbf{E}'(\mathbf{r})|^2\mathbf{E}'(\mathbf{r}), \tag{3.90}$$

where $\mathbf{E}'(\mathbf{r})$ is the local field at $\mathbf{r}$. Applying the current conservation law given by Eq. (3.4) to Eq. (3.90) gives

$$\nabla\cdot\left(\sigma^{(1)}(\mathbf{r})\left[-\nabla\phi(\mathbf{r}) + \mathbf{E}_0 + \frac{\sigma^{(3)}}{\sigma^{(1)}(\mathbf{r})}\mathbf{E}'(\mathbf{r})|\mathbf{E}'(\mathbf{r})|^2\right]\right) = 0, \tag{3.91}$$

where $\mathbf{E}_0$ is the applied electric field, and $-\nabla\phi(\mathbf{r}) + \mathbf{E}_0 \equiv \mathbf{E}'(\mathbf{r})$ is the local field. It is convenient to consider the last two terms in the square brackets in Eq. (3.91) as a renormalized external field

$$\mathbf{E}_e(\mathbf{r}) = \mathbf{E}_0 + \mathbf{E}_f(\mathbf{r}) = \mathbf{E}_0 + \frac{\sigma^{(3)}}{\sigma^{(1)}(\mathbf{r})}\mathbf{E}'(\mathbf{r})|\mathbf{E}'(\mathbf{r})|^2, \tag{3.92}$$

where the field

$$\mathbf{E}_f(\mathbf{r}) = \frac{\sigma^{(3)}}{\sigma^{(1)}(\mathbf{r})}\mathbf{E}'(\mathbf{r})|\mathbf{E}'(\mathbf{r})|^2 \tag{3.93}$$

may change over the film but its averaged value $\langle \mathbf{E}_f(\mathbf{r})\rangle$ is collinear with the applied field $\mathbf{E}_0$. Then the average current density $\langle \mathbf{j}(\mathbf{r})\rangle$ is also collinear to $\mathbf{E}_0$ in the macroscopically isotropic films considered here. Therefore, the average current can be written as

$$\langle \mathbf{j} \rangle = \frac{\mathbf{E}_0}{E_0^2} \left(\mathbf{E}_0 \cdot \langle \mathbf{j} \rangle \right) = \frac{\mathbf{E}_0}{E_0^2} \frac{1}{A} \int \mathbf{E}_0 \cdot \mathbf{j}\left(\mathbf{r} \right) \, d\mathbf{r}, \tag{3.94}$$

where A is the total area of the film and the integration is performed over the film area. Expressing the current $\mathbf{j}\left(\mathbf{r}\right)$ in Eq. (3.94) in terms of the nonlocal conductivity matrix defined by Eq. (3.63) gives

$$\langle \mathbf{j} \rangle = \frac{\mathbf{E}_0}{E_0^2} \frac{1}{A} \int \left[\mathbf{E}_0 \widehat{\Sigma}(\mathbf{r}, \mathbf{r}_1) \mathbf{E}_e \left(\mathbf{r}_1 \right) \right] \, d\mathbf{r} \, d\mathbf{r}_1. \tag{3.95}$$

If this equation is integrated over the surface coordinates $\mathbf{r}$ and the symmetry of nonlocal conductivity matrix (3.67) is used, we obtain

$$\langle \mathbf{j} \rangle = \frac{\mathbf{E}_0}{E_0^2} \frac{1}{A} \int \left(\mathbf{j}_0 \left(\mathbf{r} \right) \cdot \mathbf{E}_e \left(\mathbf{r} \right) \right) \, d\mathbf{r}, \tag{3.96}$$

where $\mathbf{j}_0\left(\mathbf{r}\right)$ is the current induced at the coordinate $\mathbf{r}$ by the constant external field $\mathbf{E}_0$. By substituting the renormalized external field $\mathbf{E}_e(\mathbf{r})$ from Eq. (3.92) into Eq. (3.96) and integrating over the coordinates, the average current is found to be

$$\langle \mathbf{j} \rangle = \mathbf{E}_0 \left[\sigma_e^{(1)} + \frac{\left\langle \sigma^{(3)} \left(\mathbf{E}(\mathbf{r}) \cdot \mathbf{E}'(\mathbf{r}) \right) \left| \mathbf{E}'(\mathbf{r}) \right|^2 \right\rangle}{E_0{}^2} \right], \tag{3.97}$$

where $\sigma_e^{(1)}$ and $\mathbf{E}(\mathbf{r})$ are the effective conductivity and local fluctuating field, respectively, obtained in the linear approximation, i.e. for $\sigma^{(3)} \equiv 0$. Comparison of Eqs. (3.97) and (3.89) allows us to establish the expression for the effective Kerr conductivity

$$\sigma_e^{(3)} = \frac{\left\langle \sigma^{(3)} \left(\mathbf{E}(\mathbf{r}) \cdot \mathbf{E}'(\mathbf{r}) \right) \left| \mathbf{E}'(\mathbf{r}) \right|^2 \right\rangle}{E_0{}^2 \left| E_0 \right|^2}. \tag{3.98}$$

We stress out that this result does not depend on the "weakness" of the nonlinearity, i.e., it holds even for a strong nonlinear case $\left|\sigma^{(3)} E^2\right| \gg \sigma_e^{(1)}$. For the case of weak nonlinearities, the local field $\mathbf{E}'(\mathbf{r})$ in Eq. (3.98) can be replaced by the linear local field $\mathbf{E}(\mathbf{r})$ resulting in the following equation:

$$\sigma_e^{(3)} = \frac{\left\langle \sigma^{(3)} E^2(\mathbf{r}) \left| \mathbf{E}(\mathbf{r}) \right|^2 \right\rangle}{E_0^2 \left| E_0 \right|^2}, \tag{3.99}$$

which gives the effective nonlinear conductivity in terms of the *linear* local field.

In the absence of the metal grains, the local electric field $\mathbf{E}(\mathbf{r}) \simeq \mathbf{E}_0$ and the effective nonlinear Kerr conductivity $\sigma_e^{(3)}$ coincides with the Kerr conductivity $\sigma^{(3)}$. Thus the enhancement of the Kerr nonlinearity G_K due to the excitation of SP is given by the following equation

$$G_K = \frac{\left\langle E^2(\mathbf{r})|\mathbf{E}(\mathbf{r})|^2 \right\rangle}{E_0^2|E_0|^2} = M_{2,2}, \tag{3.100}$$

where $M_{2,2}$ is the fourth moment of the local field [see Eq. (3.28)]. Therefore, the enhancement of the Kerr nonlinearity G_K can be ultimately expressed in terms of the local field $\mathbf{E}(\mathbf{r})$ in the linear approximation.

So far we have assumed that the nonlinear Kerr conductivity $\sigma^{(3)}$ is the same in the metal and in the dielectric. When $\sigma_m^{(3)} \neq \sigma_d^{(3)}$ the above derivations can be repeated and the following result emerges for the effective Kerr conductivity

$$\sigma_e^{(3)} = p\sigma_m^{(3)} \frac{\left\langle E^2(\mathbf{r})|\mathbf{E}(\mathbf{r})|^2 \right\rangle_m}{E_0^2|E_0|^2} + (1-p)\sigma_d^{(3)} \frac{\left\langle E^2(\mathbf{r})|\mathbf{E}(\mathbf{r})|^2 \right\rangle_d}{E_0^2|E_0|^2}, \tag{3.101}$$

where the angular brackets $\langle .. \rangle_m$ and $\langle .. \rangle_d$ stand for the averaging over the metal and the dielectric regions, respectively. The formula (3.101) for enhancement of the cubic nonlinearity in percolating composites was first obtained by Aharony [Aharony, 1987], Stroud and Hui [Stroud and Hui, 1988] and Bergman [Bergman, 1989]. Similar expression was independently obtained by Shalaev et al. to describe the Kerr enhancement in aggregates of metal particles [Shalaev, 2000], [Shalaev, 1996]. The general equation Eq. (3.98) was derived by [Shalaev and Sarychev, 1998]. The enhancement for the third and second harmonic generation was considered by [Hui and Stroud, 1997], [Hui *et al.*, 1998], and [Hui *et al.*, 2004b]. The equations for the specular emission of the high harmonics are similar to Eq. (3.98). In the next section, we will demonstrate that the diffuse-like scattering can prevail for the high harmonics.

According to Eq. (3.100), the value of the Kerr enhancement G_K is proportional to the fourth power of the local fields averaged over the sample. This is similar to the case of surface-enhanced Raman scattering with the enhancement factor G_R given by Eq. (3.81). Note, however, that Kerr enhancement G_K is a complex quantity whereas G_R is a real positive quantity.

Because the enhancement for Raman scattering is determined by the average of $|\mathbf{E}|^4$, which is phase insensitive, the upper limit for the enhancement is realized in this case.

The enhancement of the Kerr nonlinearity can be estimated analytically using the methods developed in Sec. 3.3.4. We consider first the case when the conductivity $\sigma^{(3)}(\mathbf{r})$ of the dielectric component is of the same order of magnitude or larger than in the metal component. (The opposite case of almost linear dielectric $\left|\sigma_d^{(3)}\right| \ll \left|\sigma_m^{(3)}\right|$ will be considered below). Then the Kerr enhancement G_K is approximated by

$$G_K \sim \left|\sigma_e^{(3)} / \left\langle \sigma^{(3)}(\mathbf{r}) \right\rangle \right| = \left|\varepsilon_e^{(3)} / \left\langle \varepsilon^{(3)}(\mathbf{r}) \right\rangle \right| \sim |M_{2,2}| \\ \sim \left(\frac{|\varepsilon_m|}{\varepsilon_d} \right)^{\frac{2\nu+s}{t+s}} \left(\frac{|\varepsilon_m|}{\varepsilon_m''} \right)^{3-\gamma(7-D)} \tag{3.102}$$

where Eq. (3.31) is used for the moment $M_{2,2}$ of the local field. For $\omega \ll \omega_p$, the Kerr enhancement for two-dimensional composites including semicontinuous metal films can be estimated as $G_K \sim \varepsilon_d^{-3/2} (\omega_p/\omega_\tau)^3 (\omega_\tau/\omega)^{0.7}$, where the Drude formula (3.1) is used for the metal dielectric constant ε_m, and we have substituted $t = s = \nu = 4/3$, $\gamma = 0.14$. For a silver-on-glass semicontinuous film, the Kerr enhancement is $G_K \sim 10^5 \div 10^6$.

The results of numerical simulations for G_K are shown as a function of the metal filling factor p for $D = 2$ in Fig. 3.20. The plot has a well-pronounced two-peak structure. Our simulations show that the dip at $p = p_c$ is proportional to the loss factor κ. This implies that at $p = p_c$, the enhancement is directly proportional to the local field moment $M_{2,2}$ given above: $G_K \sim \kappa M_{2,2}$. This result is the consequence of the special symmetry of a self-dual system at $p = p_c$.

As shown in Sec. 3.3.4, the local field maxima are concentrated in the dielectric gaps when $|\varepsilon_m| \gg \varepsilon_d$. Therefore the enhancement estimate in Eq. (3.102) is valid when the Kerr nonlinearity is located mainly in these gaps, which can be due to the dielectric itself or due to the adsorbed on the surface molecules). When the Kerr nonlinearity originates from the metal grains exclusively as it was done in the experiment of [Ma *et al.*, 1998] the enhancements are much smaller (also see details in [Sarychev and Shalaev, 2000]).

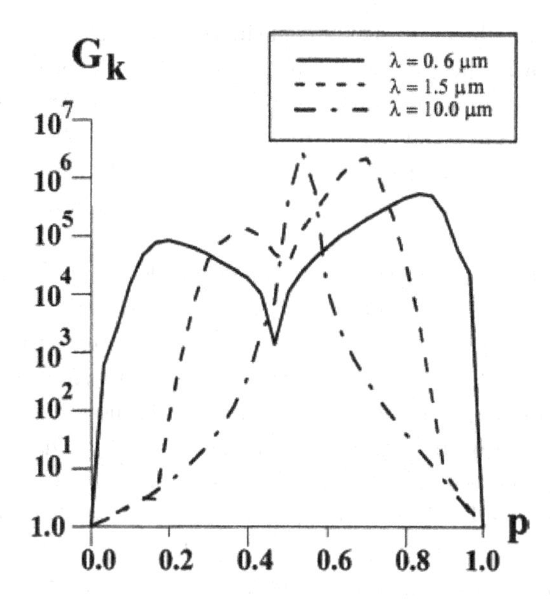

Fig. 3.20 Average enhancement of Kerr optical nonlinearity in silver semicontinuous films as function of the metal concentration p for three different wavelengths. The nonlinear Kerr permittivity $\varepsilon^{(3)}$ is the same for metal and dielectric components.

3.7 Percolation-enhanced Nonlinear Scattering: High Harmonic Generation

Nonlinear optical processes of the n-th order are proportional to the nth power of the electric field $E^n(\mathbf{r})$ and, therefore, the strong spatial fluctuations of the "nonlinear" field source, $\propto E^n(\mathbf{r})$, can result in giant nonlinear scattering from a composite material (see Fig. 3.9.) In this section, we consider percolation-enhanced nonlinear scattering (PENS) from a random metal-dielectric film (also referred to as a semicontinuous metal film) for the case when the metal filling factor p is close to the percolation threshold, $p = p_c$. Specifically, we study the enhanced nonlinear scattering, which originates from the local field oscillations at frequency $n\omega$, when a percolation metal-dielectric film is illuminated by a monochromatic EM wave of the frequency ω.

At the percolation, an infinite metal cluster spans over the entire sample and the metal-dielectric transition occurs in a semicontinuous metal film (see Sec. 1.2). Optical excitations of the self-similar fractal clusters formed by metal particles near p_c result in giant, scale-invariant, field fluctuations, which make the considered here PENS process differ from the known phenomena of surface polariton excitations and harmonic generation from smooth and rough metal surfaces (see [Liebsch and Schaich, 1989; Murphy *et al.*, 1989; Reider and Heinz, 1995; Tsang, 1995; Watts *et al.*, 1997a] and [Bozhevolnyi *et al.*, 1998; McGurn *et al.*, 1985; Sanches-Gil *et al.*, 1994; Tsang, 1996; Hecht *et al.*, 1996] and references therein).

It was shown in Sec. 3.4.1 that although the Rayleigh scattering can be strongly enhanced under certain conditions, it still contributes to the specular reflection and transmission on a small scale. In contrast, we will demonstrate that the PENS with a broad angular distribution can be a leading optical process. The plasmon localization, discussed in this chapter, leads to the concentration of the electromagnetic energy in sharp, nm-sized peaks, so called "hot spots." The local fields in such areas are enhanced enormously, with the local intensity I_{loc} exceeding the applied intensity I_0 by four to five orders of magnitude (see Figs. 3.3, 3.6, 3.7, and 3.9). The local fields gigantically fluctuate in an amplitude, a phase and a polarization over the film. The local-field enhancements associated with the hot spots are important for amplifying various optical processes such as, for example, the surface enhanced Raman scattering (Sec. 3.5).

One of most interesting theoretical predictions for the metal-dielectric composites is that nonlinear light scattering at the n-th frequency harmonic $n\omega$ of the incident light can be significantly enhanced and it is characterized by a broad, nearly isotropic angular distribution [Sarychev *et al.*, 1999a]. At the first glance, the PENS effect in semicontinuous films is surprising since a nanostructured percolation film is, on average, homogeneous on the microscopic scale. There are no wavelength-sized speckles on the film, which could result in conventional diffuse light scattering. Indeed, if the inhomogeneity scale a of local field fluctuations is much smaller than the wavelength λ, the diffuse component of the linear scattering, which is proportional to $\sim (a/\lambda)^4 \langle I_{loc} \rangle / I_0$ (the angular brackets stand for averaging over the film), remains small in comparison with the amount of specular reflected and transmitted light. In contrast, the nonlinear PENS contribution caused by the giant nm-scale field fluctuations comes from the plasmon localization. The intensity of nonlinear scattering at frequency $n\omega$ is $\sim (a/\lambda)^4 \langle (I_{loc}/I_0)^n \rangle$; thus the strongly fluctuating local sources at $n\omega$ frequency are enhanced by the factor $(I_{loc}/I_0)^n$, which is very large and increases with n as it is shown in Figs. 3.9 and 3.10. Hence, the role of field fluctuations is dramatically increased for nonlinear scattering. When

$(a/\lambda)^4$ $(I_{loc}/I_0)^n > 1$, the total diffuse component of the scattered light intensity from a semicontinuous film can be much larger than the coherent signal radiated in the specular reflected and transmitted directions. Then the non-linear PENS effect can be thought as the manifestation of the non-linear critical opalescence resulting from the localization of the plasmons on the film surface. Thus the main theoretical prediction for the PENS effect holds true, that is: The percolation metal films should be strongly scattering objects for nonlinear signals, with a broad, nearly isotropic angular intensity distribution. This is very different from the linear scattering, in which the diffuse intensity component, although somewhat enhanced, still remains a relatively small portion of the overall intensity.

In this section we discuss the experiment that proves the existence of the PENS effect [Breit *et al.*, 2001]. We show that strongly enhanced second harmonic generation (SHG), which is characterized by a broad angular distribution, is observed on the gold-glass films near the percolation threshold. The diffuselike SHG scattering contrasts with highly directional linear reflection and transmission from these nanostructured films. The observations by [Breit *et al.*, 2001] are in agreement with theoretical calculations of the PENS effect observed on the gold-glass percolation films for SHG.

Samples of the percolation metal films were prepared by the group of Prof. Gadenne (L.M.O.V., Universitede Versailles Saint Quentin) by the evaporation of gold films (10 nm thickness) on a glass substrate at room temperatures under ultrahigh vacuum (10^{-9} torr). In order to determine how close to the percolation threshold the films were, the resistivity and the deposited mass thickness were measured during the film deposition. The optical reflection and transmission of the samples were determined outside the vacuum chamber and were compared with the known optical properties of percolation samples. Transmission electron microscopy was performed by depositing the same film on a copper grid covered by a very thin SiO_2 layer. (A sample of transmission electron microscope (TEM) image of a percolation film fragment is shown in Fig. 3.2). For a base-line comparison, smooth continuous gold films were also deposited under the same conditions. To obtain SHG from the films, a pulsed laser light was used at three different wavelengths $\lambda_{exc} = 770$, 800, and 920 nm, which were corresponding to the fundamental frequencies generated by a Ti:Sapphire laser (pulse duration 150 fs). The beam was focused on a $4 \cdot 10^3$ μm^2 size-area of the film and was incident at 45° with respect to the film normal, as shown in Fig.3.21(a). Measurements were performed for both S- and $P-$ polarizations of the incident light. Linear and nonlinear light scattering from the percolation metal

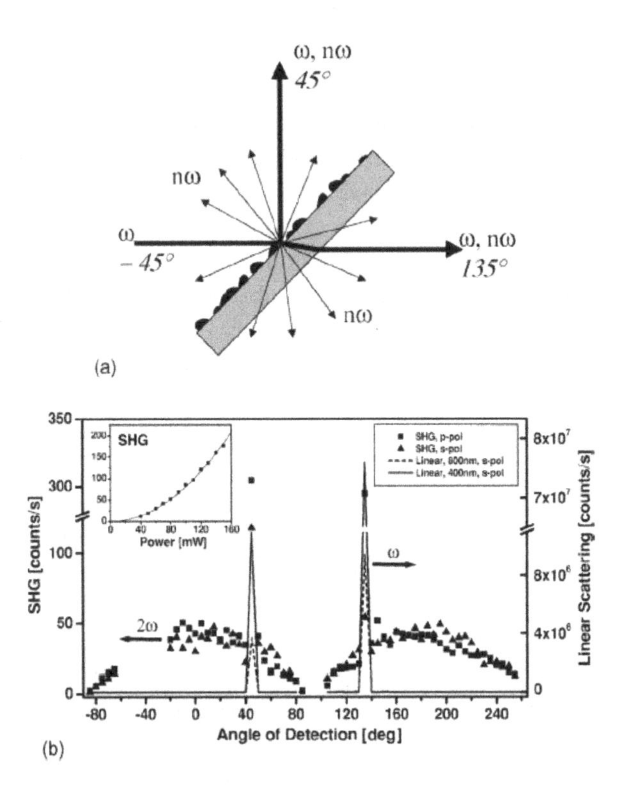

Fig. 3.21 (a) Linear reflection and transmission at frequency ω and diffuselike PENS (at frequency $n\omega$) from percolation metal-dielectric film. (b) Angular distributions of reflected and transmitted light (lines) and of SHG (symbols) from percolation gold-glass film; the polarizations of the incident light are indicated. The gap in the SHG data near 245° is due to the geometry of the excitation setup. Inset: intensity dependence for diffuse SHG detected at 0°.

film results in the coherent beams in the specular reflected and transmitted directions and also in diffuse light portion shown in Fig. 3.21. The scattered light was detected by a photomultiplier used for a photon counting. The interference filters were installed for the spectral selection. For comparison, linear scattering and reflection of incident light at 400 nm and 800 nm was also studied. For angular dependence measurements, the photomultiplier was moved around the sample. Light scattering by both the percolation and continuous gold films was compared. The experimental results for SHG from the percolation films are shown in Fig. 3.21(b). Coherent SHG signal detection was performed at $\lambda_{det} = 400$ nm (corresponding to double SHG frequency 2ω) in the specular reflected and transmitted directions, 45° and

135°, respectively. Using a small pinhole in the front of the photomultiplier and a high angular resolution camera, the SHG signals were found to have a Gaussian intensity distribution as a function of detection angle, with a narrow width of about 1°. For the P polarization of the incident beam, the coherent SHG component radiated in the specular reflected direction was about 3 times larger than for the S-polarization. This difference indicated that for the specular reflected signal, the usual dominance of the SHG from P-polarized excitation over that from S-polarized excitation, which is known from flat continuous metal surfaces, [Bloembergen *et al.*, 1968] is preserved to some extent.

In addition to the coherent SHG component at 45° and 135°, a strong diffuselike signal, characterized by the broad angular distribution, was measured at 2ω, as seen in Fig. 3.21(b). The intensity of this diffuse component exhibits the I^2 dependence expected for SHG, as illustrated in the insert of Fig. 3.21(b) for the normal direction 0°. The total (angle-integrated) intensity of this diffuse SHG component exceeds the intensity of the coherent light reflected in the specular direction by a factor of *350* for the P-polarized incident light, and by 10^3 for the S-polarized light. It is important to note that the highly diffused characteristics of the SHG component is the signature of the predicted PENS effect. Two more observations are also important: i) The diffuse SHG intensity exhibits roughly a cosine-angle dependence. ii) The diffuse SHG intensity is about the same for both the S- and P polarizations of the incident beam. The above observations point out to a nearly random orientation of the nm-scale nonlinear sources for the diffuse SHG. This conclusion is consistent with the localization of the surface plasmons in the percolation films discussed in the beginning of this chapter. All the SHG measurements described above were repeated at the fundamental wavelengths $\lambda = 770$ and 920 nm. Qualitatively, very similar results were obtained for the case of $\lambda = 770$ and 920 nm but with different ratios of the diffuselike scattered and collimated signal.

For comparison, linear scattering of incident light at $\lambda = 400$ nm and 800 nm was also measured on the same gold-glass percolation film [solid and dashed lines in Fig. 3.21(b), respectively]. In contrast to the SHG signal, the angle-integrated diffuse component of the light intensity was smaller than the components of the linear signal in the reflected and transmitted directions by the factor of 240. This observation confirms the theoretical expectation that linear scattering from percolation films has much less diffusive nature than the PENS.

Similar measurements for the linear and nonlinear light scattering were also performed on continuous (i.e., smooth) gold films. As expected, both the linearly scattered light and the SHG light components were highly collimated in the specular reflected and transmitted directions of these films [Bloembergen *et al.*, 1968], [Shen, 1984]. The directional characteristics of the SHG signal from continuous gold films are thus in a strong contrast with those from the percolation films, which are predominantly diffusive-like. In addition, the total, i.e., angle-integrated SHG intensity from a percolation gold film is greater than SHG signal from a smooth film by a factor of 60 for the P-polarization and by roughly by 10^3 for the S polarization. The total SHG enhancement originates from the huge local-field enhancement, which is stipulated by the plasmon localization in the percolation films. Similar comparisons of SHG intensities from the percolation and continuous gold films were also performed at $\lambda_{det} = 385$ and 460 nm, where we obtained enhancements of roughly 50 and 90-times increase, respectively, for the P-polarized SHG. We can then conclude that the enhancement increases toward larger wavelengths and discuss this wavelength dependence in more detail below.

It is worth noting that all experiments by [Breit *et al.*, 2001] were performed below an excitation intensity of 10^8 W/cm^2, where only the fundamental beam at $\lambda = 800$ nm and SHG at $\lambda = 400$ nm could be detected; all other spectral components were below the detection limit. However, above 10^8 W/cm^2-intensity threshold, a continuum background radiation in a broad spectral range was detected. This plasmon-enhanced white light generation was observed earlier by [Ducourtieux *et al.*, 2000]

To understand these phenomena, we use the theory of the PENS developed by [Sarychev *et al.*, 1999a]. Specifically, we calculate the diffuse component of SHG from a percolation gold film and compare it with the specular reflected SHG from both the percolation and the continuous gold films. By using the scattering theory approach, discussed in Sec. 3.4.1, we obtain that the integral scattering intensity in all but specular directions is given by

$$S \cong \frac{k^2}{c^2} \int \left(\left\langle \mathbf{j}^{(n)}_{\mathbf{r}_1} \cdot \mathbf{j}^{(n)*}_{\mathbf{r}_2} \right\rangle - \left\langle \mathbf{j}^{(n)} \right\rangle^2 \right) d\mathbf{r}_1 \, d\mathbf{r}_2, \tag{3.103}$$

where $k = \omega/c$, and $\mathbf{j}^{(n)}$ is a nonlinear surface current (oscillating at frequency $n\omega$), which serves as a local source for the scattered light. We consider subwavelength fluctuations and thus assume that the integral vanishes for the distances $|\mathbf{r}_2 - \mathbf{r}_1| \ll \lambda$, therefore, omit the $\exp(ik|\mathbf{r}_2 - \mathbf{r}_1|)$ term.

The nonlinear local fields and currents are strongly enhanced in the areas of plasmon localization (see Figs. 3.9 and 3.10). As discussed earlier, these extremely sharp peaks of nonlinear light intensity are spatially separated on the film. Because of the gigantic spatial fluctuations, we can use "local"-field approximation and estimate the diffuse component of the $n\omega$-scattering as

$$\begin{aligned}\left\langle \mathbf{j}_{\mathbf{r}_1}^{(n)} \cdot \mathbf{j}_{\mathbf{r}_2}^{(n)*} \right\rangle - \left\langle \mathbf{j}^{(n)} \right\rangle^2 &\simeq \left\langle |\mathbf{j}^{(n)}|^2 \right\rangle a^2 \delta\left(|\mathbf{r}_2 - \mathbf{r}_1|\right) \\ &\simeq \omega^2 a^4 \left\langle |\varepsilon_{n\omega}^e|^2 |E_\omega|^{2n} \right\rangle \delta\left(|\mathbf{r}_2 - \mathbf{r}_1|\right) \\ &\simeq \frac{\omega^2 a^4}{|E_{n\omega}^{(0)}|^2} \left\langle |\varepsilon_{n\omega} E_{n\omega}|^2 |E_\omega|^{2n} \right\rangle a^4 \delta\left(|\mathbf{r}_2 - \mathbf{r}_1|\right),\end{aligned} \tag{3.104}$$

where $\varepsilon_{n\omega}^e = \langle \varepsilon_{n\omega} E_{n\omega} \rangle / E_{n\omega}^{(0)}$ is the effective nonlinear permittivity, $\varepsilon_{n\omega} E_{n\omega}$ is the electric displacement vector component induced by the constant electric field $E_{n\omega}^{(0)}$ in the film. ($E_{n\omega}^{(0)}$ has the same direction as the incident wave). In transition to the final form of Eq. (3.104), we have used the approach developed in Sec. 3.4.1 (see derivation of Eq. 3.81). In our model, the local permittivity $\varepsilon_{n\omega}$ takes values of $\varepsilon_{m,n\omega}$ (with probability p) and ε_d (with probability $1-p$) for the metal and the dielectric, respectively. We also take into account the field enhancement for the fields at both the fundamental frequency ω and the generated frequency $n\omega$.

PENS can formally be written as $S = G^{(n)} I_{n\omega}$, where $I_{n\omega}$ is the signal intensity from a continuous metal film and $G^{(n)}$ is the enhancement factor for a semicontinuous metal film. The factor $G^{(n)}$ then gives the ratio of the nonlinear signals from "broken" and "perfect" mirrors, representing semicontinuous and continuous films, respectively. As it follows from Eqs. (3.103) and (3.104), the diffuse part of enhancement, $G_{dif}^{(n)}$ estimates as:

$$G_{dif}^{(n)} = (ka)^4 \frac{\left\langle |\varepsilon_{n\omega} E_{n\omega}|^2 |E_\omega|^{2n} \right\rangle}{|\varepsilon_{m,n\omega} E_{n\omega}^{(0)}|^2 |E_\omega^{(0)}|^{2n}} \tag{3.105}$$

where $E_\omega^{(0)}$ and $E_{n\omega}^{(0)}$ are the applied (probed) fields at the fundamental and generated frequencies. Note that this result does not depend on the amplitudes $E_\omega^{(0)}$ and $E_{n\omega}^{(0)}$.

Although the "Rayleigh factor" $(ka)^4$ in Eq. (3.105) is typically small [for discussed here films with $a \leq 50$ nm, where a is a typical "in-plane" size of metal grains, $(ka)^4 \sim 2 \cdot 10^{-2}$], the resultant field-enhancement for a nonlinear scattering with $n \geq 2$ can still be very large. This is in contrast to a linear scattering ($n = 1$) where the specular reflected component for semicontinuous metal films, typically, exceeds the diffuse scattering (Sec. 3.4.1)

With increase of the order of nonlinearity, enhancement factors increase for both diffuse and collimated scattering, but the diffuse component increases at much greater rate. The PENS is essentially a nonlinear effect and becomes progressively larger for higher nonlinearities. For comparison, we also derive the enhancement factor $G_c^{(n)}$ for the collimated signal at $n\omega$-frequency on a semicontinuous film:

$$G_c^{(n)} = \frac{|\langle \varepsilon_{n\omega} E_{n\omega} E_\omega^n \rangle|^2}{|\varepsilon_{m,\,n\omega} E_{n\omega}^{(0)}|^2 |E_\omega^{(0)}|^{2n}}. \tag{3.106}$$

The PENS effect can formally be expressed as $G_{dif}^{(n)} \gg G_c^{(n)}$. In the estimate (3.106) we have neglected, for simplicity, the tensor form of the high-order harmonic susceptibility. Detailed consideration of the specular scattering of third- and the second-order harmonics is presented by [Hui and Stroud, 1997], [Hui *et al.*, 1998], and [Hui *et al.*, 2004b].

We use the Drude formula for the permittivity of gold (Eqs. (3.1) and (3.3)) to estimate $G_c^{(n)}$. Since there is little correlation in the field distribution at ω and $n\omega$ frequencies, we can use the decoupling

$$\langle \varepsilon_{n\omega} E_{n\omega} E_\omega^n \rangle = \langle \varepsilon_{n\omega} E_{n\omega} \rangle \langle E_\omega^n \rangle \tag{3.107}$$

and estimate

$$\langle \varepsilon_{n\omega} E_{n\omega} \rangle = \varepsilon_{e,n\omega} E_{n\omega}^{(0)} = \sqrt{\varepsilon_{m,n\omega} \varepsilon_d} E_{n\omega}^{(0)}, \tag{3.108}$$

where the Dykhne formula [Dykhne, 1971] for the effective permittivity ε_e in $2D$ percolation system: $\varepsilon_e = \sqrt{\varepsilon_m \varepsilon_d}$ was used. We recall that we consider the semicontinuous gold films at the percolation threshold. To estimate $\langle E_\omega^n \rangle$ in Eq. (3.107), we consider the resonance composite where $\varepsilon_d = -\operatorname{Re}(\varepsilon_m) \simeq 1$ at the percolation threshold $p = p_c \simeq 0.5$. Since the phase of the field local E_ω^n changes over the composite whose components give equal "plus" (ε_d) and "minus" (ε_m) contributions, the lower estimate is $|\langle E_\omega^n \rangle| \simeq |E_\omega^{(0)}|$ is valid. When we go to the lower frequency limit so that $|\varepsilon_m(\omega)| \gg 1$, the local fields increase and, correspondingly, the moment

$|\langle E_\omega^n\rangle|$ increases. Using Eq. (3.44) for the field increasing factor Ξ, we obtain

$$|\langle E_\omega^n\rangle| / |E_\omega^{(0)}|^n \sim \Xi \sim \left|\frac{\varepsilon_m}{\varepsilon_d}\right|^{\frac{n-1}{2}}. \tag{3.109}$$

We substitute Eqs. (3.107), (3.108) and (3.109) into Eq. (3.106) to obtain the enhancement for the collimated signal at $n\omega$ frequency:

$$G_c^{(n)} \sim \frac{|\varepsilon_m|^{n-1}}{|\varepsilon_{m,\,n\omega}|\,\varepsilon_d^{n-2}}. \tag{3.110}$$

To further estimate the frequency dependence for $G_c^{(n)}$, we use a free-electron approximation for the metal permittivity $\varepsilon_m \sim -(\omega_p/\omega)^2$ obtaining

$$G_c^{(n)} \sim n^2 (\omega_p/\omega)^{2(n-2)} / \varepsilon_d^{n-2}. \tag{3.111}$$

This estimate shows that the collimated nonlinear signal from a semicontinuous metal film is enhanced in comparison with a continuous metal film, for third- and higher-order harmonics, except the SHG. Indeed, experimentally we find $G_c^{(2)} \sim 0.1$, i.e., the collimated signal is not enhanced for the second harmonic.

By applying the same analysis to Eq. (3.105), we obtain the following estimate for the diffuse PENS:

$$G_{dif}^{(n)} \sim (ka)^4 \frac{\langle|\varepsilon_{n\omega}E_{n\omega}|^2\rangle}{|\varepsilon_{m,n\omega}E_{n\omega}^{(0)}|^2} M_{2n}(\omega) \tag{3.112}$$

where $M_{2n}(\omega)$ are the 2nd-moment of the local field given by Eq. (3.45). Note that $\langle|\varepsilon_{n\omega}E_{n\omega}|^2\rangle$ can be rewritten as

$$\langle|\varepsilon_{n\omega}E_{n\omega}|^2\rangle \simeq G(a)\,|\langle\varepsilon_{n\omega}E_{n\omega}\rangle|^2,$$

where $G(a)$ is the correlation function is given by Eq. (3.60). For the film with $D = 2$, $s = t = \nu \simeq 1.3$, Eqs. (3.60) and (3.45) give $G(a) \simeq M_2$. We invoke the Dykhne formula (3.108) once again, obtaining

$$\langle|\varepsilon_{n\omega}E_{n\omega}|^2\rangle \simeq M_2(n\omega)\,|\varepsilon_{m,n\omega}|\,\varepsilon_d \left|E_{n\omega}^{(0)}\right|^2.$$

Substituting this estimate into Eq. (3.112) gives

$$G_{dif}^{(n)} \sim (ka)^4 (\varepsilon_d/|\varepsilon_{m,\,n\omega}|)\,M_2(n\omega)\,M_{2n}(\omega) \tag{3.113}$$

where $M_2(n\omega)$ and $M_{2n}(\omega)$ are the moments for the local field oscillating at frequencies $n\omega$ and ω.

For SHG ($n = 2$), formula (3.113) together with Eq. (3.45) gives

$$\frac{G_{dif}^{(2)}}{(ka)^4} \sim \left|\frac{\varepsilon_{m,2\omega}}{\varepsilon_d}\right|^{-1/2} \left|\frac{\varepsilon_{m,2\omega}}{\varepsilon''_{m,2\omega}}\right|^{1-\gamma} \left|\frac{\varepsilon_{m,\omega}}{\varepsilon_d}\right|^{3/2} \left|\frac{\varepsilon_{m,\omega}}{\varepsilon''_{m,\omega}}\right|^{3-5\gamma}, \tag{3.114}$$

which with the use of the Drude formula (3.1), transforms into

$$G_{dif}^{SHG}/(ka)^4 \sim \left(2^{2-\gamma}/\varepsilon_d\right)(\omega_p/\omega)^2 (\omega/\omega_\tau)^{4-6\gamma}, \tag{3.115}$$

for $\omega \ll \omega_p$.

The normalized PENS factor for SHG, $G_{dif}^{SHG}/(ka)^4$, calculated from the scaling analysis according to Eq. (3.114) is shown in Fig. 3.22 as a function of the wavelength. As seen, the PENS factor increases with wavelength. The physical explanation for this behavior is that the local fields responsible for generation of the SHG enhancement increase towards longer wavelengths. In addition to the results from the scaling analysis, the enhancement factors obtained from the SHG measurements at fundamental wavelengths $\lambda_{exc} = 770$, 800, 920 nm are plotted for the $P-$ polarization.

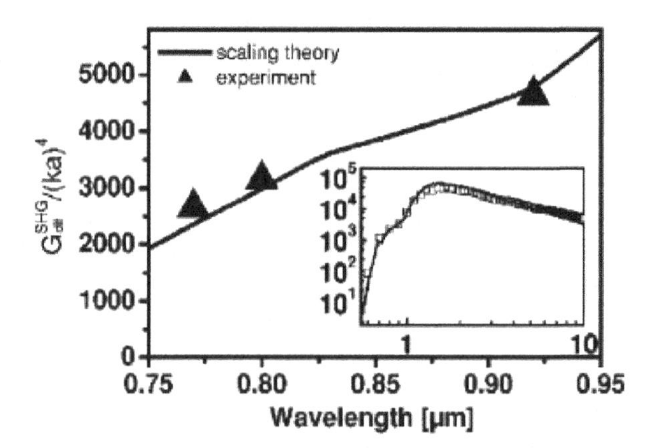

Fig. 3.22 Normalized PENS enhancement factor for SHG, $G_{dif}^{SHG}/(ka)^4$, as function of wavelength for percolation gold-glass film. The solid line is the result of theoretical formula (3.114). The solid triangles are the enhancement factor determined from SHG measurements at the fundamental wavelengths 770, 800, and 920 nm. Inset: theoretical data of the normalized PENS enhancement factor for SHG over a broad spectral range. The solid line is the result of theoretical formula (3.114), and the squares are the results of numerical simulations based on Eq. (3.105).

We see from the graphs that the predictions of the scaling theory are in a good agreement with the experimental data. The increase of the enhancement factor with wavelength confirms experimentally the wavelength dependence of the local fields predicted by the theory. The insert in Fig. 3.22 shows the theoretical results over a wider wavelength range. In addition to the results of the scaling analysis, we also show the results of the numerical simulations, which were performed using the real-space renormalization group procedure described in Sec. 3.2.2. As seen in Fig. 3.22, the results of the two calculations agree very well. The calculations predict a flattening of the wavelength dependence for $G_{dif}^{SHG}/(ka)^4$ beyond ~ 1.5 μm, and a subsequent slight deviation from Eq. (3.112).

It is worth noting that PENS seems be a robust effect. The enhanced nonlinear scattering resulting from the localized plasmons occurs provided that the field correlation length is smaller than the percolation length characterizing the mean size of fractal metal clusters on the film. This condition results in the estimate (3.47) for the metal concentration range $\Delta p = |p - p_c|$, where the effect occurs. As follows from this estimate, Δp shrinks when the frequency decreases. However, in the visible range, the PENS occurs within a broad range of p, even relatively far from p_c. Thus, it may be possible that an "unusually" large diffuse SHG component reported in earlier studies of SHG from metal-dielectric films, placed far from the percolation [Chen *et al.*, 1981], [Aktsipetrov *et al.*, 1993], [Kuang and Simon, 1995] is related to the PENS effect. The robustness of the PENS effect was checked in [Breit *et al.*, 2001] by performing additional angle-resolved SHG experiments on semicontinuous metal films with $p > p_c$ and $p < p_c$ conditions

Chapter 4

Optical Properties of Metal-dielectric Films: Beyond Quasistatic Approximation

The formalism developed in Sec. 3 cannot be used to describe the optical properties of the semicontinuous films in an important case of the strong skin effect in metals. In an attempt to expand the theoretical recipes beyond the quasistatic approximation, the approach based on the full set of Maxwell's equations has been proposed [Sarychev *et al.*, 1995], [Lagarkov *et al.*, 1997c], [Levy-Nathansohn and Bergman, 1997a], [Levy-Nathansohn and Bergman, 1997b]. It does not invoke the quasistatic approximation as the fields are not assumed to be curl-free in the film. Although the theory was developed mainly for the metal-insulator thin films, it is in fact quite general and can be applied to any kind of inhomogeneous film under appropriate conditions. The theory can be called the "generalized Ohm's law" and we will present it here and explaining the meaning.

We will restrict ourselves to the case where all the external fields are parallel to the film plane of symmetry. This means that all the waves (incident, reflected, or transmitted) are travelling in the direction perpendicular to the film plane. The consideration is focused on the electric- and magnetic-field magnitudes at certain distances *away* from the film. We relate the outside fields to the *inside* currents. We then assume that the size of film inhomogeneity is much smaller than the incident wavelength λ (but not necessarily smaller than the skin depth) so that the fields away from the film are curl-free and can be expressed as the gradients of potential fields. The electric displacement and magnetic induction, averaged over the film thickness, obey the usual two-dimensional continuity equation. Therefore the equations for the fields (e.g., $\nabla \times \mathbf{E} = 0$) and the equations for the currents (e.g., $\nabla \cdot \mathbf{j} = 0$) are the *same* as in the quasistatic case. The only difference is that the fields and the currents are now related by new material equations and that there exist the magnetic currents along with the electric currents.

To determine these new material equations, we first find the electric and magnetic field distributions inside the conductive and dielectric regions of the film. The boundary conditions determine the solutions of the Maxwell's equations for the fields inside bulk metal (the metal grain) completely when the frequency is fixed. Therefore the internal fields, which change very rapidly in a "z"-direction (perpendicular to the film surface), depend linearly on the electric and magnetic fields outside of the film. The currents inside the film are linear functions of the local internal fields given by the usual local constitutive equations. Therefore the currents flowing *inside* the film also depend linearly on the electric and magnetic fields *outside* the film. However, the electric current averaged over the film thickness now depends not only on the external electric field, but also on the external magnetic field. The same is true for the average magnetic induction current. This brings us to two linear equations that connect the two types of the average internal currents to the external fields. These equations can be considered as a generalized Ohm's law (GOL) in the non-quasistatic case [Levy-Nathansohn and Bergman, 1997a]. Thus the GOL forms the basis of a new approach to calculating the electromagnetic properties of inhomogeneous films.

4.1 Generalized Ohm's Law (GOL) and Basic Equations

We consider, for simplicity, the case where all the external fields are parallel to "x, y"-plane, meaning that the plane wave propagates normal to the film's surface. The metallic film has thickness d. The external electromagnetic wave is incident on the back film surface at $z = -d/2$ and is transmitted from the front surface at $z = d/2$ as shown in Fig. 4.1. A typical spatial scale of the metallic grains a is assumed to be much smaller than the free–space wavelength λ, $a \ll \lambda$.

First, we consider the electric and magnetic fields in the close vicinity to the film's front reference plane $z = d/2 + h$ and to the back reference plane $z = -d/2 - h$ (both are sketched in Fig. 4.1). The electric and magnetic fields are considered at the distance h in front of the film are designated as

$$\mathbf{E}_1(\mathbf{r}) = \mathbf{E}(\mathbf{r}, d/2 + h), \quad \mathbf{H}_1(\mathbf{r}) = \mathbf{H}(\mathbf{r}, d/2 + h), \tag{4.1}$$

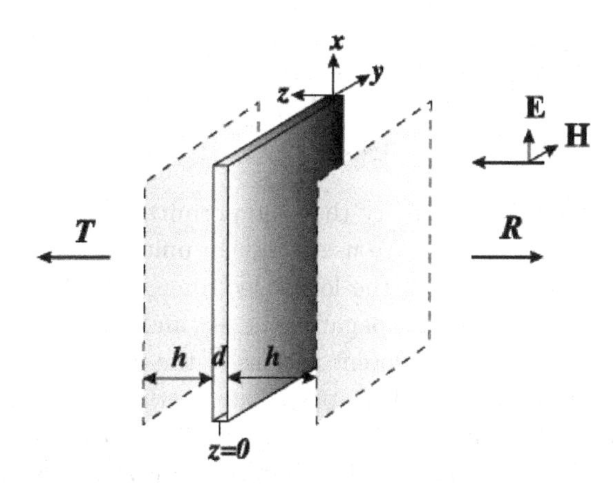

Fig. 4.1 Electromagnetic field of wavelength λ is incident onto a metal-dielectric film. The field is partially reflected and absorbed, with the reminder being transmitted through the film. Electric and magnetic fields are considered at $z = -a/2 - h$ and at $z = a/2 + h$ (the front and the back planes of the film).

and those at a distance h behind the film are denoted by

$$\mathbf{E}_2(\mathbf{r}) = \mathbf{E}(\mathbf{r}, -d/2 - h), \quad \mathbf{H}_2(\mathbf{r}) = \mathbf{H}(\mathbf{r}, -d/2 - h). \tag{4.2}$$

All the fields are monochromatic with the usual $\exp(-i\omega t)$ type dependence; the vector $\mathbf{r} = \{x, y\}$ is a two-dimensional vector in the (x, y)-plane.

We introduce the average electric displacement

$$\mathbf{D}(\mathbf{r}) = k \int_{-d/2-h}^{d/2+h} \mathbf{D}(\mathbf{r}, z)\, dz = k \int_{-d/2-h}^{d/2+h} \varepsilon(\mathbf{r}, z)\, \mathbf{E}(\mathbf{r}, z)\, dz, \tag{4.3}$$

and the average magnetic induction

$$\mathbf{B}(\mathbf{r}) = k \int_{-d/2-h}^{d/2+h} \mathbf{B}(\mathbf{r}, z)\, dz = k \int_{-d/2-h}^{d/2+h} \mu(\mathbf{r}, z)\mathbf{H}(\mathbf{r}, z)\, dz, \tag{4.4}$$

where $k = 2\pi/\lambda = \omega/c$ is the free-space wavevector. In the case of laterally inhomogeneous films, the vectors $\mathbf{D}$ and $\mathbf{B}$ are dependant of $\mathbf{r}$. Vectors $\mathbf{D}$ and $\mathbf{B}$, introduced by Eqs. (4.3) and (4.4) have the same dimension as electric displacement and magnetic induction.

The permittivity $\varepsilon(\mathbf{r}, z)$ and permeability $\mu(\mathbf{r}, z)$ take different values for dielectric regions of the film:

$$\varepsilon(\mathbf{r}, z) = \{\varepsilon_d, \ |z| < d/2; \ 1, \ |z| > d/2\}, \tag{4.5}$$

$$\mu(\mathbf{r}, z) = \{\mu_d, \ |z| < d/2; \ 1, \ |z| > d/2\}, \tag{4.6}$$

and for metal regions:

$$\varepsilon(\mathbf{r}, z) = \{\varepsilon_m, \ |z| < d/2; \ 1, \ |z| > d/2\}, \tag{4.7}$$

$$\mu(\mathbf{r}, z) = \{\mu_m, \ |z| < d/2; \ 1, \ |z| > d/2\}. \tag{4.8}$$

We assume hereafter for simplicity that the permittivity ε is a scalar and the permeability $\mu_d = \mu_d = 1$. We use Gaussian units.

In the GOL approximation, the local electromagnetic field is a superposition of two plane waves propagating in $+z$ and $-z$ directions. This superposition is different for different regions of the film. We neglect scattered and evanescent waves, which propagate in the $(x, y)-$ plane since their amplitudes are proportional to $(a/\lambda)^2 \ll 1$. This allows us to use a two-plane wave approximation when both electric $\mathbf{E}(\mathbf{r}, z)$ and magnetic $\mathbf{H}(\mathbf{r}, z)$ fields have components in the $(x, y)-$ plane only. After that, the Maxwell equations $\operatorname{curl} \mathbf{E}(\mathbf{r}, z) = ik\mathbf{B}(\mathbf{r}, z)$ and $\operatorname{curl} \mathbf{H}(\mathbf{r}, z) = -ik\mathbf{D}(\mathbf{r}, z)$ take the following forms:

$$\frac{d}{dz}\left[\mathbf{n} \times \mathbf{E}(\mathbf{r}, z)\right] = ik\mathbf{H}(\mathbf{r}, z), \tag{4.9}$$

$$\frac{d}{dz}\left[\mathbf{n} \times \mathbf{H}(\mathbf{r}, z]\right) = -ik\varepsilon\left(\mathbf{r}, z\right)\mathbf{E}(\mathbf{r}, z). \tag{4.10}$$

Unit vector $\mathbf{n} = \{0, 0, 1\}$ is normal to the film's surface. The permittivity $\varepsilon(\mathbf{r}, z)$ is defined above. The solution to Eqs. (4.9) and (4.10) is the superposition of two plane waves, which are determined either by the fields $\mathbf{E}_1$ and $\mathbf{H}_1$ or by fields $\mathbf{E}_2$ and $\mathbf{H}_2$, defined on the front or back reference planes, respectively (see Eqs. (4.1) and (4.2)). Therefore in GOL approximation, the fields $\mathbf{E}(\mathbf{r}, z)$ and $\mathbf{H}(\mathbf{r}, z)$ are completely determined by either fields $\mathbf{E}_1(\mathbf{r})$ and $\mathbf{H}_1(\mathbf{r})$ or by $\mathbf{E}_2(\mathbf{r})$ and $\mathbf{H}_2(\mathbf{r})$ or, in general, by a linear combination of any pair (E and H) of these fields.

The equations (4.9) and (4.10) transform after the integration over the coordinate "z" within the boundaries $z = -d/2 - h$ (back) and $z = d/2 + h$ (front):

$$[\mathbf{n}\times(\mathbf{E}_1(\mathbf{r}) - \mathbf{E}_2(\mathbf{r}))] = i\mathbf{B}(\mathbf{r}), \quad [\mathbf{n}\times(\mathbf{H}_1(\mathbf{r}) - \mathbf{H}_2(\mathbf{r}))] = -i\mathbf{D}(\mathbf{r}). \tag{4.11}$$

Since the fields $\mathbf{E}_{1,2}(\mathbf{r})$ and $\mathbf{H}_{1,2}(\mathbf{r})$ have components only in the $(x, y)-$ plane, they are curl–free (otherwise they would have non-zero "z"-components according to the Maxwell equations). It is convenient to introduce the linear combinations of the fields $\mathbf{E}_{1,2}(\mathbf{r})$ and $\mathbf{H}_{1,2}(\mathbf{r})$, namely,

$$\mathbf{E}(\mathbf{r}) = \mathbf{E}_1(\mathbf{r}) + \mathbf{E}_2(\mathbf{r}), \quad \mathbf{H}(\mathbf{r}) = \mathbf{H}_1(\mathbf{r}) + \mathbf{H}_2(\mathbf{r}), \tag{4.12}$$

which are also curl–free:

$$\operatorname{curl}\mathbf{E}(\mathbf{r}) = 0\,, \quad \operatorname{curl}\mathbf{H}(\mathbf{r}) = 0\,. \tag{4.13}$$

The vectors $\mathbf{E}(\mathbf{r})$ and $\mathbf{B}(\mathbf{r})$ completely determine the fields $\mathbf{E}(\mathbf{r}, z)$ and $\mathbf{H}\,(\mathbf{r}, z)$. For example the electric and magnetic fields inside the film equal to

$$\begin{aligned}\mathbf{E}(\mathbf{r},z) = &\frac{\mathbf{E}(\mathbf{r})\cos(k\,n\,z)}{2\cos(X)\cos(Y) - 2\,n\sin(X)\sin(Y)}\\ &+\frac{i\,\mathbf{H}(\mathbf{r})\sin(k\,n\,z)}{2\,n\cos(X)\cos(Y) - 2\sin(X)\sin(Y)},\end{aligned} \tag{4.14}$$

and

$$\begin{aligned}\mathbf{H}(\mathbf{r},z) = &\frac{\mathbf{H}(\mathbf{r})\,n\cos(k\,n\,z)}{2\,n\cos(X)\cos(Y) - 2\sin(X)\sin(Y)}\\ &+\frac{i\,\mathbf{E}(\mathbf{r})\,n\sin(k\,n\,z)}{2\cos(X)\cos(Y) - 2\,n\sin(X)\sin(Y)}\end{aligned} \tag{4.15}$$

correspondingly, where $X = h\,k$, $Y = d\,k\,n/2$, and the refractive index n takes the values $n_m = \sqrt{\varepsilon_m}$ and $n_d = \sqrt{\varepsilon_d}$ for the metallic and the dielectric regions of the film. The average electric displacement $\mathbf{D}(\mathbf{r})$ and average magnetic induction $\mathbf{B}(\mathbf{r})$ are obtained from Eqs. (4.3) and (4.4). The continuity equations for the average "electric displacement" $\mathbf{D}(\mathbf{r})$ and for the average "magnetic induction" $\mathbf{B}(\mathbf{r})$ in Eqs. (4.11) are obtained by the integration of the local (3D) equations $\operatorname{div}\mathbf{D}\,(\mathbf{r}, z) = 0$, $\operatorname{div}\mathbf{B}\,(\mathbf{r}, z) = 0$ between the reference planes $z = -d/2 - h$ and $z = d/2 + h$, which results in

$$\operatorname{div}\mathbf{D}(\mathbf{r}) = 0\,, \quad \operatorname{div}\mathbf{B}(\mathbf{r}) = 0\,, \tag{4.16}$$

where we take into account the boundary conditions $D_z = 0$ and $B_z = 0$ at $z = -d/2 - h$ and $z = d/2 + h$.

Equations (4.11), (4.13) and (4.16) represent the system of generalized Ohm's equations, which connect electric and magnetic fields, determined in two-dimensional reference planes, to the average electric displacement (electric current) and average magnetic induction (magnetic current), which flow in the film. Thus the entire physics of 3D inhomogeneous film, which is described by the full set of Maxwell equations, has been reduced to a simple set of quasistatic equations (4.13) and (4.16). To solve these equations, we only need the material equations, which connect the fields $\mathbf{E}$ and $\mathbf{H}$ to the "currents" $\mathbf{D}$ and $\mathbf{B}$.

Since the fields $\mathbf{E}(\mathbf{r})$ and $\mathbf{H}(\mathbf{r})$ completely determine the fields inside the film in the GOL (the two-wave approximation), the average two-dimensional electric displacement vector $\mathbf{D}(\mathbf{r})$ and the average magnetic induction vector $\mathbf{B}(\mathbf{r})$ can be presented as linear combinations $\mathbf{D} = u\,\mathbf{E}+g_1\mathbf{H}$ and $\mathbf{B} = v\,\mathbf{H}+g_2\mathbf{E}$, where u , v, g_1, and g_2 are dimensionless *Ohmic* parameters. For simplicity, we consider films having a mirror symmetry with respect to the reflection from the $z = 0$ plane. For such films, parameters $g_1 = 0$, and $g_2 = 0$ as it follows from the mirror–like symmetry [Levy-Nathansohn and Bergman, 1997a], [Lagarkov *et al.*, 1997c]. Therefore, we write

$$\mathbf{D}(\mathbf{r}) = u(\mathbf{r})\mathbf{E}(\mathbf{r})\,, \quad \mathbf{B}(\mathbf{r}) = v(\mathbf{r})\mathbf{H}(\mathbf{r})\,, \tag{4.17}$$

The equations (4.17) have the form, typical for the material equations in electromagnetism [Weiglhofer, 2003]. They include parameters u and v, which incorporate the local geometry of the film. To find the dimensionless Ohmic parameters, $u(\mathbf{r})$ and $v(\mathbf{r})$, we fix electric $\mathbf{E}(\mathbf{r})$ and magnetic $\mathbf{H}(\mathbf{r})$ fields and solve Eqs. (4.9) and (4.10). The obtained fields $\mathbf{E}(\mathbf{r},z)$ and $\mathbf{H}(\mathbf{r},z)$ are then substituted into Eqs. (4.3) and (4.4), which together with Eqs. (4.17), results in

$$u = \frac{\tan(a\,k/4) + n\,\tan(d\,k\,n/2)}{1 - n\,\tan(a\,k/4)\,\tan(d\,k\,n/2)}, \tag{4.18}$$

$$v = \frac{n\,\tan(a\,k/4) + \tan(d\,k\,n/2)}{n - \tan(a\,k/4)\,\tan(d\,k\,n/2)}, \tag{4.19}$$

where a is the lateral size of the metal grains; the refractive index n takes the values $n_m = \sqrt{\varepsilon_m}$ and $n_d = \sqrt{\varepsilon_d}$ for the metallic and the dielectric regions of the film, respectively. In Eqs. (4.18) and (4.19), the distance to the reference planes was set to

$$h = a/4, \tag{4.20}$$

[Sarychev *et al.*, 1995], [Levy-Nathansohn and Bergman, 1997a], [Sarychev *et al.*, 2000a], [Shubin *et al.*, 2000], and [Sarychev and Shalaev, 2003]. Note that this inter-reference plane distance is very close to the exact result $h = 2a/(3\pi)$, obtained from the diffraction on a small aperture in a perfectly conducting screen [Bethe, 1944].

In order to characterize the film transmittance and reflectance, we average Eq. (4.11) over the film (x, y) – plane, obtaining

$$[\mathbf{n}\times(\langle\mathbf{E}_1\rangle-\langle\mathbf{E}_2\rangle)]=i\,\langle v\mathbf{H}\rangle\ ,\quad [\mathbf{n}\times(\langle\mathbf{H}_1\rangle-\langle\mathbf{H}_2\rangle)]=-i\,\langle u\mathbf{E}\rangle\ , \tag{4.21}$$

where we have used Eqs. (4.17) for the vectors **D** and **B**. We then introduce the effective film parameters u_e and v_e, through the spatial averaging:

$$u_e\,\langle\mathbf{E}\rangle=\langle u\mathbf{E}\rangle\ ,\quad v_e\,\langle\mathbf{H}\rangle=\langle v\mathbf{H}\rangle\ , \tag{4.22}$$

thereby, obtaining

$$[\mathbf{n}\times(\langle\mathbf{E}_1\rangle-\langle\mathbf{E}_2\rangle)]=\ iv_e\,(\langle\mathbf{H}_1\rangle+\langle\mathbf{H}_2\rangle)\ , \tag{4.23}$$

$$[\mathbf{n}\times(\langle\mathbf{H}_1\rangle-\langle\mathbf{H}_2\rangle)]=-iu_e\,(\langle\mathbf{E}_1\rangle+\langle\mathbf{E}_2\rangle) \tag{4.24}$$

through the definitions of the vectors **E** and **H** (4.12). Resultant equations (4.23) and (4.24) relate the spatially averaged fields on both sides of the film.

Suppose that the plane wave is incident from the right half-space $z<0$ as seen from Fig. 4.1) so that its electric field depends on z as $\exp(i\,k\,z)$. The incident wave is partially reflected and partially transmitted. The combine field amplitude in the right half-space can be written as

$$\tilde{\mathbf{E}}_2\,(z)\ =\mathbf{E}_0\{\exp\left[i\,k\,\left(z+d/2+h\right)\right]+r\exp\left[-i\,k\,\left(z+d/2+h\right)\right]\},$$

where r is the reflection coefficient and $\mathbf{E}_0$ is amplitude of the incident wave. On the front film surface, the electric field acquires the form $\tilde{\mathbf{E}}_1\,(z)=t\mathbf{E}_0\exp\left[i\,k\,\left(z-d/2-h\right)\right]$, where t is the transmission coefficient. At the reference planes $z=d/2+h$ and $z=-d/2-h$, the average field equals to $\langle\mathbf{E}_1\rangle$ and $\langle\mathbf{E}_2\rangle$, respectively; therefore, $\langle\mathbf{E}_1\rangle=\tilde{\mathbf{E}}_1\,(d/2+h)=t\,\mathbf{E}_0$ and $\langle\mathbf{E}_2\rangle=\tilde{\mathbf{E}}_2\,(-d/2-h)=\ (1+r)\,\mathbf{E}_0$. The same matching for the magnetic fields gives $\langle\mathbf{H}_1\rangle=t\,[\mathbf{n}\times\mathbf{E}_0]$ and $\langle\mathbf{H}_2\rangle=(1+r)\ [\mathbf{n}\times\mathbf{E}_0]$. Substitution of these expressions for $\langle\mathbf{E}_{1,2}\rangle$ and $\langle\mathbf{H}_{1,2}\rangle$ in Eqs. (4.23) and (4.24) results in two linear equations for t and r. Solving of those gives film's reflectance

$$R\equiv|r|^2=\left|\frac{u_e-v_e}{(i+u_e)(i+v_e)}\right|^2\ , \tag{4.25}$$

transmittance

$$T\equiv|t|^2=\left|\frac{1+u_e v_e}{(i+u_e)(i+v_e)}\right|^2\ , \tag{4.26}$$

and absorptance

$$A = 1 - T - R. \tag{4.27}$$

This means that effective Ohmic parameters u_e and v_e completely determine the *observable* optical properties of inhomogeneous films.

Since the fields $\mathbf{E}(\mathbf{r})$ and $\mathbf{H}(\mathbf{r})$ are curl–free (Eqs. (4.13)), they can be represented as gradients of certain scalar potentials as it follows:

$$\mathbf{E}(\mathbf{r}) = -\boldsymbol{\nabla}\phi(\mathbf{r}), \qquad \mathbf{H}(\mathbf{r}) = -\boldsymbol{\nabla}\psi(\mathbf{r}). \tag{4.28}$$

By substituting these expressions, first in Eq. (4.17) and then into Eq. (4.16), we obtain the equations

$$\nabla \cdot [u(\mathbf{r}) \nabla \varphi(\mathbf{r})] = 0\,, \quad \nabla \cdot [v(\mathbf{r}) \nabla \psi(\mathbf{r})] = 0\,, \tag{4.29}$$

which can be solved independently for the potentials φ and ψ. The solutions are defined by the incident field $\mathbf{E}_0$ through the following additional conditions

$$\begin{aligned} \langle -\nabla \varphi \rangle &= \langle \mathbf{E} \rangle \equiv (1 + r + t)\, \mathbf{E}_0, \\ \langle -\nabla \psi \rangle &= \langle \mathbf{H} \rangle \equiv (1 - r + t)\, [\mathbf{n} \times \mathbf{E}_0]\,. \end{aligned} \tag{4.30}$$

When the Poisson's equations (4.29) are solved, the effective parameters u_e and v_e are obtained from the equations $\langle u(\mathbf{r}) \nabla \varphi(\mathbf{r}) \rangle = u_e \langle \nabla \varphi(\mathbf{r}) \rangle$ and $\langle v(\mathbf{r}) \nabla \psi(\mathbf{r}) \rangle = v_e \langle \nabla \psi(\mathbf{r}) \rangle$. Note that the effective parameters do not depend on the averages of $\langle \mathbf{E} \rangle$ and $\langle \mathbf{H} \rangle$. Thus the problem of the field distribution and optical properties of the metal-dielectric films reduces to non-coupled quasistatic conductivity problem [Eq. (4.29)], to which an extensive theory exists. Thus numerous analytical as well as numerical methods developed in the percolation theory (see Ch. 3) can be applied to find the effective parameters u_e and v_e of the film.

4.2 Transmittance, Reflectance, and Absorptance

When the skin depth (penetration depth), $\delta = 1/\left(k \operatorname{Im} n_m\right)$, is much smaller than the grain size a, the electromagnetic field does not penetrate in metal grains. The Ohmic parameters u_m and v_m take values

$$u_m = -\frac{4}{a\,k}, \; v_m = \frac{ka}{4}\,. \tag{4.31}$$

When the film is thin $k\,n_d \ll 1$ (d is film thickness) and $\varepsilon_d \sim 1$, Eqs. (4.18) and (4.19) give

$$u_d = \frac{k\varepsilon_d'}{4}a, \; v_d = \frac{k}{2}\left(d + a/2\right), \tag{4.32}$$

where we have introduces the reduced dielectric constant $\varepsilon_d' = 1 + 2\varepsilon_d d/a$. Note that in the limit of the strong skin effect, the parameters u_m and v_m are real and the parameter u_m is of inductive character, i.e., it has the sign opposite to the dielectric parameter u_d. In contrast, the Ohmic parameter v remains essentially the same ($v \sim ka$) for the dielectric and for the metal regions regardless of the value of the skin effect.

It is instructive to trace the evolution of the film optical properties when the surface density of metal p is increasing. When $p = 0$, the semicontinuous film is purely dielectric and the effective parameters u_e and v_e coincide with the dielectric Ohmic parameters given by Eqs. (4.32). By substituting $u_e = u_d$ and $v_e = v_d$ in Eqs. (4.25), (4.26) and (4.27) and assuming that the dielectric film has no loss and is optically thin ($dk\varepsilon_d \ll 1$), we obtain the reflectance $R = d^2\left(\varepsilon_d - 1\right)^2 k^2/4$, transmittance $T = 1 - d^2\left(\varepsilon_d - 1\right)^2 k^2/4$, and the absorbance $A = 0$ that coincide with the well known results for a thin dielectric film (see, for example, [Landau *et al.*, 1984; Jackson, 1998]).

There is no surprise that a film without loss has zero absorbance. When the ratio of the penetration length (skin depth) $\delta = 1/\,k\,\mathrm{Im}\,n_m$ is negligible in comparison with the film thickness d and $|n_m| \gg 1$, the loss is also absent in the limit of a full metal coverage of the film surface when the metal concentration $p = 1$. In this case, the film acts as a perfect metal mirror. Indeed, the substitution of the Ohmic parameters $u_e = u_m$ and $v_e = v_m$ from Eqs. (4.31) into Eqs. (4.25), (4.26) and (4.27) gives $R = 1$ for the reflectance, while the transmittance T and absorbance A are both equal to zero. Note that the optical properties of the semicontinuous film do not depend on the particle (grain) size a for the metal concentrations $p = 0$ and $p = 1$ since the properties of the purely dielectric and purely metal films do not depend on the shape of the metal grains.

We consider now the semicontinuous metal film at the percolation threshold $p = p_c$. For the strong skin effect, the film can be thought of as a mirror, which is broken onto small pieces with typical size a, each is much smaller than the incident wavelength λ. At the percolation threshold the exact formulas $u_e = \sqrt{u_d u_m}$ and $v_e = \sqrt{v_d v_m}$ hold [Dykhne, 1971]. Thus following equations for the effective Ohmic parameters are obtained from Eqs. (4.32) and (4.31):

$$u_e(p_c) = i\sqrt{\varepsilon'_d}, \qquad v_e(p_c) = \frac{ka}{4}\sqrt{1+\frac{2d}{a}}. \tag{4.33}$$

It follows from these equations that $|v_e/u_e| \sim a\,k \ll 1$ so that the effective Ohmic parameter v_e can be neglected in comparison with u_e. By substituting the effective Ohmic parameter $u_e(p_c)$ given by Eq. (4.33) in Eqs. (4.25), (4.26) and (4.27), the film optical properties at the percolation are obtained:

$$R(p_c) = \frac{\varepsilon'_d}{\left(1+\sqrt{\varepsilon'_d}\right)^2}, \tag{4.34}$$

$$T(p_c) = \frac{1}{\left(1+\sqrt{\varepsilon'_d}\right)^2}, \tag{4.35}$$

$$A(p_c) = \frac{2\sqrt{\varepsilon'_d}}{\left(1+\sqrt{\varepsilon'_d}\right)^2}. \tag{4.36}$$

We notice that when the metal grains are oblate enough so that $\varepsilon_d d/a \ll 1$, the reduced dielectric function $\varepsilon'_d = 1 + 2\varepsilon_d d/a$ transforms to unity $\varepsilon'_d \to 1$ and the above expressions simplify to the universal result

$$R = T = 1/4, \qquad A = 1/2. \tag{4.37}$$

This means that the effective adsorption exists in a semicontinuous metal film even if neither dielectric nor metal grains absorb the light energy. An original mirror (the film), which is broken onto small pieces, still effectively absorbs energy from the electromagnetic field. The effective absorption in a loss-free film means that the electromagnetic energy is stored in the system and that the amplitudes of the local electromagnetic field increase infinitely. In a real semicontinuous metal film, the local field will saturate due to non-zero loss but significant field fluctuations take place on the film surface when the loss is small.

To find the optical properties of semicontinuous films for arbitrary metal concentrations p, the effective medium theory can be implemented. The theory has been originally developed to provide a semi-quantitative description of the transport properties of percolating composites [Bergman and Stroud, 1992] and was discussed in Sec. 2.2. In the considered here the two dimensional case, we substitute the depolarization factors $g_{\parallel} = g_{\perp} = 1/2$ in Eq. (2.9) and replace σ_m, σ_d, and σ_e either by u_m, u_d, and u_e or by v_m,

v_d, and v_e obtaining the following equations for the effective parameters u_e and v_e:

$$p\frac{u_m - u_e}{u_m + u_e} + (1-p)\frac{u_d - u_e}{u_d + u_e} = 0, \tag{4.38}$$

$$p\frac{v_m - v_e}{v_m + v_e} + (1-p)\frac{v_d - v_e}{v_d + v_e} = 0. \tag{4.39}$$

The equations can be rewritten as:

$$u_e^2 - \Delta p\, u_e\,(u_m - u_d) - u_d\, u_m = 0, \tag{4.40}$$

$$v_e^2 - \Delta p\, v_e\,(v_m - v_d) - v_d\, v_m = 0, \tag{4.41}$$

where the reduced concentration $\Delta p = (p - p_c)/p_c$ is introduced. It follows from Eq. (4.41) that for the strong skin effect when v_m and v_d are given by Eqs. (4.31) and (4.32), the effective Ohmic parameter $|v_e| \ll 1$ for all the metal concentrations p. Therefore v_e can neglected in Eqs. (4.25) and (4.26). We also omit, for further simplification, the Ohmic parameter u_d in comparison with u_m in the second term of Eq. (4.40) [cf. Eqs. (4.31) and (4.32)] and introduce the parameter $\Delta = 2\Delta p/(ak)$ to simplify the Eq. (4.40) to

$$u_e^2 + 2\Delta u_e + \varepsilon_d' = 0. \tag{4.42}$$

At the percolation threshold $p = p_c = 1/2$, the reduced concentration $\Delta p = 0$, Eq. (4.42) gives the effective Ohmic parameter $u_e'(p_c) = \sqrt{\varepsilon_d'}$ that coincides with the exact Eq. (4.33) and results in reflectance, transmittance, and absorbance given by Eqs. (4.34), (4.35) and (4.36) respectively.

For concentrations different from the percolation threshold, Eq. (4.42) gives

$$u_e = -\Delta + \sqrt{\Delta^2 - \varepsilon_d'}, \tag{4.43}$$

which becomes real for $|\Delta| > \sqrt{\varepsilon_d'}$. Then Eqs. (4.34), (4.35) and (4.36) results in the zero absorption $A = 1 - R - T = 1 - |u_e|^2/|i + u_e|^2 - 1/|i + u_e|^2 = 0$ (recall that the effective Ohmic parameter v_e is neglected). On the other hand the effective Ohmic parameter u_e has a non-vanishing imaginary part in the vicinity of a percolation threshold

$$|\Delta| < \sqrt{\varepsilon'_d} \tag{4.44}$$

and, therefore, the absorption is non-zero, defined as

$$A = \frac{2\sqrt{-\Delta^2 + \varepsilon'_d}}{1 + \varepsilon'_d + 2\sqrt{-\Delta^2 + \varepsilon'_d}}. \tag{4.45}$$

It has a well-defined maximum at the percolation threshold. The film adsorbs the electromagnetic energy for the metal concentration in the absorption band $p_c\left(1 - \pi a\sqrt{\varepsilon'_d}/\lambda\right) < p < p_c\left(1 + \pi a\sqrt{\varepsilon'_d}/\lambda\right)$, which is inversely proportional to the wavelength. As discussed above, the effective absorption in a loss-free metal-dielectric film means that local electromagnetic fields strongly fluctuate in the system.

In a real semicontinuous metal film the skin depth takes a finite value. We observe a well-defined absorption band near the percolation threshold. The experimental results and the calculations in the GOL approximation are presented in Fig. 3.4 for gold semicontinuous film [Yagil *et al.*, 1991]. The effective parameters u_e and v_e were found from Eqs. (4.40) and (4.41). The absorptance reaches its maximum at the percolation threshold and decreases with the metal concentration p steering away from p_c. The anomalous absorption maximum at the percolation threshold has been first discovered in the experiments by [Yagil *et al.*, 1992; Gadenne *et al.*, 1988; Gadenne *et al.*, 1989; Yagil *et al.*, 1991].

The strong skin effect influence on the film's reflectance, transmittance, and absorption was also investigated in the microwave region [Lagarkov *et al.*, 1997c]. In this experiment, the macroscopic copper-dielectric films were prepared by the lithographic method. The samples were made of the circular copper spots cut off from a foil and arranged on a plastic substrate. The spot diameter was $a = 2$ mm and the thickness was $d = 0.04$ mm. The film microwave properties were studied at $\lambda = 2.5$ cm so that the skin depth $\delta \simeq 1.0\ \mu$m was much smaller than the metal thickness $\delta \ll d$. In the case of a regular (periodical) spot array, the absorption A was less than 5% for investigated surface concentrations of the copper. For the random arrangement of the spots, a wide absorption band was observed around the percolation threshold p_c where A achieved 40% as it is shown in Fig. 4.2. The local field fluctuations observed in Fig. 4.3 as well as effective absorption in the nearly loss-free film allow us to speculate that the localization of the surface plasmons was observed in this experiment [Lagarkov *et al.*, 1997c].

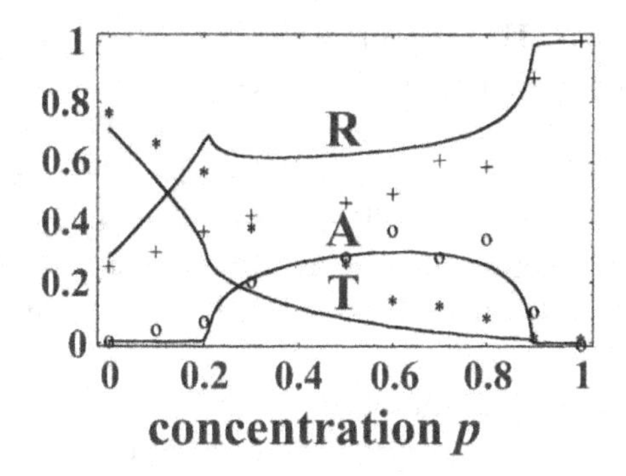

Fig. 4.2 **R**eflectance – "+", **T**ransmittance –"*", and **A**bsorptance – "○" as function of the metal concentration p; lines – theoretical results obtained from GOL.

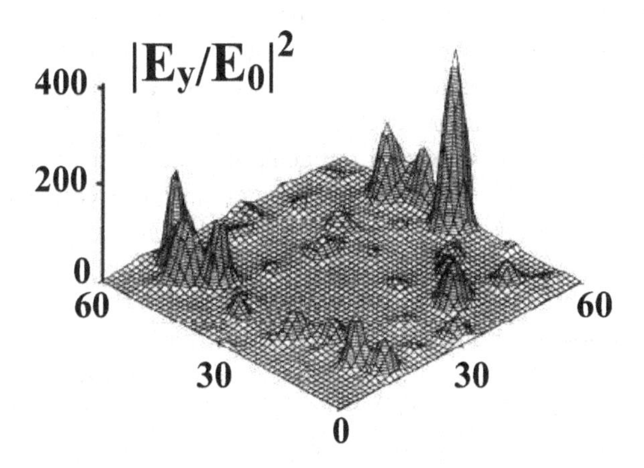

Fig. 4.3 The surface plasmon localization in a microwave region. An electric field distribution on the surface of a macroscopic copper-dielectric film is shown for $\lambda = 2.5$ cm at the metal's concentration $p = 0.6$. A copper thickness $h = 0.04$ mm.

4.3 Numerical Simulations of Local Electric and Magnetic Fields

To find the local electric $\mathbf{E}(\mathbf{r})$ and magnetic $\mathbf{H}(\mathbf{r})$ fields, we solve Eqs. (4.29), which we first rewrite is for the convenience:

$$\nabla \cdot \left[u(\mathbf{r})\nabla\phi'(\mathbf{r})\right] = \mathcal{E}_E, \qquad \nabla \cdot \left[v(\mathbf{r})\nabla\psi'(\mathbf{r})\right] = \mathcal{E}_H, \tag{4.46}$$

where $\phi'(\mathbf{r})$ and $\psi'(\mathbf{r})$ are the fluctuating parts of the potential fields $\phi(\mathbf{r})$ and $\psi(\mathbf{r})$, describing $\nabla\phi(\mathbf{r}) = \nabla\phi'(\mathbf{r}) - \mathbf{E}_a$, $\langle\phi'(\mathbf{r})\rangle = 0$, $\langle\nabla\phi(\mathbf{r})\rangle = \mathbf{E}_a$

and $\nabla\psi(\mathbf{r}) = \nabla\psi'(\mathbf{r}) - \mathbf{H}_a$, $\langle\psi'(\mathbf{r})\rangle = 0$, $\langle\nabla\psi(\mathbf{r})\rangle = \mathbf{H}_a$. By using the above, $\mathcal{E}_E = \nabla\cdot[u(\mathbf{r})\mathbf{E}_a]$ and $\mathcal{E}_H = \nabla\cdot[v(\mathbf{r})\mathbf{H}_a]$; which allows us to find the solution field of Eq. (4.46) and calculate the effective Ohmic parameters u_e and v_e, which are clearly independent of $\mathbf{E}_a$ and $\mathbf{H}_a$. Then we can calculate reflection r and transmission t coefficients (see Eqs. (4.25) and (4.26)) and use the Eqs. (4.30) to express the vectors $\mathbf{E}_a$, $\mathbf{H}_a$ and, therefore, the local fields $\mathbf{E}(\mathbf{r}) = \nabla\phi(\mathbf{r})$, $\mathbf{H}(\mathbf{r}) = \psi(\mathbf{r})$ in terms the amplitude $\mathbf{E}_0$ of the incident wave.

In the case of the strong skin effect, when the penetration depth δ is smaller than the film thickness d, the Ohmic parameter u_m is inductive according to Eq. (4.31) for all the spectral ranges regardless of the metal properties. If so, the percolation metal-dielectric film represents a set of randomly distributed $L-$ and $C-$ elements for *all* the spectral ranges when the skin effect is strong in the metal grains. Note that Ohmic parameter v takes the same sign and rather close absolute values for the metal and for the dielectric inclusions according to Eqs. (4.31), and (4.32). The film then can be thought as the collection of capacitive (C-) elements in "v"-space. Therefore, the resonance phenomena are absent in the solution to the second Eq. (4.46). The fluctuations of the potential ψ' can be neglected in comparison to the φ' fluctuations. For this reason, we concentrate on the solution to the first set of Eq. (4.46).

Equations (4.46) have the same form as Eq. (3.5) so that we can use the familiar square lattice discretization algorithm. Then numerical methods discussed in Sec. 3.2.2 can be used to find the local field distributions. The real space renormalization method was employed to solve Eq. (3.6) and to calculate the local field $\mathbf{E}(\mathbf{r})$ in terms of the average field $\mathbf{E}_a$. The effective Ohmic parameter u_e is determined from Eq. (4.22). In a similar manner, the magnetic field $\mathbf{H}(\mathbf{r})$ and the effective parameter v_e can be found. Note that the *same* lattice should be used to determine the fields $\mathbf{E}(\mathbf{r})$ and $\mathbf{H}(\mathbf{r})$. The directions of the external fields $\mathbf{E}_a$ and $\mathbf{H}_a$ may be chosen arbitrary when the effective parameters u_e and v_e are calculated since the effective parameters does not depend on the direction of the field for the film, which is macroscopically isotropic.

Though the effective parameters do not depend on the external field, the local electric $\mathbf{E}_1(\mathbf{r})$ and magnetic and $\mathbf{H}_1(\mathbf{r})$ fields do depend on the incident wave. The local fields $\mathbf{E}_1(\mathbf{r})$ and $\mathbf{H}_1(\mathbf{r})$ are defined in the reference plane $z = d/2+h$, which corresponds to the front surface of the film (see Fig. 4.1). Note that the field $\mathbf{E}_1(\mathbf{r})$ is observed in a typical near-field experiment (see, e.g., [Markel and George, 2000]). It follows from Eqs. (4.11), (4.12), and (4.17) that the fields $\mathbf{E}_1(\mathbf{r})$ and $\mathbf{H}_1(\mathbf{r})$ are expressed in terms of the fields $\mathbf{E}(\mathbf{r})$ and $\mathbf{H}(\mathbf{r})$, which are determined by the amplitude of the incident EM wave $\mathbf{E}_0$:

$$\mathbf{E}_1(\mathbf{r}) = \frac{1}{2}\left(\mathbf{E}(\mathbf{r}) - iv(\mathbf{r})\left[\mathbf{n} \times \mathbf{H}(\mathbf{r})\right]\right), \tag{4.47a}$$

$$\mathbf{H}_1(\mathbf{r}) = \frac{1}{2}\left(\mathbf{H}(\mathbf{r}) + iu(\mathbf{r})\left[\mathbf{n} \times \mathbf{E}(\mathbf{r})\right]\right). \tag{4.47b}$$

The local electromagnetic fields obtained from the numerical solution of Eqs. (4.46) are plotted in Figs. 4.4 and 4.5. The local electric and magnetic fields are larger than the intensity of the incident plane wave by 4–5 orders of magnitude. For the considered wavelength $\lambda = 1\ \mu$m, the silver permittivity approximates as $\varepsilon_m \simeq -\left(\omega_p/\omega\right)^2$ (see Eq. (3.1)). Then the skin depth $\delta = 1/(k\,\mathrm{Im}\,n_m) \simeq -c/\omega_p$ is independent of the frequency and estimates as $\delta \simeq 20$ nm for the silver. The local fields are shown for the film thickness $d = 50$ nm, which corresponds to the intermediate skin effect when the ratio $d/\delta \sim 2$. Yet, the fluctuations of the magnetic field exceed the local electric field. With the possible future development of optical magnetic nano-antennas discussed in Sec. 2.5, one could measure both the local electric and the local magnetic fields in the near-field optics experiments. The giant local field fluctuations were first experimentally observed for the metal-dielectric films with a strong skin effect (see Fig. 4.3) in the microwave region [Lagarkov *et al.*, 1997c].

4.4 Spatial Moments of Local Electric and Magnetic Fields

The results obtained in the previous section allow to find the spatial moments of the local electric field $\mathbf{E}_1$ distributed on the reference plane $z = d/2 + h$ (see Figs. 4.4 and 4.5). The electric field $\mathbf{E}_1$ is expressed in terms of the $\mathbf{E}$ and $\mathbf{H}$ fields by the means of Eqs. (4.47a) and (4.47b). The fluctuations of the local "magnetic induction" $\mathbf{B}(\mathbf{r}) = v(\mathbf{r})\mathbf{H}(\mathbf{r})$ can be neglected in the Eq. (4.47a) as it was discussed after the Eq. (4.46). Therefore, the moment $M^E_{n,m} = \langle|\mathbf{E}_1(\mathbf{r})|^n \mathbf{E}_1^m(\mathbf{r})\rangle / \left(|\langle\mathbf{E}_1\rangle|^n \langle\mathbf{E}_1\rangle^m\right)$, where $|\mathbf{E}_1(\mathbf{r})|^n \equiv \left(\mathbf{E}_1^* \cdot \mathbf{E}_1\right)^{n/2}$ and $\mathbf{E}_1^m \equiv \left(\mathbf{E}_1 \cdot \mathbf{E}_1\right)^{m/2}$, approximately equals to the $M^E_{n,m} \simeq \langle|\mathbf{E}(\mathbf{r})|^n \mathbf{E}^m(\mathbf{r})\rangle / \left(|\langle\mathbf{E}\rangle|^n \langle\mathbf{E}\rangle^m\right)$. Since the fluctuating electric field $\mathbf{E}(\mathbf{r}) = -\nabla\phi'(\mathbf{r})$ is obtained from the Poisson differential equation (4.46) its moment $M^E_{n,m}$ estimates from Eq. (3.45), where the metal and dielectric

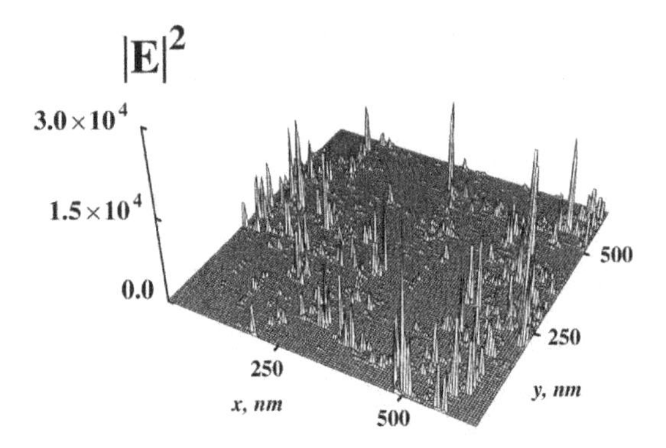

Fig. 4.4 The distribution of the local electric fields $|\mathbf{E}|^2 = |\mathbf{E}_1(\mathbf{r})|^2$ on a silver semicontinuous film for the wavelength $\lambda = 1\ \mu m$ and thickness $h = 50$ nm. The concentration of silver particles is at the percolation threshold (silver surface concentration $p \approx 0.5$). The amplitude of the incident field is taken as the unity ($|\mathbf{E}_0| = 1$).

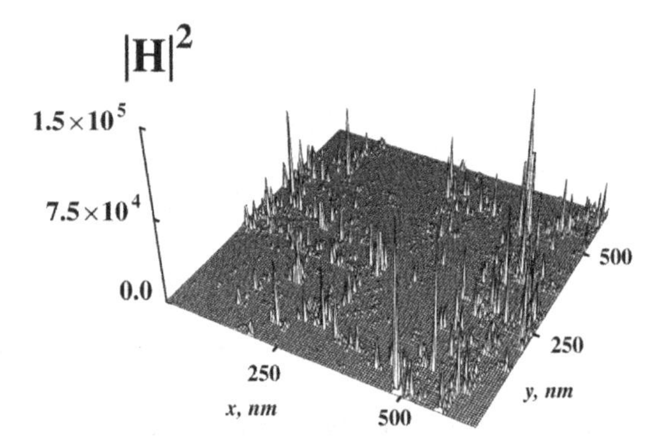

Fig. 4.5 The distribution of the local magnetic field intensity $|\mathbf{H}|^2 = |\mathbf{H}_1(\mathbf{r})|^2$ in a silver semicontinuous film for the wavelength $\lambda = 1\ \mu m$ and the thickness $h = 50$ nm, The concentration of silver particles is at the percolation threshold (silver surface concentration $p \approx 0.5$). The amplitude of the incident wave is unity $|\mathbf{E}_0| = 1$.

permittivities should be replaced by the Ohmic parameters u_m and u_d respectively.

Consider the moments $M^E_{n,m}$ for arbitrary strong skin effect assuming that the real part of the metal dielectric constant ε_m is negative, large in absolute value and can be approximate by the Drude formula (3.1). The Drude formula is then substituted in Eq. (4.18) to obtain the Ohmic parameter u_m in the limit $\omega \ll \omega_p$ and $\omega \gg \omega_\tau$. If the Ohmic parameters u_m and u_d are substituted into Eq. (3.45), we obtain

$$M^E_{n,m} \sim \left(\frac{|u_m|}{u_d}\right)^{\frac{n+m-1}{2}} \left(\frac{|u_m|}{u''_m}\right)^{(n+m-1)(1-2\gamma)+\gamma} \tag{4.48}$$

$$\sim \left(\frac{\omega_p^2}{\varepsilon_d\,\omega_\tau^2} f_1\right)^{(n+m-1)/2} \left(\frac{\omega}{\omega_\tau} f_2\right)^{\gamma(3-2n-2m)},$$

$$f_1 = \frac{4\tanh(x)^3\,\left[1+\frac{a}{2d}\tanh(x)\right]}{x\left\{\tanh(x)+x\left[1-\tanh(x)^2\right]\right\}^2} > 1, \tag{4.49}$$

$$f_2 = \frac{2\tanh(x)\,\left[1+\frac{a}{2d}\tanh(x)\right]}{\tanh(x)+x\left[1-\tanh(x)^2\right]^2} > 1,$$

where $u''_m = \operatorname{Im} u_m$ and $x = d/2\delta = d\omega_p/2c$ is the ratio of the film thickness d and the skin depth $\delta \approx c/\omega_p$. When the shape of the metal inclusions have a shape of oblate spheroids, the moments $M^E_{n,m}$ increase significantly with increasing the skin effect, i.e., increasing the parameter x.

Let consider now the far-infrared, microwave and radio-frequency bands, where the metal conductivity σ_m acquires its static value, i.e., it is positive and does not depend on the frequency. Then it follows from Eq. (4.18) that Eq. (3.45) for the moments acquires the following form

$$M^E_{n,m} \sim F^{(n+m-1)/2}\left(\frac{ak}{4}F\right)^{\gamma(3-2n-2m)/2} \gg 1, \tag{4.50}$$

where $F = (4\pi^2/\varepsilon_d)\,(a/\lambda)\,(\sigma_m/\omega) = 2\pi a\sigma_m/\,(c\varepsilon_d)$. Since the typical metal conductivity is much lager than frequency ω in the microwave and radio bands, the moments remain large at these frequencies. For example, the factor $F \sim 10^6$ for the copper particles of the size $a = 100$ μm so that the forth field moment $M^E_{4,0} \sim 10^7/(ak)^{0.35}$.

We proceed with describing the fluctuations of the local magnetic field $\mathbf{H}_1(\mathbf{r})$ in the reference plane $z = d/2 + h$. The fluctuations of the field $\mathbf{H}(\mathbf{r})$ can be still neglected. Then it follows from Eq. (4.47b) that the local magnetic field moments $M^H_{n,m} = \langle|\mathbf{H}_1(\mathbf{r})|^n \mathbf{H}_1^m(\mathbf{r})\rangle/\left(|\langle\mathbf{H}_1\rangle|^n \langle\mathbf{H}_1\rangle^m\right)$ transform immediately into

$$M^H_{n,0} \equiv M^H_n \simeq \langle\,|\mathbf{D}(\mathbf{r})|^n\rangle/\,|\langle\mathbf{E}_1\rangle|^n, \tag{4.51}$$

as $\mathbf{D}(\mathbf{r}) = u(\mathbf{r})\mathbf{E}(\mathbf{r})$. Thus, the external electric field induces the electric currents in a semicontinuous metal film and these currents, in their turn, generate the strongly fluctuating local magnetic field.

We then generalize the approach suggested by Dykhne [Dykhne, 1971] for the nonlinear case [Baskin *et al.*, 1997] to estimate the moments $\langle\, |\mathbf{D}(\mathbf{r})|^n \rangle$ of the "electric displacement" in semicontinuous metal films. Since the electric displacement vector is $\mathbf{D}(\mathbf{r}) = u(\mathbf{r})\mathbf{E}(\mathbf{r})$ (see 4.17), the following equation is written

$$\langle |\mathbf{D}(\mathbf{r})|^n \rangle = \alpha(u_m, u_d) \langle | \ \mathbf{E}(\mathbf{r})|^n \rangle \,, \tag{4.52}$$

where the coefficient $\alpha(u_m, u_d)$ is a function of variables u_m and u_d given by Eq. (4.18). We again consider the concentrations corresponding to the percolation threshold $p = p_c$ and set $p_c = 1/2$. We assume that the statistical properties of the system do not change with the spatial inter-change of the metal and the dielectric. If all the conductivities are increased by the factor k, the average nonlinear displacement $\langle |\mathbf{D}|^n \rangle$ increases by the factor $|k|^n$ so that the coefficient $\alpha(u_m, u_d)$ increases by $|k|^n$ times as well. This means that the coefficient $\alpha(u_m, u_d)$ has an important scaling properties, namely,

$$\alpha(ku_m, ku_d) = |k|^n \, \alpha(u_m, u_d). \tag{4.53}$$

By taking $k = 1/u_m$ the following equation is obtained

$$\alpha(u_m, u_d) = |u_m|^n \, \alpha_1(u_m/u_d). \tag{4.54}$$

Now we perform the Dykhne transformation:

$$\mathbf{D}^*(\mathbf{r}) = [\mathbf{n} \times \mathbf{E}(\mathbf{r})] \,, \tag{4.55}$$

$$\mathbf{E}^*(\mathbf{r}) = [\mathbf{n} \times \mathbf{D}(\mathbf{r})] \,. \tag{4.56}$$

It is easy to verify that the introduced above field $\mathbf{E}^*$ is still a potential field, that is $\nabla \times \mathbf{E}^* = 0$, whereas the displacement field $\mathbf{D}^*$ is conserved: $\nabla \cdot \mathbf{D}^* = 0$. The fields $\mathbf{D}^*$ and $\mathbf{E}^*$ are coupled by the GOL: $\mathbf{D}^* = u^*\mathbf{E}^*$, where the Ohmic parameter u^* takes values $1/u_m$ and $1/u_d$. Therefore, the equation $\langle | \ \mathbf{D}^*|^n \rangle = \alpha(1/u_m, 1/u_d) \langle |\mathbf{E}^*|^n \rangle$ holds, from which it follows that

$$\alpha(1/u_m, 1/u_d)\alpha(u_m, u_d) = 1. \tag{4.57}$$

We have assumed that at the percolation threshold $p_c = 1/2$, the statistical properties of the system do not change if we interchange the metal and the dielectric. Then the arguments in the function $\alpha(1/u_m, 1/u_d)$ can be exchanged to obtain $\alpha(1/u_d, 1/u_m)\alpha(u_m, u_d) = 1$. This equation, in turn, can be rewritten using Eq. (4.54) as $\alpha_1(u_m/u_d) = |u_d/u_m|^{n/2}$, and the final result becomes

$$\alpha(u_m, u_d) = |u_m\, u_d|^{n/2}; \tag{4.58}$$

so that the following generalization of the Dykhne's formula is obtained:

$$\langle|\mathbf{D}|^n\rangle = |u_d u_m|^{n/2} \langle|\mathbf{E}|^n\rangle. \tag{4.59}$$

The last expression for $\langle|\mathbf{D}|^n\rangle$ is substituted in Eq. (4.51) to get

$$M^H_{n,m} = |u_d u_m|^{n/2} M^E_{n,m}. \tag{4.60}$$

In the optical and in the infrared spectral ranges, it is possible to simplify Eq. (4.60) in the same way as it has been done for Eq. (4.48). Using again the Drude formula (3.1) and assuming that $\omega_\tau \ll \omega \ll \omega_p$, the following estimate is obtained

$$M^H_{n,m} = \left[\varepsilon_d' \frac{x \tanh x}{(2d/a) + x \tanh x}\right]^{n/2} M^E_{n,m}, \tag{4.61}$$

where the moment $M^E_{n,m}$ is given by Eq. (4.48) and $x = d/2\delta = d\omega_p/2c$ has the same meaning as in Eq. (4.51).

As it turns out from Eq. (4.61), the spatial moments of the local magnetic field $M^H_{n,m}$ have the same order of magnitude as the moments of the local electric field $M^E_{n,m}$ in the limit of the strong skin effect, i.e., when $x \gg 1$.

At frequencies much smaller than the typical metal relaxation rate $\omega_\tau \simeq 3.2 \times 10^{13}\, sec^{-1}$, the silver conductivity acquires its static value $\omega_p^2/4\,\pi\,\omega_\tau \simeq 10^{18}\, sec^{-1}$. In this case, the moments are given by Eq. (4.50). For $\lambda = 3$ cm ($\omega/2\pi = \nu = 10\, GHz$), the moments are $M^H_{n,m} \sim M^E_{n,m} \gg 1$, which allows us to conclude that the local electric and the local magnetics field strongly fluctuate in a wide frequency range: from the optical down to the microwave and the radio- spectral ranges (cf. Figs. 3.3 and 4.3). The fluctuations are typically stronger in the microwave and radio- bands. This is because for the strong skin effect (when the penetration depth is much smaller than the film thickness), loss is small in comparison with the electromagnetic field energy, which is accumulated in the vicinity of the film. This opens a fascinating possibility to observe the localization of the surface plasmons in the microwave experiments with localization length to be on a centimeter scale.

4.5 Extraordinary Optical Transmittance (EOT)

A flat metal surface is almost a perfect reflector for the electromagnetic waves in the visible region, and the application of metal films as mirrors has a long history. In this section, we show that even a small periodic corrugation could make a metal film semi-transparent. Essentially, a metal film becomes semi-transparent at resonant wavelengths allowing the excitation of electromagnetic waves propagating on the surface of the film.

We begin our consideration with a flat metal surface. In the optical and infrared spectral ranges, the collective excitation of the electron density coupled to the electromagnetic field results in surface plasmon polaritons (SPPs) travelling on the metal surface, as it is shown in Fig. 4.6 (see e.g., [Landau *et al.*, 1984], [Raether, 1988], [Freilikher *et al.*, 1997]). The SPPs can be excited when the real part of the metal permittivity, $\varepsilon_m = \varepsilon'_m + i\varepsilon''_m$, is negative, $\varepsilon'_m < 0$, and loss is relatively small, $\kappa = \varepsilon''_m / |\varepsilon'_m| \ll 1$, which is typical for metals in the optical range. First, for simplicity, we set $\kappa = 0$.

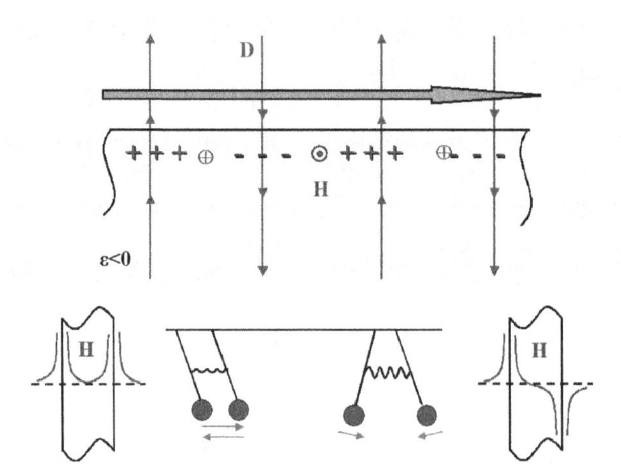

Fig. 4.6 Top diagram: Surface plasmon-polariton propagating over the metal ($\varepsilon_m < 0$) interface; electric displacement **D** has same direction on the both sides of the interface. Bottom diagrams: Symmetric and antisymmetric surface plasmon-polaritons propagating in metal film.

We also denote at the rest of this chapter the negative metal permittivity ε_m as $-n^2$, where n is the magnitude of the refractive index; the role of loss will be considered later.

At the metal-dielectric interface, the SPP is essentially an H-wave, with the direction of the magnetic field $\mathbf{H}$ parallel to the metal surface [Landau *et al.*, 1984]. In the direction perpendicular to the interface, SPPs exponentially decay in the both media (the metal and the dielectric). The relation between frequency ω and wavevector k_p of SPP can be found from the following consideration: We set the metal-dielectric interface at the $(x, y)-$plane and assume that the SPP propagates in the "x"-direction, with the field $\mathbf{H}$ directed in the "y"-direction ($\mathbf{H} = \{0, H, 0\}$). For simplicity, we also assume that the half-space $z < 0$ is vacuum, with the dielectric constant $\varepsilon = 1$, and neglect, as mentioned, the metal loss for the $z > 0$ half-space. We seek solutions for the magnetic field H in the following form

$$\begin{aligned} H &= H_1 = H_0 \exp\left(ik_p x + \Lambda_1 z\right), &\quad \text{for } z < 0; \\ H &= H_2 = H_0 \exp\left(ik_p x - \Lambda_2 z\right), &\quad \text{for } z > 0; \end{aligned} \tag{4.62}$$

where $\Lambda_1 = \sqrt{k_p^2 - k^2}$, $\Lambda_2 = \sqrt{k_p^2 + (kn)^2}$, $k = \omega/c$ is the wavevector and $n = \sqrt{-\varepsilon_m}$. Thus, the boundary conditions, requiring that the tangential components of the magnetic field are continuous at $z = 0$, are satisfied: $H_1(x, z = 0) = H_2(x, z = 0)$. The electric field $\mathbf{E}$, found from the Maxwell equation $\operatorname{curl}\mathbf{H} = -ik\varepsilon\mathbf{E}$ with $\varepsilon = 1$ for $z < 0$ and $\varepsilon = \varepsilon_m = -n^2$ for $z > 0$, has components E_{x1} and E_{x2}. The continuity requirement for the tangential component $E_{x1} = E_{x2}$ results in the following equation

$$\frac{\partial H_1}{\partial z} = -\frac{1}{n^2}\frac{\partial H_2}{\partial z}, \tag{4.63}$$

for $z = 0$. At $n > 1$, this equation can be satisfied and it yields the dispersion relation for the wavevector k_p of SPP [Landau *et al.*, 1984]

$$k_p = \frac{kn}{\sqrt{n^2 - 1}}. \tag{4.64}$$

Note, that for $n > 1$, the wavevector k_p is real and $k_p > k$ so that the H-field (see Eq. (4.62)) decays exponentially in the metal and in vacuum.

The component E_z (perpendicular to the propagation plane) of the electric field in the SPP takes the following values on the metal interface: $E_z(-0) = -\left(k_p/k\right) H_0 \exp\left(ik_p x\right)$ on the vacuum side of the film interface and $E_z(+0) = \left(k_p/n^2 k\right) H_0 \exp\left(ik_p x\right) \neq E_z(-0)$ on the opposite (metal) side. This electric field discontinuity occurs due to the surface charge density

$$\rho(x) = \frac{1}{4\pi}\left[E_z(+0) - E_z(-0)\right] = \frac{\left(1+n^2\right)}{4\pi n\sqrt{n^2-1}} H_0 \exp\left(ik_p x\right), \qquad (4.65)$$

which propagates together with the electric and magnetic fields along the film surface.

Thus, the SPP is a wave, which consists of the electromagnetic field coupled to surface charges. Since the SPP propagation includes rearrangement of the electron density, it is not surprising that its speed $c_p = \omega/k_p = c\sqrt{n^2-1}/n$ is always slower than the speed of light c. This explains why the surface plasmon polaritons cannot be excited by an external electromagnetic wave on a perfectly flat metal surface.

When the refractive index n approaches unity from above (that is the metal dielectric constant $\varepsilon_m \to -1$), the SPP velocity c_p vanishes so that the SPP "stops" on the metal surface. In this case, the surface charge diverges as $\left(n^2-1\right)^{-1/2}$; so does the normal component of the electric field. This phenomenon is known as the plasmon resonance in a thin metal plate. We note that the SPPs can propagate not only on the metal surface but also on the surfaces of artificial electromagnetic crystals, for example, wire-mesh crystals [Sievenpiper *et al.*, 1996], [Pendry *et al.*, 1996], [Sarychev *et al.*, 2000b], [Sarychev and Shalaev, 2000], (also see discussion in Ch. 4.5.3), in specially organized metal-dielectric layers [Sievenpiper *et al.*, 1998], [Sievenpiper *et al.*, 1999] or over perfect conductor surface with an array of holes or grooves [Pendry *et al.*, 2004]. This propagation occurs because the real part of the *effective* dielectric constant can be negative in these mesostructures. Such excitation of SPPs was observed in *2D* superconducting wire networks deposited onto a dielectric substrate [Parage *et al.*, 1998] and over tube structures [Hibbins *et al.*, 2005]. Exact analytical solution for the surface waves propagated on the ribbed metal surfaces was obtained in the beginning of fifties and can be found, for example, in the book by [Vainshtein, 1969]. The dissertation by Daniel Sievenpiper [Sievenpiper, 1999] contains a very good review of the high-impedance electromagnetic surfaces that can support "artificial" SPP. The propagation of SPP on the surfaces that mimic LHM was considered, for example, by [Caloz *et al.*, 2005] and [Caloz *et al.*, 2004].

We consider now SPP propagation on a metal film of the finite thickness d. The film is placed at the $z = 0$. Then two kinds of SPPs occur: those correspond to the symmetric and antisymmetric solutions to the Maxwell equations defined with the respect to the reflection in the film's principal plane $z = 0$. Hereafter, we will use the "refractive index" defined as $n = \sqrt{-\varepsilon'_m} > 1$ and neglect the loss. We are interested in the consequences of a strong skin effect $\exp(-dkn) \ll 1$ so that the field decays exponentially in the bulk of the film. Then, by applying the same approach as it was demonstrated above for the film with the finite thickness, we find that the propagation of SPPs is determined by the following expressions for the SPP wavevectors:

$$k_{1,2} = k_p \left[1 \pm \frac{2n^2}{n^4 - 1} \exp(-dk_p n) \right], \tag{4.66}$$

where k_1 and k_2 correspond to the symmetric and the antisymmetric wavevectors of the SPPs, respectively, and k_p is defined by Eq. (4.64). It is important to note that the symmetric and the antisymmetric SPPs propagate on the both sides of the film as shown in Fig. 4.7. Moreover, since these SPPs represent the film eigenmodes, the magnitudes of the electric and magnetic fields are the same on the both interfaces. This consideration holds for arbitrarily thick film, although the difference in speeds for the two SPPs becomes exponentially small for the optically thick films $dk_p n \gg 1$. The velocities of both the symmetric and the antisymmetric SPPs are less than the speed of light. These SPPs cannot be excited by an external

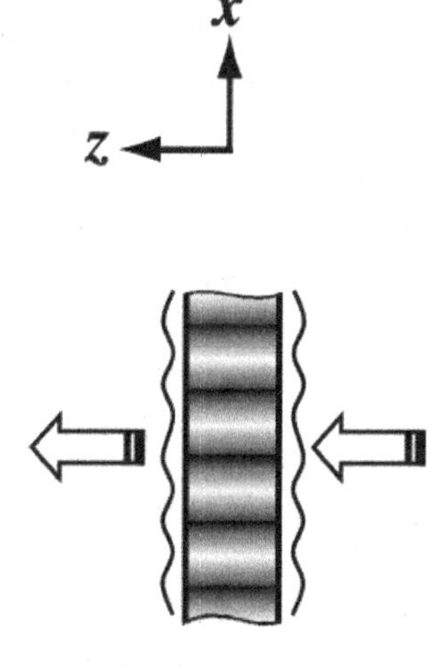

Fig. 4.7 The light incident on a modulated film. It first excites surface plasmon-polariton (SPP) on the front interface of the film, which then couples to the SPP on the back interface; the back-side SPP is eventually converted into the transmitted light.

electromagnetic wave incident from vacuum since that would violate the momentum conservation.

The situation changes dramatically when the film has a periodically modulated material properties, say, a refractive index. In this case, the EM field inside the film also gets modulated. When the spatial period of the modulation coincides with the wavelength of a SPP, it can be excited by an incident EM wave. An example of such spatial modulation is a square array of nanoholes punched in a film as it was done in Refs. [Ebbesen *et al.*, 1998], [Ghaemi *et al.*, 1998], [Kim *et al.*, 1999], [Thio *et al.*, 1999], [Martin-Moreno *et al.*, 2001]. Another example of a refractive-index modulation (which we will study) is the light-induced modulation because of the optical Kerr nonlinearity [Dykhne *et al.*, 2003], [Smolyaninov *et al.*, 2005].

The problem of light interaction with periodically modulated metal surfaces and films (see illustration in Fig. 4.7) has a long history, starting with a discovery of the Wood's anomalies. In 1902 Robert Wood reported [Wood, 1902] that the reflection spectrum from a metal diffraction grating could drastically change at some frequencies, which are now known as the excitation SPP frequencies [Agranovich and Mills, 1982], [Boardman, 1982], [Kawata, 2001]. Earlier experimental studies, as well as theories based on the perturbation approach, are reviewed in a book by [Raether, 1988]. The first full-scale computer simulations of SPPs that propagate on a metal surface with sinusoidal and sawtooth profiles was performed by [Laks *et al.*, 1981]. Although the material profile variation in this work was relatively small, yet the authors obtained almost a flat dispersion curve $\omega(k_p)$ for the SPP which, as we understand now, indicates the EM field localization in the grooves. SPP propagation arrest was observed in semicontinuous films by [Seal *et al.*, 2005]. The problem of the electromagnetic wave interaction with corrugated metal surfaces was extensively studied in the microwave-band literature. In particular, ribbed metal surfaces and waveguides (called also septate waveguides) are shown to support slow electromagnetic modes with the properties similar to SPPs in optics [Mahmoud, 1991], [Ilyinsky *et al.*, 1993], [Hibbins *et al.*, 1999]. The possibility to decrease loss in the waveguides by corrugating their walls (the gliding effect) was also considered in [Vainshtein, 1969], [Schill and Seshadri, 1988] and [Schill and Seshadri, 1989]).

A long-standing problem of the existence of SPP modes localized on the subwavelength grooves has attracted a lot of attention (see, e.g., [McGurn and Maradudin, 1985], [Seshadri, 1986], [Barnes *et al.*, 1995], [Pendry *et al.*,

1996], [Barnes *et al.*, 1996], [Madrazo and Nieto-Vesperinas, 1997], [Watts *et al.*, 1997b], [Freilikher *et al.*, 1997], [Sobnack *et al.*, 1998], [Lopez-Rios *et al.*, 1998]). Experiments reported by [Lopez-Rios *et al.*, 1998] have demonstrated well-pronounced minima in the specular reflection spectra, which can be explained by localization of SPPs on the groves, whose widths were much smaller than the radiation wavelength. The plasmon localization could also happen in the inter-grain gaps of metal films or in metal nanocavities as reported in [Moresco *et al.*, 1999] and [Coyle *et al.*, 2001]. Localization of the surface plasmons in the semicontinuous metal films were discussed in Ch. 3 (for recent development see also [Seal *et al.*, 2005], [Seal *et al.*, 2006], [Genov *et al.*, 2005].) The excitation of SPPs in a long chain of gold nanoparticles was observed by [Krenn *et al.*, 1999]. It is interesting to note that the plasmon excitations were localized in the inter-particle gaps in according with the theory presented in Sec. 1.1. Localization of SPPs in a subwavelength cavity was observed in the near-field experiments by [Bozhevolnyi and Pudonin, 1997] (see also [Bozhevolnyi and Coello, 2001] and references therein), the waveguiding of SPPs in an array of metal nanostructures was also investigated [Bozhevolnyi *et al.*, 2001].

The electromagnetic surface-plasmon resonances continue to be subject of research interest because the enhanced local fields associated with SPP resonances play an important role in the surface-enhanced Raman scattering and in nonlinear optical processes on rough metal surfaces (see, for example, [Pendry *et al.*, 1996], [Sarychev and Shalaev, 2000]). Another motivation for studying the electromagnetic properties of these nanoscale structures is the possibility of using them for near-field Superresolution readout of the optical disks and the subwavelength lithography [Tominaga *et al.*, 2001], [Peng, 2001], [Shi and Hesselink, 2004]. The confinement of the plasmon between a flat metal surface and metal nanoparticle resulted in the enhancement of the local field intensity by orders of magnitude; see for example [Leveque and Martin, 2006] and [Drachev *et al.*, 2006].

The growing interest to the subwavelength optics of metal films was further boosted by Ebbesen, Lezec, Ghaemi, Thio and Wolff when they discovered the extraordinary optical transmission (EOT) through nanohole arrays in optically thick metal films [Ebbesen *et al.*, 1998], [Ghaemi *et al.*, 1998], [Kim *et al.*, 1999], [Thio *et al.*, 1999], [Grupp *et al.*, 2000], [Martin-Moreno *et al.*, 2001]. A review of earlier studies for light interaction with subwavelength holes can be found in Ref. [Wannemacher, 2001]. The possibility that an opaque metallic film can be transparent, provided that an incident electromagnetic wave excites the coupled surface plasmons, was demonstrated in early work of [Dragila *et al.*, 1985]. It is also interesting to point out that as far as in the beginning of the last century, Lord Rayleigh predicted that a perfectly flat rigid surface with cylindrical holes could still have acoustic resonances at some special depths of the holes [Rayleigh, 1920].

The electromagnetic fields in subwavelength holes, grooves, or slits of metal films were simulated numerously after discovering EOT (see, e.g. [Wannemacher, 2001], [Schroter and Heitmann, 1998], [Porto *et al.*, 1999], [Treacy, 1999], [Astilean *et al.*, 2000], [Tan *et al.*, 2000], [Popov *et al.*, 2000], [Salomon *et al.*, 2001], [Lalanne *et al.*, 2000], [Cao and Lalanne, 2002], [Krishnan *et al.*, 2001]). In nearly all the simulations, the local EM field was strongly enhanced in subwavelength apertures for certain frequencies. This enhancement is considered to be the reason for the enhanced resonant transmittance. Yet, the distributions for the local EM fields vary significantly in the simulations performed by different authors. For example, the resonant field concentrates exactly inside deep grooves and slits, according to simulations of [Pendry *et al.*, 1996], [Watts *et al.*, 1997b], [Sobnack *et al.*, 1998], [Astilean *et al.*, 2000], [Tan *et al.*, 2000], while the simulations of [Madrazo and Nieto-Vesperinas, 1997], [Schroter and Heitmann, 1998] predict that the resonant field is strongly enhanced in close vicinity but outside the subwavelength grooves. According to [Pendry *et al.*, 1999], the field is concentrated in deep slits but distributed all over the film for more shallow slits. Computer simulations in [Wannemacher, 2001] predict a rather broad maximum for the local-field intensity, which centers in a nanohole but extends over distances much greater than the hole diameter. According to [Treacy, 1999] and [Treacy, 2002], where a so-called dynamic diffraction was studied for thin metallic gratings, the local magnetic field is strongly enhanced in some regions outside the slits of the metal film, and in the slits themselves, the local field has a clear minimum. As it is stressed in Ref. [Treacy, 1999], the obtained field maxima were different from the SPPs. Analysis of [Ebbesen *et al.*, 1998], [Ghaemi *et al.*, 1998], [Kim *et al.*, 1999], [Thio *et al.*, 1999], [Grupp *et al.*, 2000] and [Martin-Moreno *et al.*, 2001] experiments was performed by [Vigoureux, 2001] in terms of the diffraction theory, which also did not invoke the SPP excitation. In the near-field experiments of [Sonnichsen *et al.*, 2000], the EM field intensity (at $\lambda = 0.9$ μm) was measured around a single nanohole and a pair of nanoholes, having diameter $a = 0.3$ μm and separated by 2 μm distance from each other. For a single hole, the field was concentrated inside a hole, whereas for the pair of holes, a well-defined trace of the field was observed between the holes, which was interpreted as a SPP.

A theory of EOT was presented by [Martin-Moreno *et al.*, 2001]. The theory considers the SPPs on both sides of an optically thick metal film, which are connected through the evanescent modes in the nanoholes, which are treated as the subwavelength waveguides. The theory qualitatively reproduces the long-wave peak in EOT obtained in one of the Ebbesen's *et al.* experiments [Grupp *et al.*, 2000]. The theory of [Martin-Moreno *et al.*, 2001] also predicts that the transmittance of a thin lossless metal film has two asymmetric maxima that merge together with an increase of the film thickness; eventually, there is a single peak, which becomes progressively smaller with a further increase of the film thickness. A similar theory was developed by [Popov *et al.*, 2000] for the light transmission through subwavelength array of holes. This theory also attributes the resonant transmittance to the coupling of SPPs through the nanoholes. However, it predicts that this coupling exists only for a metal with a finite conductivity. In computer simulations and in qualitative descriptions performed in Refs. [Lalanne *et al.*, 2000], [Cao and Lalanne, 2002], the extraordinary transmittance through a periodical array of subwavelength slits was considered. The authors attribute the transmittance to the *internal* plasmon modes of the slits and arrive to the conclusion that the SPP excitation "is not the prime mechanism responsible for the extraordinary transmission of subwavelength metallic gratings with very narrow slits."

The narrow slits support the propagating TEM mode, where the electric field is perpendicular to the walls. As result the Fabry-Perot resonances can be excited in the slit and, therefore, the transmittance has maxima. This behavior was first theoretically considered by Takakura [Takakura, 2001] in one mode approximation and then obtained in the experiment [Suckling *et al.*, 2004]. Computer simulations indicates that the resonance transmittance is sensitive to the details of the em field distribution at the entrance (exit) of the slit. Detailed consideration of the field in the slit (beyond one mode approximation) gives, for example, the strong blueshift of the resonance frequency with increasing the width of the slit [Kukhlevsky *et al.*, 2005]. Periodic array of the slits can have Fano type of the resonance instead of symmetrical, Lorentz shape [Lee and Park, 2005]. The transmittance through two dimensional array of the parallel metal rods was considered by [Gippius *et al.*, 2005] (see also references therein.) Such system is also known as photonic crystal slab. The transmittance depends on the distance between the rods but also on the cross-section of a rod.

The transmittance almost vanishes in the forbidden band of the *dielectric* photonic crystal. When the slab of the silicon photonic crystal was cut at both interfaces with suitable truncations of the first and last layers, the 100% peak transmittance were obtained due to excitation the surface waves [Laroche *et al.*, 2005].

It was suggested that the extraordinary transmittance could result from the excitation of internal modes in the holes punched in a metal film [Shubin *et al.*, 2000], [Sarychev *et al.*, 2002]. The importance of the SPP excitation around a hole was demonstrated in the experiments by [Thio *et al.*, 2001], [Thio *et al.*, 2002], [Lezec *et al.*, 2002], [Garcia-Vidal *et al.*, 2003], [Degiron and Ebbesen, 2004], [Lezec and Thio, 2004], where the light transmittance through a hole surrounded by a system of periodic rings with grooves was investigated. When the period of the rings was approximately equal to the wavelength of SPP, the electric field in a vicinity of the hole increased significantly. It is interesting to note that direct experimental measurements by [Lezec *et al.*, 2002] have shown that the local EM field concentrates in a close proximity to the hole at the resonance when the hole is surrounded by the system of rings. It has been also shown that a periodic texture on the exit side can give rise to beam with a small angular divergence, very different from the expected quasi-isotropic diffraction of small hole [Lezec *et al.*, 2002], [Martin-Moreno *et al.*, 2003], [Garcia-Vidal *et al.*, 2003], [Barnes *et al.*, 2003]. The extraordinary transmittance through the textured hole can be observed in the microwave [Hibbins and Sambles, 2002] and [Lockyear *et al.*, 2004], where metal itself does not support SPP (see discussion below.)

The resonance transmittance though a metal film with a periodically modulated refractive index was first theoretically considered by [Shubin *et al.*, 2000]. In this case, the increase in the transmittance was attributed to the excitation of SPPs, which propagated on the both interfaces of the film. The computer simulations were performed for a similar system by [Avrutsky *et al.*, 2000] and [Tan *et al.*, 2000]. The authors considered an optically thick metal film with a periodical set of symmetric conical groves [Tan *et al.*, 2000] and sinusoidal grooves [Avrutsky *et al.*, 2000] on both sides of the film. The grooves were placed in such a way that their bases were opposite to each other (apexes away). [Avrutsky *et al.*, 2000] also considered the gratings on the opposite sides shifted by a half of the period with respect to each other. At certain frequencies, the incident electromagnetic field causes the resonant enhancement for the field inside the grooves. The excitation, which authors [Tan *et al.*, 2000] treat as standing (localized)

SPPs, tunnels through the metal film from the bottom of the illuminated grooves to the opposite side of the film and excites the standing SPPs on the surface. Thus, the resonance transmittance through the otherwise optically thick metal film occurs. The transmittance vanishes exponentially when the groove pitch exceeds 100 μm in the film. The results of simulations presented by [Tan *et al.*, 2000] and [Avrutsky *et al.*, 2000] give qualitatively a similar physical picture. Experiments performed by [Avrutsky *et al.*, 2000] for corrugated gold films, had indeed shown a maximum in the transmittance. In the experiment and computer simulations by [Schroter and Heitmann, 1999], the optical transmittance and reflectance were investigated for a modulated silver film with the average thickness 50 μm and the modulation amplitude of 30 to 35 μm. The observed transmittance (reflectance) exhibits a double-peak maximum (minimum), which corresponds to the excitation of the SPP on the surface of the film. These results agree with the theory [Dykhne *et al.*, 2003], which we present below. Similar approach was developed by Darmanyan and Zayats [Darmanyan and Zayats, 2003], however they did not take into account loss in the metal that limited the extraordinary transmittance.

Recently Henri Lezec and Tineke Thio [Lezec and Thio, 2004], [Thio, 2006] reconsidered the results of their original experiment [Ebbesen *et al.*, 1998] that led to the discovering the extraordinary optic transmittance. Recall that very large transmittance enhancement up to factor 10^3 were reported and subsequently quoted. In the work [Lezec and Thio, 2004] the authors compared the transmittance of the hole array not to a theoretical result but to the transmittance of a single, isolated hole. The lone hole they used was in every way identical to ones in the array, and it was fabricated in the same metal film. The only difference was that it had nothing around it but the smooth metal surface. Lezec and Thio also accounted for the fact that an array of holes sends light in well collimated beam (see e.g., [Bravo-Abad *et al.*, 2006]) whereas a single hole projects light into a hemisphere. As a consequence the fraction of the incident power that gets transmitted appears to depend on the size of the lens used to collect the light for measurements on the far (front) side. One needs to correct for this effect to extract the true transmission coefficient, i.e., the total power radiating from the holes divided by the power incident on the total area of the holes. The authors obtained the experimental enhancement G by dividing the transmission coefficient for the array of the holes by transmission coefficient of a single hole. In most of the experiments [Lezec and Thio, 2004] the enhancement was on the order unit, increasing in the optimal configuration to $G \approx 5 \div 7$. This result is much less than the "orders of magnitude" reported in Nature [Ebbesen *et al.*, 1998]. To understand the discrepancy the authors reexamined their Nature results and found that the hole diameter a, which was quoted for the original experiment, was smaller than the actual diameter. The diameter enters as a^{-6} in the estimate of the enhancement G. Therefore the underestimate of the hole diameter a resulted in huge overestimate of the transmittance enhancement. Thus the works [Lezec and Thio, 2004], [Thio, 2006] resolve the problem of EOT through the array of the subwavelength holes.

On the other hand, the bull's-eye structure, consisting of a single hole surrounded by 16 rings provides the enhancement $G \simeq 30$ when parameters are optimized [Lezec and Thio, 2004]. According to [Thio, 2006] G can be as large as 400 in the optimized bull's-eye structure. In this sense EOT still exists. Note, however, that the bull's-eye enhancement was at most six: $G < 6$ in the experiment by Degiron and Ebbesen [Degiron and Ebbesen, 2004].

It was also proclaimed by [Lezec and Thio, 2004] that SPP has nothing to do with EOT since this phenomenon is due to the excitation of the special evanescent waves that are generated by interaction of the incident light with holes, rings, and other nonuniformities on the metal surface. This result does not depend on the material of the film and holds for the metal films ($\varepsilon_m' < 0$) as well as for perfect electrical conductor (PEC) films, where $\sigma_m = \infty$ and $\varepsilon_m = \infty$. However, the extensive computer simulations, performed by [Chang *et al.*, 2005] for the gold film and for PEC films, give quite different results for the transmittance through a nanohole. The transmittance for a single hole in the gold film, which supports evanescent modes as well as SPP, is much larger than the transmittance of the hole in PEC, which supports evanescent modes only. The simulations [Chang *et al.*, 2005] give well defined set of maxima and minima for the transmittance through the regular array of the holes in the gold film. The maxima are much more narrow and less in amplitude in the case of PEC. In both cases (gold, PEC) the enhancement G was $G \simeq 1 \div 2$ for the most of the maxima.

The local field was also detailed investigated in the simulations [Chang *et al.*, 2005]. The component of the electric field, which is perpendicular to the film, has maximum at the rim of the holes. This component of the local electric field corresponds to SPP and/or evanescent modes. The electric field which is parallel to the film has maximum inside the holes.

This maximum shifts to front (back) interface of the film for maximum (minimum) of the transmittance.

The dispersion of the transmittance was measured by [Barnes *et al.*, 2004]. In this experiment light impinges on the film at various angles that is with various in-plane wavevectors $\mathbf{k}_{||}$. The position of a transmittance and absorptance maximum is measured as a function of the frequency ω and the vector $\mathbf{k}_{||}$. For P polarization of the impingement light thus obtained dispersion $\omega\left(\mathbf{k}_{||}\right)$ corresponds to the dispersion of the $SPPs$ propagating at metal-air and metal-glass interfaces. It is interesting to note that there is no dispersion for S polarization: with a good accuracy the resonance frequency does not depend on $\mathbf{k}_{||}$. The authors conclude that "results provide strong experimental evidence for transmission based on diffraction, assisted with SPP". This conclusion was supported by computer simulations [Lalanne *et al.*, 2005] where authors identify the SPP modes responsible for the extraordinary transmission. They showed that the near-field has a hybrid character combining hole periodicity and localized effects relying on the fine structure of the aperture geometry. EOT is unambiguously attributes to the excitation of SPP on the periodically corrugated surface. Detailed investigation of the near-field showed that the component of the magnetic field, which is parallel to the film has maxima – hot spots near the outer boundary of the hole. On the other hand the component of the magnetic field, which is perpendicular to the film, has maxima inside the holes. This local field distribution corresponds to that by [Chang *et al.*, 2005]. The transmittance enlacement $G \leq 50$ could be obtained by optimizing the system according to [Lalanne *et al.*, 2005].

SPP is dumped out in the terahertz (THz) band, where the loss factor $\kappa \gg 1$. Yet the experiments [Rivas *et al.*, 2003] and [Cao and Nahata, 2004] revealed well defined peaks in the transmittance that are similar to EOT. The THz transmittance trough the square array of the square holes was investigated by [Rivas *et al.*, 2003]. The plate was made of doped silicon with large imaginary permittivity ($\varepsilon' = -18.1$, $\varepsilon'' = 91.8$ at 1 THz). The thick film with thickness $d = 280$ μm and size of the hole $a = 70$ μm has no any resonance transmittance. The thin film, which thickness $d = 100$ μm is almost equal to the size a of the hole, has well defined maximum in the transmittance. The first resonance wavelength (frequency 0.8 THz) corresponds to the period of 400 μm of the hole array.

In another THz experiment [Cao and Nahata, 2004] the transmittance of the stainless still plate with regular array of the holes was investigated for the frequencies $0.1 \div 0.5$ THz. The diameter of a hole $a = 0.4$ mm was larger than the thickness $d = 0.075$ mm. Yet one can see in the time-domain spectroscopy long lasting resonance state, which corresponds to the peak in the transmittance.

The computer simulation by [de Abajo *et al.*, 2005a], performed for the regular array of the holes in the infinitely thin plate of PEC, are in agreement with the THz experiments. The EOT is obtained for the wavelength very close to the period of the array, i.e., very close to the onset of the diffraction. At the resonance the transmittance is shown to complete (100%) even for arbitrary narrow holes. For qualitative discussion of the transmittance the authors suggested the model, where each hole in PEC palate is considered as a dipole and the resonance transmittance corresponds to the collective excitations in the system of the dipoles. Note that results of [de Abajo *et al.*, 2005a] disagrees with the simulations by [Chang *et al.*, 2005], discussed above.

Another approach was adopted by [Pendry *et al.*, 2004] who proposed that a surface of perfect conductor pierced by array of subwavelength holes supports surface waves. Pendry, Martin-Moreno and Garcia-Vidal speculate that the structured metal surface has negative effective permittivity, which frequency dependence equivalent to the optic permittivity of loss-free metal (see Eq. (3.1), where $\omega_\tau = 0$.) The cutoff frequency, i.e., frequency where the effective permittivity of the pierced metal becomes negative coincides with the cutoff frequency of a single hole. Than Pendry, Martin-Moreno, and Garcia-Vidal extend the consideration to any structured metal surface that is supposed to have negative effective permittivity and supports SPP. They wrote "Although a flat PEC surface supports no bound states, the presence of holes, however small, produces a SPP like bound surface state. Indeed, almost any disturbance of the flat surface will bind a state. Bound states are found in the following circumstances: a surface with array of holes of finite depth, however shallow, a surface with an array of grooves of finite depth; and surface covered by a layer of dielectric, however thin."

The existence of the "artificial" SPP was confirmed in the experiment [Hibbins *et al.*, 2005]. The bound states were detected in microwave on the metal surface densely pierced by square holes. To bound the impinging em wave with artificial SPP an additional corrugation was made on the top of the holes. The excitation of SPP is seen as a minimum of the reflectance at certain frequencies and angels of incidence. The dispersion of the artificial SPP was close to the light cone and somehow different from a simple low proposed in Ref. [Pendry *et al.*, 2004]. The results similar to the experiment [Hibbins *et al.*, 2005] were obtained in computer simulations by [de Abajo and Saenz, 2005].

The importance of the connection between resonance transmittance and any kind of bound states was stressed by [Borisov *et al.*, 2005], [Qu and Grischkowsky, 2004], and [Selcuk *et al.*, 2006]. Let us consider a frequency selective surface and broad band pulse impinging on the surface. Suppose that the band of the pulse is much larger than the transmittance window. It follows from the uncertainty principle that in order for the transmittance to occur the radiation should last much longer than the duration of the initial pulse. This implies that the corresponding electromagnetic mode have to be trapped by structured material for sufficiently long time. Such modes are known in the scattering theory as scattering resonances. The example of the trapped mode is the plasmon resonances in the array of the dielectric spheres buried in the metal film ([de Abajo *et al.*, 2005b], [Sugawara *et al.*, 2006] and references therein.)

We have presented a short introduction to the problem of EOT. However, the field develops so fast that it is literally impossible to suggest a complete and fully applicable list of references. We can recommend the references in the following review and original papers [Barnes *et al.*, 2003], [Zayats *et al.*, 2005], and [Kukhlevsky *et al.*, 2005]. Yet, it follows from above consideration that after almost ten years of intense research and despite a convergence of views, no global consensus exists. We cannot even give unambiguous answers to the very basic questions:

How large can be the enhancement G of the resonance transmittance?

Does EOT exist indeed?

In this situation an analytical solution of EOT can shed a light on the problem. Having the generic, analytical solution a reader can take a pencil (or, say, Wolfram's Mathematica) and reproduce all the calculations to answer above questions by (him/her)self.

We present a simple model, which a) reproduces main features of EOT and b) provides with an explicit analytical solution, which can become a starting point in considering behaviors of more complicated systems. We adopt the approach, in which the film modulation is small. This allows us to develop an *analytical theory* for EOT, which has certain advantages over numerical simulations because of its generic nature. A detailed analysis allows us to find the resonance conditions under which a modulated film becomes semi-transparent. Our approach can be used to find the resonances for nonspecular incidence [Kats and Nikitin, 2004]. We also extend our consideration to the case when nonlinear optical effects are essential. Specifically, we consider a dependence of the metal dielectric function on the intensity of SPP (which is strongly enhanced at the resonance) so that the SPP can further increase the initial (blended "seed") modulation of the film's refractive index. This ultimately leads to the increase of the film transmittance. In the original paper [Dykhne *et al.*, 2003] and than in the work by [Porto *et al.*, 2004] the bistable behavior of the transmittance was predicted. The bistability of EOT was recently obtained in the very important experiment by [Wurtz *et al.*, 2006].

When analyzing the transmittance behavior, we will assume for simplicity, that the electromagnetic wave incidences normally to the film surface as sketched in Fig. 4.7. The refractive index of the film has a spatial profile in the "x"-direction with the period of $a = 2\pi/q$, where q is spatial wavevector of the film's modulation. The modulation can be either intrinsic (pre-fabricated) or external (induced by the incident light due to the optical nonlinearity in the film's material). When the frequency of an incident wave is such that SPP wavelengths $\lambda_{1,2} = 2\pi/k_{1,2}$ (where $k_{1,2}$ are given by Eq. (4.66)) coincide with the period of the modulation a, the SPPs are excited in the film. Since the film is optically thick, the SPP first gets excited on the back interface of the film. Eventually it "spreads" over the other side of the film, so that SPPs on both front and back interfaces of the film are eventually excited. There is a straightforward analogy between the two SPPs on the opposite sides of the film and two identical, coupled oscillators as it is illustrated in Fig. 4.6. The coupling between the two oscillators can be arbitrary weak. Nevertheless, if we push the first oscillator then, after some time (which depends on the oscillator's coupling strength), the second oscillator starts to oscillate with the same amplitude as the first one. Similarly, the two SPPs on different sides of the film will have eventually the same amplitude. When a SPP propagates on the front-side of the film, it interacts with the film modulation index and, as a result, converts its energy back to the plane wave, which was re-emitted from the film. Therefore, at the resonance, the film becomes almost transparent, regardless of its thickness; however, the width of the transmittance resonance shrinks when the film thickness d increases [Shubin *et al.*, 2000].

Note that the profile amplitude $g \simeq \Delta n/n$ of the film modulation does not play any role. The modulation g can be arbitrary small, yet the SPPs are excited, and the film becomes transparent. Moreover, one does not even need the holes in order for the resonant transmittance to occur. The only thing required is such that the both sides of a metal film are modulated with the same spatial period (the pitch). The minimum of the modulation pitch for the extraordinary transmittance depends on the combined loss in the metal. The transmittance maxima typically have a double-peak form, indicating the excitation of symmetric and antisymmetric SPPs.

The amplitude g of the modulation and the film thickness d are very important for the temporal dynamics of EOT. The smaller g and the larger d the more time it takes for SPP to percolate from backside to the front of the film. During the penetration from back to the front interface of the film the developing SPP can be seen as a trapped mode (see e.g., [Borisov *et al.*, 2005], [Qu and Grischkowsky, 2004], and [Selcuk *et al.*, 2006].)

In the next two subsections, we consider a detailed theory of the resonant light transmittance through linear and nonlinear metal films.

4.5.1 *Resonant transmittance*

A periodic modulation of the film's index (or dielectric permittivity) can be represented by the Fourier series, where the spatial harmonics of Fourier decomposition are described by wavevector q. The resonant transmittance in this model is described by the resonance equality between a SPP wavevector $k_{1,2}$ and the wavevector q of the qth spatial harmonics (the resonance interaction of the SPP mode with the qth Fourier harmonics). Since other spatial harmonics are off the resonance, their amplitudes are not enhanced and small so that we pay attention to the interaction of the incident field with the resonant mode only. Suppose that the magnetic field $\mathbf{H}$ in the incident light has only "y"-component $\mathbf{H} = \{0, H, 0\}$. We consider the variation of the dielectric constant of the film as $\varepsilon = -n^2 [1 + g\cos(qx)]$, where the amplitude of the modulation is small, $g \ll 1$ (see Fig. 4.7); recall that "x" coordinate is along the metal film, which is placed in "x, y" plane. In actual calculations, it is convenient to use a slightly different, but an equivalent to the above, equation:

$$\varepsilon(\mathbf{r}) = -n^2 [1 - g\cos(qx)]^{-1}, \tag{4.67}$$

with $g \ll 1$. An amplitude of the plane electromagnetic wave propagating normal to the film's surface (along the z axis), depends only on a "z"-coordinate. In the course of the field interaction with the film (4.67), an electromagnetic harmonic, which is proportional to $g\cos(qx)$, is generated. Its amplitude is proportional to the film modulation g. This harmonic, in its turn, also interacts with the film and excites a harmonics at double spatial frequency, $2q$, with the amplitude proportional to $g^2\cos(2qx)$. As a result of the cascade process, the whole spectrum of the electromagnetic waves is excited in the film. The amplitude of the lq-th harmonics ($g^l\cos(qlx)$) is proportional to g^l. The resonant transmittance occurs when these harmonics are converted back into a plane wave, which is transmitted through the film. Since $g \ll 1$, the transmitted electromagnetic harmonics of the lowest possible decomposition order give the maximum contribution to the spectrum. We thus restrict our consideration to the lowest $\cos(qx)$-harmonics and neglect all the others (e.g. all $\cos(lqx)$- type harmonics with $l > 1$).

We consider the magnetic field in the form $H_y(x,z) = H(z) + H_q(z)\cos(qx)$ where $H(z)$ and $H_q(z)$ are the two functions, which determine the electromagnetic field inside and outside of the film. To find electric **E** and magnetic **H** fields inside the film, we substitute the magnetic field

$$\mathbf{H} = \{0,\ H(z) + H_q(z)\cos qx,\ 0\}, \tag{4.68}$$

and modulated dielectric constant (4.67) into the Maxwell equations and equate the terms that have the same dependence on the "x" coordinate [Rayleigh, 1896]. Neglecting higher harmonics, we find that

$$\begin{aligned} H(z)'' - (kn)^2 H(z) - \frac{g}{2} H_q(z)'' &= 0, \\ H_q(z)'' - \left[(kn)^2 + q^2\right] H_q(z) - gH(z)'' &= 0. \end{aligned} \tag{4.69}$$

The solution to this system of the differential equations is

$$\begin{aligned} \left\{ \begin{matrix} H(z) \\ H_q(z) \end{matrix} \right\} &= \mathbf{A}\ \mathrm{sech}(\frac{d\,k\,n\Lambda_1}{2})\,[X_1\cosh(k\,n\,z\,\Lambda_1) - X_2\sinh(k\,n\,z\,\Lambda_1)] \\ &+ \mathbf{B}\ \mathrm{sech}(\frac{d\,n\,k\Lambda_2}{2})\,[Y_1\cosh(k\,n\,z\,\Lambda_2) - Y_2\sinh(k\,n\,z\,\Lambda_2)], \end{aligned} \tag{4.70}$$

where X_1, X_2, Y_1, and Y_2 are constants, and Λ_1,Λ_2 are the dimensionless eigenvalues.

$$\Lambda_1^2 = \frac{2 - Q + {q_1}^2}{2 - g^2}, \qquad \Lambda_2^2 = \frac{2 + Q + {q_1}^2}{2 - g^2}, \tag{4.71}$$

$$Q = \sqrt{{q_1}^4 + 2g^2\,(1 + {q_1}^2)}, \quad q_1 = \frac{q}{kn}\,. \tag{4.72}$$

Vectors **A** and **B** in Eq. (4.70) are the eigenvectors for Eqs. (4.69), which are equal to

$$\mathbf{A} = A = \left\{ \begin{array}{c} \Lambda_1^2 \\ \dfrac{2\,(1 - \Lambda_1^2)}{g} \end{array} \right\}, \ \mathbf{B} = \left\{ \begin{array}{c} \dfrac{1 + q_1^2 - \Lambda_2^2}{g\,(1 + q_1^2)} \\ \dfrac{\Lambda_2^2}{1 + q_1^2} \end{array} \right\}, \tag{4.73}$$

with Q and q_1 given by Eq. (4.72).

The electric field **E** inside the film is obtained by the back substitution of the magnetic field given by Eqs. (4.68) and (4.70) into the Maxwell equation. Neglecting harmonics $\sim \cos{(lqx)}$ with $l > 1$, we then obtain:

$$\mathbf{E} = \{E\,(z) + E_q\,(z)\cos qx,\, 0,\, E_z\,(z)\sin qx\}\,, \tag{4.74}$$

with $E\,(z)$, and $E_q\,(z)$given by

$$\begin{aligned} &\left\{ \begin{array}{c} E\,(z) \\ E_q\,(z) \end{array} \right\} \\ &\quad = \mathbb{U}\cdot\mathbf{A}\ \ \Lambda_1 \,\mathrm{sech}(\frac{d\,k\,n\Lambda_1}{2})\ [X_1\ \sinh(k\,n\,z\,\Lambda_1) - X_2\ \cosh(k\,n\,z\,\Lambda_1)] \\ &\qquad + \mathbb{U}\cdot\mathbf{B}\ \ \Lambda_2 \,\mathrm{sech}(\frac{d\,kn\,\Lambda_2}{2})\ [Y_1\ \sinh(k\,n\,z\,\Lambda_2) - Y_2\ \cosh(k\,n\,z\,\Lambda_2)]\,, \end{aligned} \tag{4.75}$$

the matrix U defined by

$$\mathbb{U} = \frac{i}{n} \left\{ \begin{array}{cc} 1 & g/2 \\ g & 1 \end{array} \right\}, \tag{4.76}$$

and the vectors **A**, **B** given by Eq. (4.73). The component E_z of the electric field

$$\begin{aligned} E_z = \ &\frac{\sin(q)\,q_1}{g\,n} [\, \frac{2\,(\Lambda_1^2 - 1)}{\cosh(d\,k\,n\,\Lambda_1/2)} \\ &* (-X1\ \cosh(k\,n\,z\,\Lambda_1) + X2\ \sinh(k\,n\,z\,\Lambda_1)) \\ &+ \frac{g\,\Lambda_2^2}{(1 + q_1^2)\cosh(d\,k\,n\,\Lambda_2/2)} \\ &* (Y1\ \cosh(k\,n\,z\,\Lambda_2) - Y2\ \sinh(k\,n\,z\,\Lambda_2))\ ], \end{aligned} \tag{4.77}$$

which is perpendicular to the film surface, changes its sign along the film surface as it should be for SPP em field.

It follows from Eqs. (4.68), (4.70), (4.74), and (4.75) that the fields inside the film are fully determined by the two pairs of constants, namely, $\{X_1, X_2\}$ and $\{Y_1, Y_2\}$. By their physical meaning, the constants $\{X_1, X_2\}$ in Eqs. (4.70) and (4.75) describe the fundamental (incident) beam, whereas the other two constants, $\{Y_1, Y_2\}$, describe the $\cos(qx)$ (generated harmonic) mode. For a flat film, $g = 0$ so that the eigenvalues in Eqs. (4.70) and (4.75) equal to $\Lambda_1 = 1$ and $\Lambda_2 = (kn)^{-1}\sqrt{k^2n^2 + q^2}$, while the eigenvectors (Eq. (4.73)) are $\mathbf{A}= \{1, 0\}$, $\mathbf{B}= \{0, 1\}$, and the matrix $\mathbb{U}$ (see Eq. (4.76)) is $\mathbb{U} = i\{\{1, 0\}, \{0, 1\}\}/n$.

The magnetic field of the incident and reflected waves can be represented as

$$H(z) = H_0 \exp(ikz) + rH_0 \exp(-ikz), \tag{4.78}$$

for $z \leq -d/2$, where H_0 is the amplitude of the incident wave, r is the reflection coefficient and $R = |r|^2$ is the film reflectance. The transmitted wave field has the form

$$H(z) = tH_0 \exp(ikz), \tag{4.79}$$

for $z \geq d/2$, where $T = |t|^2$ is the transmittance. For the qth Fourier mode, we use the radiation boundary conditions such that

$$H_q(z) = Y_3 H_0 \exp(-iq_z z_1) \tag{4.80}$$

for $z \leq -d/2$, where $z_1 = z + d/2$;

$$H_q(z) = Y_4 H_0 \exp(iq_z z_2), \tag{4.81}$$

for $z \geq d/2$, where $z_2 = z - d/2$,

$$q_z^2 = k^2 - q^2, \tag{4.82}$$

Y_3 and Y_4 are some constants. Note that for the resonant transmittance, the wavevector k obeys $|k| < |q|$ so that q_z is imaginary quantity and the field H_q decays exponentially outside of the film.

The combined electromagnetic field in the free space is completely determined by the vector set $\mathrm{X} = \{X_1, X_2, X_3, X_4\}$ and $\mathrm{Y} = \{Y_1, Y_2, Y_3, Y_4\}$, where we choose

$$X_3 \equiv r \exp(ikd/2), \qquad X_4 \equiv t \exp(ikd/2). \tag{4.83}$$

We set the amplitude of the incident wave $H_0 = \exp(ikd/2)$ so that the electric, $\mathbf{E}_{rf}$, and magnetic, $\mathbf{H}_{rf}$, fields right before the film entering ($z \leq -d/2$) are described by

$$\mathbf{E}_{rf} = \left\{ \begin{array}{c} \exp(i\,k\,z_1) - X_3 \exp(-i\,k\,z_1) - Y_3 \dfrac{q_z}{k} \cos(q\,x) \exp(-i\,k\,z_1) \\ 0 \\ -iY_3 \dfrac{q}{k} \sin(q\,x) \exp(-i\,k\,z_1) \end{array} \right\}, \tag{4.84}$$

$$\mathbf{H}_{rf} = \left\{ \begin{array}{c} 0 \\ \exp(i\,k\,z_1) + X_3 \exp(-i\,k\,z_1) + Y_3 \cos(q\,x) \exp(-i\,k\,z_1) \\ 0 \end{array} \right\}, \tag{4.85}$$

where $z_1 = z + d/2$.The electric $\mathbf{E}$ and magnetic $\mathbf{H}$ fields inside the film $(-d/2 \leq z \leq d/2)$ are given by the Eqs. (4.74), (4.75), and (4.68), (4.70), respectively. The fields leaving the film $(z \geq d/2)$, i.e., transmitted fields, are equal to

$$\mathbf{E}_{tr} = \left\{ \begin{array}{c} X_4 \exp(i\,k\,z_2) + Y_4 \dfrac{q_z}{k} \cos(q\,x) \exp(i\,k\,z_2) \\ 0 \\ -iY_4 \dfrac{q}{k} \sin(q\,x) \exp(i\,k\,z_2) \end{array} \right\}, \tag{4.86}$$

$$\mathbf{H}_{tr} = \left\{ \begin{array}{c} 0 \\ X_4 \exp(i\,k\,z_2) + Y_4 \cos(q\,x) \exp(i\,k\,z_2) \\ 0 \end{array} \right\}, \tag{4.87}$$

where $z_2 = z - d/2$.

We now match the "x"-components of the electric (magnetic) fields $\mathbf{E}_{rf}$, $(\mathbf{H}_{rf})$ and the corresponding components of the inside fields $\mathbf{E}$, $(\mathbf{H})$ at $z = -d/2$ $(z_1 = 0)$. We first equate the terms with the same dependence on the "x"-coordinate to obtain four linear equations connecting X_1, X_2, Y_1, Y_2 and X_3, Y_3. Matching fields $\mathbf{E}_{tr}$, $\mathbf{H}_{tr}$ and $\mathbf{E}$, $\mathbf{H}$ in the $z = d/2$ $(z_2 = 0)$ plane gives other four linear equations connecting, this time connecting, X_1, X_2, Y_1, Y_2 and X_4, Y_4. The eight combined equations for the components of vectors X and Y can be written in a matrix form as

$$\widehat{\mathrm{H}} \cdot \mathrm{X} + g\, \widehat{\mathrm{G}}_1 \cdot \mathrm{Y} = \mathrm{Z}, \qquad \widehat{\mathrm{H}}_q \cdot \mathrm{Y} + g\, \widehat{\mathrm{G}}_2 \cdot \mathrm{X} = 0, \tag{4.88}$$

where vector Z is proportional to the amplitude of the incident field H_0, chosen to be $H_0 = \exp(ikd/2)$ so that

$$\mathrm{Z} = \{1, 1, 0, 0\}. \tag{4.89}$$

The matrices $\widehat{\mathrm{H}}$, $\widehat{\mathrm{H}}_q$, $\widehat{\mathrm{G}}_1$, and $\widehat{\mathrm{G}}_2$ are 4×4 matrices, whose explicit forms are discussed below. At this point, we note that all the matrices remain finite as $g \to 0$.

In the limit of vanishing modulation the matrix $\widehat{\mathrm{H}}$ is represented by

$$\widehat{\mathrm{H}} = \begin{Bmatrix} 1 & \tanh(\frac{dkn}{2}) & -1 & 0 \\ -\frac{i}{n}\tanh(\frac{dkn}{2}) & -\frac{i}{n} & 1 & 0 \\ 1 & -\tanh(\frac{dkn}{2}) & 0 & -1 \\ \frac{i}{n}\tanh(\frac{dkn}{2}) & -\frac{i}{n} & 0 & -1 \end{Bmatrix}. \tag{4.90}$$

When $g = 0$, the first of Eq. (4.88) reduces to $\widehat{\mathrm{H}} \cdot \mathrm{X} = \mathrm{Z}$, which gives the well-known results for the reflectance and the transmittance from the uniform metal films. Indeed, $R = |X_3|^2$, $T = |X_4|^2$ so that

$$R = \left| \frac{1 + n^2}{m^2 - 2i\,n\coth(d\,k\,n)} \right|^2, \tag{4.91}$$

$$T = \left| \frac{2\,n}{2\,n\cosh(d\,k\,n) + im^2\sinh(d\,k\,n)} \right|^2,$$

where

$$m = \sqrt{n^2 - 1} \tag{4.92}$$

(see, e.g., [Landau *et al.*, 1984]).

When $g = 0$, the second of Eq. (4.88) reduces to $\widehat{\mathrm{H}}_q^{(0)}(k) \cdot \mathrm{Y} = 0$, where $\widehat{\mathrm{H}}_q^{(0)}(k) = \widehat{\mathrm{H}}_q(k,\, g = 0)$ and we have shown the explicit dependence of matrix $\widehat{\mathrm{H}}_q^{(0)}$ on the wavevector k. The equation $\widehat{\mathrm{H}}_q^{(0)}(k) \cdot \mathrm{Y} = 0$ has a non-trivial solution when the determinant of $\widehat{\mathrm{H}}_q^{(0)}(k)$ is zero. The condition $\det\left[\widehat{\mathrm{H}}_q^{(0)}(k)\right] = 0$ gives the dispersion equation $q(k)$ for the surface plasmon polariton in a flat metal film that coincides with the Eq. (4.66). Therefore, Eq. (4.88) describes simultaneously transmittance T, reflectance R, and the surface plasmon-polariton excitation.

We investigate now the transmittance, reflectance and the surface plasmon excitation in the modulated film. For a finite film modulation g, the solution to Eqs. (4.88) can be written in the form of

$$\mathrm{X} = \left(\widehat{\mathrm{H}} - g^2\,\widehat{\mathrm{G}}_1 \cdot \widehat{\mathrm{H}}_q^{-1} \cdot \widehat{\mathrm{G}}_2\right)^{-1} \mathrm{Z}. \tag{4.93}$$

Although the second term in the brackets is proportional to $\propto g^2$ $(g \ll 1)$, it cannot be neglected, because $\widehat{\mathrm{H}}_q^{-1}$ is a singular matrix whose eigenvalues can be very large at the resonance. The matrix $\widehat{\mathrm{H}}_q$ has two small eigenvalues D_1 and D_1 that vanish in the limit of zero g. Therefore, the matrix $\widehat{\mathrm{H}}_q^{-1}$ can be decompose as

$$\widehat{\mathrm{H}}_q^{-1} = \frac{|D_1\rangle\,\langle D_1|}{D_1} + \frac{|D_2\rangle\,\langle D_2|}{D_2} + \widehat{\mathrm{H}}_{q,reg}^{-1}, \tag{4.94}$$

where $\langle D_1|$ and $\langle D_2|$ are the eigenvectors. In the considered case of the strong skin effect, when $\exp(-dkn) \ll 1$, the matrix $\widehat{\mathrm{H}}_q^{-1}$ can be written in the following form

$$\widehat{\mathrm{H}}_q^{-1} = \frac{1}{D_1}\left\{\begin{matrix} 1 & -im & 1 & im \\ 0 & 0 & 0 & 0 \\ 1 & -im & 1 & im \\ 1 & -im & 1 & im \end{matrix}\right\}$$

$$+\frac{1}{D_2}\left\{\begin{matrix} 0 & 0 & 0 & 0 \\ 1 & -im & -1 & -im \\ 1 & -im & -1 & -im \\ -1 & im & 1 & im \end{matrix}\right\} + \widehat{\mathrm{H}}_{q,reg}^{-1}, \tag{4.95}$$

where

$$D_{1,2} = \frac{2m\left(1+n^2\right)\Delta}{n} \pm 4\zeta - \frac{g^2 n^2}{2}, \tag{4.96}$$

$$\Delta = k/q - m/n,$$

$$\zeta = \exp\left(-d\sqrt{(kn)^2 + q^2}\right) \simeq \exp\left(-dnq\left(1 + \frac{m}{n}\Delta\right)\right), \tag{4.97}$$

where we assume the detuning $\Delta \ll 1$ and approximate the matrixes $|D_1\rangle\,\langle D_1|$ and $|D_2\rangle\,\langle D_2|$ in Eq. 4.95 by their values for the flat thick film, i.e., for $\Delta = 0$, $g = 0$, $\zeta = 0$. Equations $D_{1,2} = 0$ for the singularities of the matrix $\widehat{\mathrm{H}}_q^{-1}$ give the dispersion equations for the surface plasmons in the modulated film. When the modulation g vanishes Eq. (4.96) coincides with

Eq. (4.66) for the flat metal film. Note, that Eq. (4.97) for the dimensionless resonance Δ can be written as

$$\Delta = \frac{a}{\lambda} - \frac{\sqrt{n^2-1}}{n}, \tag{4.98}$$

where λ is the wavelength of the incident light and a is the period of the film permittivity modulation.

Since the term $g^2\ \widehat{\mathrm{G}}_1 \cdot \widehat{\mathrm{H}}_q^{-1} \cdot \widehat{\mathrm{G}}_2$ of Eq. (4.93) is proportional to g^2 we can neglect the regular part $\widehat{\mathrm{H}}_{q,reg}^{-1}$ of matrix $\widehat{\mathrm{H}}_q^{-1}$ as well as other terms proportional to g^2 in matrices $\widehat{\mathrm{H}}$, $\widehat{\mathrm{G}}_1$ and $\widehat{\mathrm{G}}_2$. The latter are given by Eq. (4.90) and by the following equations:

$$\widehat{\mathrm{G}}_1 = \left\{ \begin{matrix} -n^2/2 & -n^2/2 & 0 & 0 \\ im/2 & im/2 & 0 & 0 \\ -n^2/2 & n^2/2 & 0 & 0 \\ -im/2 & im/2 & 0 & 0 \end{matrix} \right\}, \tag{4.99}$$

$$\widehat{\mathrm{G}}_2 = \left\{ \begin{matrix} m^2 & m^2 & 0 & 0 \\ -in & -in & 0 & 0 \\ m^2 & m^2 & 0 & 0 \\ in & -in & 0 & 0 \end{matrix} \right\},$$

where m is given by Eq. (4.92). (In derivation of Eq. (4.99) we have assumed that $\zeta \ll 1$.)

Substitution of the explicit forms for the vector Z [Eq. (4.89)] and the matrixes $\widehat{\mathrm{H}}$, $\widehat{\mathrm{H}}_q^{-1}$, $\widehat{\mathrm{G}}_1$, $\widehat{\mathrm{G}}_2$ [Eqs. (4.90), (4.95), and (4.99)] into Eq. (4.93) fully determines the vector X and its components: the transmittance $T = |X_3|^2$ and the reflectance $R = |X_4|^2$ (see Eq. (4.83)). To simplify further consideration, we neglect the off-resonant (direct) transmittance, which is proportional $\sim \zeta^2$, i.e., we set $\tanh(dkn/2) = 1$ in the matrix $\widehat{\mathrm{H}}$ given by Eq. (4.90). Direct transmittance accounting results in small corrections for the effective transmittance obtained below [Genchev and Dosev, 2004]. We obtain from Eqs. (4.93), (4.90), (4.95), and (4.99) the simplified expressions for the resonant transmittance,

$$T(\tilde{\Delta}) = \frac{4\tilde{g}^4}{\left[\left(\tilde{\Delta}-1\right)^2 + (\tilde{g}^2+\tilde{\kappa})^2\right]\left[\left(\tilde{\Delta}+1\right)^2 + (\tilde{g}^2+\tilde{\kappa})^2\right]}, \tag{4.100}$$

the reflectance,

$$R(\tilde{\Delta}) = \frac{\left(\tilde{g}^4 + \tilde{\Delta}^2 - 1\right)^2 - 2\left(\tilde{g}^4 - \tilde{\Delta}^2 - 1\right)\tilde{\kappa}^2 + \tilde{\kappa}^4}{\left[\left(\tilde{\Delta} - 1\right)^2 + (\tilde{g}^2 + \tilde{\kappa})^2\right]\left[\left(\tilde{\Delta} + 1\right)^2 + (\tilde{g}^2 + \tilde{\kappa})^2\right]}, \quad (4.101)$$

and the absorptance,

$$\begin{aligned} A(\tilde{\Delta}) &= 1 - T(\tilde{\Delta}) - R(\tilde{\Delta}) \\ &= \frac{4\,\tilde{g}^2\,\tilde{\kappa}\left(1 + \tilde{\Delta}^2 + (\tilde{g}^2 + \tilde{\kappa})^2\right)}{\left[\left(\tilde{\Delta} - 1\right)^2 + (\tilde{g}^2 + \tilde{\kappa})^2\right]\left[\left(\tilde{\Delta} + 1\right)^2 + (\tilde{g}^2 + \tilde{\kappa})^2\right]}. \quad (4.102) \end{aligned}$$

Note that all three expressions depend on the renormalized detuning from the SPP frequency,

$$\tilde{\Delta} = g^2 \frac{n\,(n - m)^2\,(n + n^3 + 2m)}{8\,(1 + n^2)\,\zeta} - \frac{\Delta}{\zeta}\frac{m\,(1 + n^2)}{2n}, \quad (4.103)$$

renormalized modulation,

$$\tilde{g}^2 = \frac{g^2 n^2 m\,(n - m)^2}{4\,(1 + n^2)\,\zeta}, \quad (4.104)$$

and the renormalized loss in the system,

$$\tilde{\kappa} = \frac{(1 + n^2)\,\kappa}{4n^2\zeta}. \quad (4.105)$$

Recall $\kappa = \varepsilon''_m / |\varepsilon_m|$, parameters m, Δ, and ζ are given by Eqs. (4.92), (4.97), and (4.98), respectively.

In Eqs. (4.100)–(4.105), we do take into account the loss by writing the metal dielectric constant ε_m in the form of $\varepsilon_m = -n^2(1 - i\kappa)$, where $n \gg 1$ while $\kappa \ll 1$. Note also that the transmittance T, reflectance R, and absorptance A also depend on the frequency detuning from the incident field, Δ. The detuning is proportional to $\propto \omega - \omega_r$, where ω is the incident field frequency, and $\omega_r = qcm/n$ is the resonance frequency, at which the wavelength of SPP excited on a flat metal surface coincides with spatial period of the film modulation.

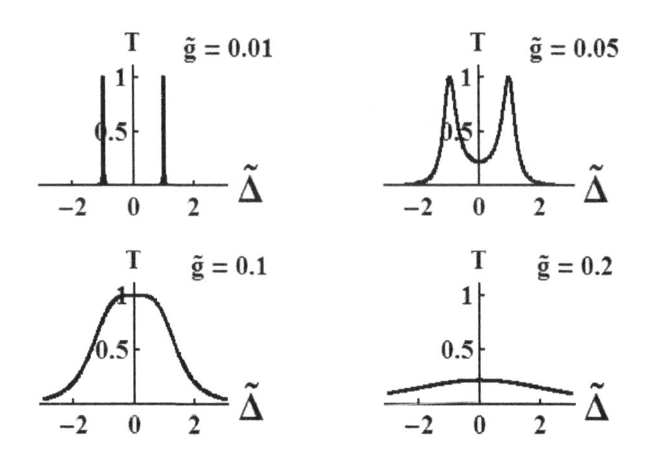

Fig. 4.8 Resonance transmittance as function of detuning $\tilde{\Delta}$, given by Eq. (4.103). The different graphs correspond to different film modulations $\tilde{g}$.

We first analyze the resonant transmittance (4.100) by ignoring the loss. We set $\tilde{\kappa} = 0$ in Eq. (4.100), which then simplifies to

$$T(\tilde{\Delta}) = \frac{4\tilde{g}^4}{\left[\left(\tilde{\Delta} - 1\right)^2 + \tilde{g}^4\right]\left[\left(\tilde{\Delta} + 1\right)^2 + \tilde{g}^4\right]}. \tag{4.106}$$

For the renormalized film modulation, $\tilde{g} < 1$, as follows from Eq. (4.106), the resonance transmittance $T(\tilde{\Delta})$ has two absolute maxima as the function of the SPP detuning $\tilde{\Delta} : T(\tilde{\Delta}_1) = T(\tilde{\Delta}_2) = 1$ at $\tilde{\Delta}_{1,2} = \pm\sqrt{1 - \tilde{g}^4}$. This means that lossless, optically thick metal film becomes fully transparent at the resonance, regardless of its thickness. It is instructive to consider how the transmittance changes when the modulation $\tilde{g}$ increases.

As seen in Fig. 4.8, the film is completely transparent at the resonance when $\tilde{g} < 1$. When $\tilde{g}$ increases and becomes larger than one, the two maxima merge together to one, with the amplitude $T_m = 4\tilde{g}^4/\left(1 + \tilde{g}^4\right)^2 < 1$. T_m *decreases* at further increase of $\tilde{g}$. This result can be easily understood if we recall that the field interaction with the film results in the radiation decay of the surface plasmon polaritons and the process of conversion into a plane-wave radiation. The radiative loss [term $\sim \tilde{g}^4$ in the denominator of Eq. (4.106)] lead to the increasing damping of SPPs. As a result, the resonant transmittance decreases with a increase of the renormalized modulation $\tilde{g}$.

The renormalized modulation $\tilde{g}$, given by Eq. (4.104), exponentially increases with film thickness $d : \tilde{g} \sim g \exp(dnq/2)$. Similar to the discussion above, the transmittance maxima merge when the thickness d increases. When d becomes larger than $d > d_c \sim 2\ln(1/g)/(nq)$, there is only one maximum in the transmittance. The maximum decays exponentially with the further increase of the film thickness d. This resonant transmittance behavior is in a qualitative agreement with the results of [Martin-Moreno *et al.*, 2001] and [Tan *et al.*, 2000].

In a real metal film, loss decreases the resonant transmittance. Yet, the illustrated above effect may remain rather profound. As it follows from Eq. (4.100), the transmittance reaches its extreme value in loss film when the renormalized detuning $\tilde{\Delta}$ is

$$\tilde{\Delta}_{0,1,2} = \{0, -\sqrt{1 - (\tilde{g}^2 + \tilde{\kappa})^2}, \sqrt{1 - (\tilde{g}^2 + \tilde{\kappa})^2}\}. \tag{4.107}$$

Thus for $\tilde{g}^2 + \tilde{\kappa} < 1$, the transmittance has two maxima $T_{\max} = \tilde{g}^4/\left(\tilde{g}^2 + \tilde{\kappa}\right)^2$at $\tilde{\Delta}_{1,2} = \pm\sqrt{1 - (\tilde{g}^2 + \tilde{\kappa})^2}$. By substituting the renormalized modulation $\tilde{g}$ (given by Eq. (4.104)) and the renormalized loss $\tilde{\kappa}$ (Eq. (4.105)) into $T_{\max} = \tilde{g}^4/\left(\tilde{g}^2 + \tilde{\kappa}\right)^2$, we obtain the equation for the resonant transmittance, which does not depend on the film thickness d:

$$T_{\text{res}} = \left(1 + \frac{(m+n)^2 \left(1+n^2\right)^2 \kappa}{g^2 \, m \, n^4}\right)^{-2}, \tag{4.108}$$

Thus, at the resonance, the metal film becomes semi-transparent, even for a rather large thicknesses, provided that

$$\tilde{g}^2 + \tilde{\kappa} = \frac{e^{d\,n\,q}\left(g^2 \, m \, n^4 \, (n-m)^2 + \left(1+n^2\right)^2 \kappa\right)}{4 n^2 \left(1+n^2\right)} < 1, \tag{4.109}$$

where we approximate $\zeta \simeq \exp(-dnq)$. The transmittance decays exponentially with increasing d for the film thickness $d > d_c$ where the critical thickness d_c is obtained from the condition $\tilde{g}^2 + \tilde{\kappa} = 1$ (see Eq. (4.109)).

The resonance transmittance T_{res} in Eq. (4.108) reaches its maximum at $n = n^* = 1.78$, where it can be approximated by $T_{\text{res}} \simeq 1/(1 + 12.4\kappa/g^2)^2$. It is interesting to note that the value of the optimal "refractive index" $n^* = 1.78$ depends neither on the film nodulation g nor on loss κ.

The resonant transmittance is also accompanied by large internal fields due to the excitation of SPPs. These fields (discussed in more details in the next section) are in charge of the anomalous light absorption that occurs at the resonance ($\tilde{\Delta} = \tilde{\Delta}_{1,2}$) (also see Eq. (4.107)), when the film becomes semi-transparent):

$$A_{\text{res}} = 1 - T_{\text{res}} - R_{\text{res}} = \frac{2\,(m+n)^2\,(1+n^2)^2\,\kappa}{g^2\,m\,n^4\left(1+\dfrac{(m+n)^2\,(1+n^2)^2\,\kappa}{g^2\,m\,n^4}\right)^2}. \tag{4.110}$$

It follows from Eqs. (4.102) – (4.105) that for the optimal value $n^* = 1.78$, the resonant absorptance A_{res} is given by

$$A_{\text{res}} \simeq \left(\kappa/g^2\right) / \left(0.2 + 2.5\kappa/g^2\right)^2. \tag{4.111}$$

A_{res} depends on the ratio of loss-factor κ and the square of the film modulation, g^2. It then reaches the maximum $A_{\text{max}} = 1/2$ for the modulation $g = 3.53\sqrt{\kappa}$ and remains the same, even when the loss and the modulation vanish in metal i.e., when $\kappa \to 0$, $g \to 0$; $g = 3.53\sqrt{\kappa}$. In this case, the amplitudes of the EM fields increase infinitely but the absorptance remains finite. This is because the radiative damping of SPPs, which is proportional to the modulation g, decreases with decreasing the loss factor κ in order to keep the overall absorptance at the same level.

In Figs. 4.9 and 4.10, we show the transmittance and absorptance of a rather thick silver film, with $d = 0.12$ μm, the spatial period $a = 0.5$ μm and the modulation given by Eq. (4.67). As seen in the figures, the transmittance and the absorptance have resonances at the incident wavelength $\lambda \simeq 0.53$ μm. (In our estimates, we used the Drude formula (3.1) for the dielectric constant of silver; which approximates the experimental data

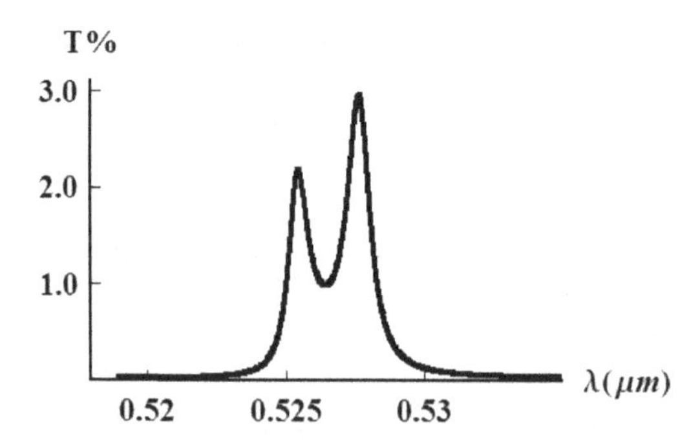

Fig. 4.9 Transmittance of modulated silver film; sample thickness $d = 0.12$ μm, spatial period of modulation is 0.5 μm, modulation amplitude $g = 0.2$.

rather well for $\lambda > 0.4$ μm, see, for example, [Johnson and Christy, 1972] and [Kreibig and Volmer, 1995].) In the absence of the modulation, the film acts as a nearly perfect mirror, with the reflectance $R = 99\%$ and the minimal transmittance $T < 0.02\%$. The situation changes dramatically when the film's dielectric constant is modulated. For the spatial period of the modulation $a = 2\pi/q = 0.5$ μm, we obtain from the direct solution of Eq. (4.93) the absorption levels of 30% at $\lambda \simeq 0.53$ μm. The transmittance also increases by two orders of magnitude: $T \simeq 4\%$. The absorptance and transmittance have a double-peak structure corresponding to the resonant excitation of symmetric and antisymmetric SPPs. Therefore the proper modulated metal film reveals EOT with the enhancement $G > 10^2$.

For real metal films, the transmittance typically increases when loss decreases, as it follows from Eq. (4.108). It is instructive to consider a

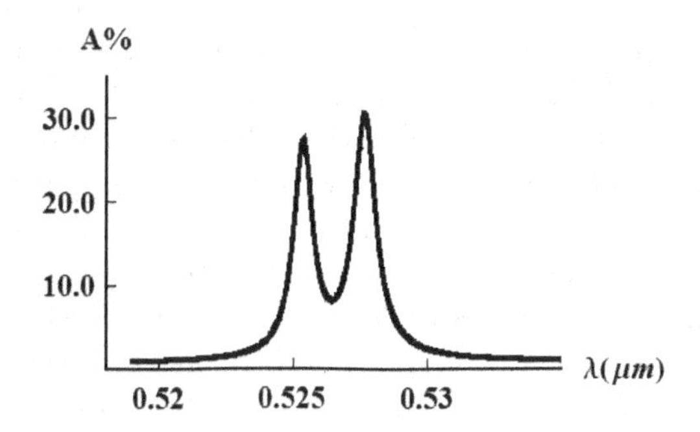

Fig. 4.10 Absorptance of modulated silver film; film thickness $d = 0.12$ μm, spatial period of modulation 0.5 μm, amplitude of modulation $g = 0.2$.

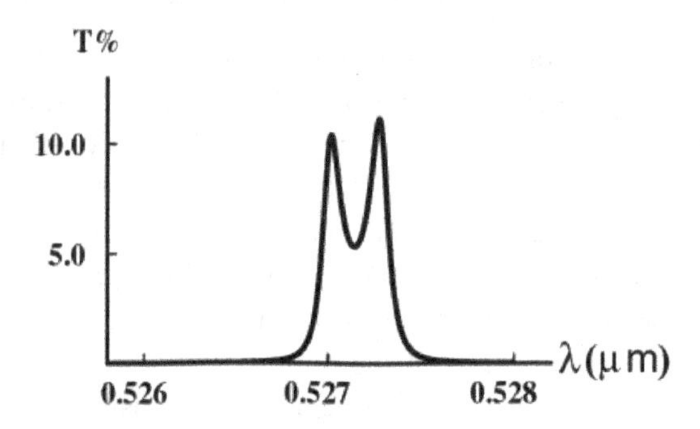

Fig. 4.11 Film transmittance at "cryogenic" temperatures. The film is silver; with the thickness $d = 0.17$ μm and the modulation $g = 0.1$.

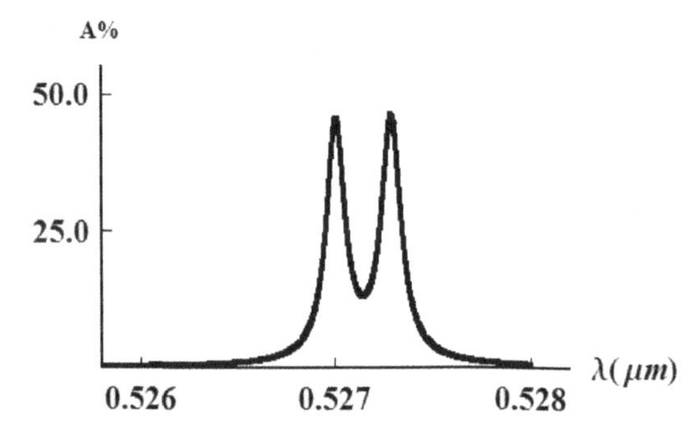

Fig. 4.12 Absorptance for "cryogenic" silver film; film thickness is $d = 0.17$ μm, and modulation is $g = 0.1$.

silver film of high (atomic) quality at cryogenic temperatures (such that the electron mean-free pass is restricted by the film thickness itself). The resonant transmittance and absorptance for such a film with thickness $d = 0.18$ μm are shown in Figs. 4.11 and 4.12, respectively. In the calculations, we assumed that loss-factor κ is ten times smaller than that one in Figs. 4.9 and 4.10. We see that the resonant absorptance increases up to $A_{\mathrm{res}} = 40\%$, while the resonance transmittance exceeds $T_{\mathrm{res}} \gtrsim 10\%$. The width of the resonance shrinks to about 1 nm. Note that without the modulation, the film has a negligible transmittance $T < 10^{-3}\,\%$ both in the optical and infrared spectral ranges ($\lambda > 0.4$ μm). Therefore, the transmittance T increases by four orders of magnitude from 10^{-3} % to 10% due to the SPP excitation in the film.

4.5.2 *Light-induced and light-controlled transmittance*

In the previous subsection, we assumed that the film's modulation was pre-fabricated. Here, we consider the case of nonlinear films, at which the film's modulation is induced, or modified (and sometimes controlled) by the light as it can be, for example, in the case of the Kerr nonlinearity. Again, the dielectric permittivity of a film has a small "seed" modulation given by Eq. (4.67), where the amplitude of the modulation $g \ll 1$. We then investigate how the index-profile (dielectric permittivity profile) modulation increases due to the nonlinearity of the film.

Exactly at the resonance, the transmitted intensity, $I_t = TI_0$, is of the same order of magnitude as the intensity of the incident wave, $I_0 = (c/8\pi)\left|E_0\right|^2$, where E_0 is the incident electric field. The transmitted wave is generated by the surface plasmon polariton, which propagates on the front (output) film interface ($z = d/2$). (As the SPP "senses" the film's modulation g while propagating, it converts its EM energy back to the light, which is emitted from the front surface.) The amplitude of SPP-emitted-light conversion is proportional to the film modulation g. Therefore, the SPP intensity $I_q \equiv |E_q|^2$ can be estimated from the equation $I_t \sim g^2 I_q$, as $I_q \sim I_t/g^2 \sim I_0/g^2 \gg I_0$. At the back (input) interface ($z = -d/2$), the SPP intensity I_q is on the same order of magnitude (see the discussion following Eq. (4.66)). The electric field E_q of the SPP is spatiality modulated, with the resonant wavevector $k_p = q$. Note, that enhanced field E_q is responsible for the anomalous resonance absorption, which could be many orders of magnitude larger than the absorption in flat metal films (see Figs. 4.9–4.12).

Since the electric field E_q of the SPP is strongly enhanced it makes sense to account for a possible nonlinearity in the optical response of the metal film. To be specific, we consider the Kerr optical nonlinearity. We assume that the electric displacement $\mathbf{D}_{in}$ in the film equals to

$$\mathbf{D} = \varepsilon_m \mathbf{E} + 12\pi\chi^{(3)}\mathbf{E}\,|\mathbf{E}|^2\,, \tag{4.112}$$

where $\chi^{(3)}$ is the Kerr susceptibility discussed in Sec. 3.6 (see also [Boyd, 1992]). We substitute the electric field $\mathbf{E}$ in the film given by Eqs. (4.74) and (4.75) into the electric displacement vector equation and obtain that the nonlinear dielectric susceptibility:

$$\varepsilon \cong -n^2\left[1 + g_i(E)\cos qx\right]^{-1}, \tag{4.113}$$

where g_i is the induced modulation,

$$g_i = g + \frac{24\pi}{n^2}\chi^{(3)}\,\mathrm{Re}(E^* E_q). \tag{4.114}$$

Equation (4.113) is similar to Eq. (4.67) except the film modulation now depends on the internal electric field.

We restrict our consideration to the nonlinearity of the film modulation only and neglect small nonlinear corrections in the numerator of Eq. (4.113). To further simplify the consideration, we replace the field term $\mathrm{Re}(E^* E_q)$ in Eq. (4.114) by its average over the film thickness $\langle \mathrm{Re}(E^* E_q)\rangle$. We use the average because the fundamental field as well as the SPP has nearly the same amplitude on both interfaces of the film at resonance. Exactly at the resonance $\tilde{\Delta} = \tilde{\Delta}_{1,2}$ (see Eq. (4.107)), the term $\langle \mathrm{Re}(E^* E_q)\rangle$ acquires the following form:

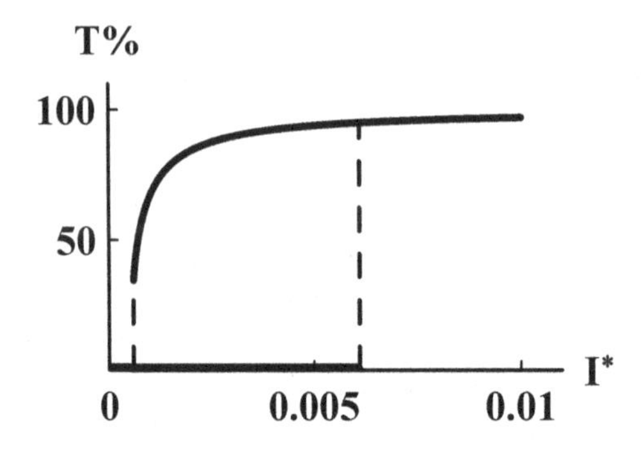

Fig. 4.13 Transmittance through "nonlinear" silver film as function of the dimensionless light intensity; film thickness is $d = 0.17\mu$m, and the "seed" modulation $g = 10^{-2}$.

$$\langle \mathrm{Re}(E^* E_q) \rangle_{\mathrm{res}} = \frac{64\pi}{c} \frac{\left(m^2 + n^4 + m\,\left(n + n^3\right)\right)}{d\,g\,m\,n^2\,q} T_{res} I_0 \tag{4.115}$$

where I_0 is the intensity of the incident light, T_{res} is the resonant transmittance given by Eq. (4.108), parameter m is given by Eq. (4.92), and d and k are the film thickness and the incident field wavevector, respectively.

It follows from Eqs. (4.114) and (4.115) that the SPP field induces the refractive-index modulation g_i via the Kerr optical nonlinearity. The induced modulation increases the transmittance and, therefore, the SPP's field amplitude. This positive feedback may eventually result in a bistability phenomenon, shown in Fig. 4.13.

We substitute Eqs. (4.108) and (4.115) in Eq. (4.114) to obtain the following equation for the induced refractive-index modulation:

$$g_i = g + \frac{g_i^3\,\left(m^2 + n^4 + m\,\left(n + n^3\right)\right)}{d\,m\,n^2\,q\,\left(g_i^2 + \frac{(m+n)^2\,\left(1+n^2\right)^2\,\kappa}{m\,n^4}\right)^2} I^* \tag{4.116}$$

which holds at the resonance $\tilde{\Delta} = \tilde{\Delta}_{1,2}$ (see Eq. 4.107). Here, we have introduced the dimensionless incident light intensity, $I^* = 768\,\pi^2\,\chi^{(3)} I_0/c$. Now we solve Eq. (4.116) with the same parameters n and κ, which we have used before to obtain Figs. 4.11 and 4.12. Thus we calculate the light-induced modulation g_i as a function of dimensionless intensity I^*. Being acquainted with modulation g_i we calculate the transmittance shown in Fig. 4.13.

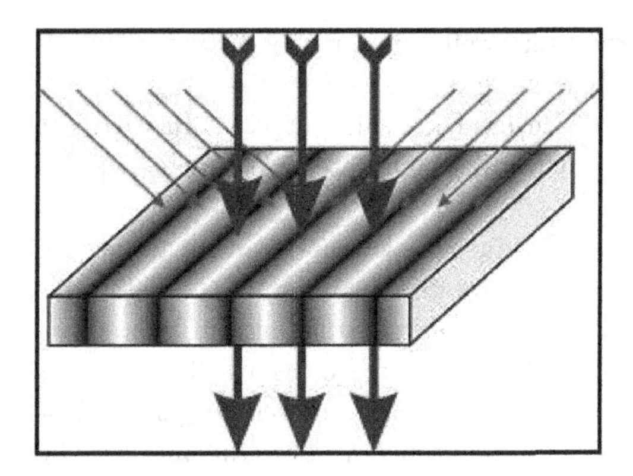

Fig. 4.14 Extraordinary optical transmittance created by two control laser beams, shown in red.

When the dimensionless incident light intensify I^* becomes larger than $I_1^* \simeq 6 \cdot 10^{-3}$, the transmittance jumps from a nearly zero up to 100%, i.e., the film suddenly becomes transparent (see Fig. 4.13). If we now decrease the incident light intensity, the film remains transparent, even for $I^* < I_1^*$ simply because the SPP has been already excited in the film. The transmittance decreases steeply only for $I^* \sim 10^{-3} < I_1^*$. Thus, the optical bistability phenomenon occurs in the modulated metal film.

It is well known that the susceptibility $\chi^{(3)}$ is rather large for noble metals, $\chi^{(3)} > 10^{-8}$ *esu*, (see, for example, [Flytzanis, 1996], [Debrus *et al.*, 2000], [Bennett *et al.*, 1999], [Ma *et al.*, 1998], [Liao *et al.*, 1998]) so that the intensity I_0 required for the bistability can be easily achieved with conventional lasers. [Note that the upper curve in Fig. 4.13 should be considered as the extrapolation since we have restricted the expansion of the nonlinear permittivity $\varepsilon\left(|E|^2\right)$ in the series of $|E|^2$ to the first two terms only (up to $O(|E|^4)$.]

We also note that the seed modulation g can be created by two additional, control laser beams (see Fig. 4.14), which are incident on the front surface from the different sides with respect to the normal. The interference between the controlling beams would result in the modulation of the film refractive index (and the film susceptibility) through the film's optical nonlinearity in the film. Thus, these beams can control the transmittance of the fundamental beam, propagating normal to the film. To provide a modulation in the refractive index, one can use a thin layer of highly nonlinear (dielectric) material placed on top of the metal film.

Photorefractive quantum-well structures (see, e.g., [Nolte, 1999]), which are known to produce large refractive-index grating at very low intensities (below 1 mW), can be used for this purpose. In this case, the required modulation at the metal-dielectric interface can be accomplished at very low intensities.

Recently, the photoinduced resonance transmittance was demonstrated in the exciting experiment by [Smolyaninov *et al.*, 2005], where the authors reported the observation of the transmittance of light through a gold film deposited on a chalcogenide glass surface. This effect is caused by formation of photoinduced diffraction grating in the chalcogenide glass near the gold film surface by an optical pump beam. The transmittance of probe beam is resonantly enhanced due to grating-induced coupling to surface electromagnetic excitations on the gold film surface. This observation demonstrates the feasibility of all-optical signal processing using the effect of the extraordinary light transmittance through thin metal films.

After the prediction of the EOT bistability was published [Dykhne *et al.*, 2003] the bistability was obtained in the computer simulations [Porto *et al.*, 2004] of the em transmittance through the metal plate with regular array of the slits that were filled with a nonlinear medium. Later the EOT bistability was detected by [Wurtz *et al.*, 2006] in real experiment for the periodically nanostructured metal film (SPP crystal) coated with a nonlinear polymer.

4.5.3 *Discussion*

We have shown that the excitation of surface plasmon-polaritons in modulated metal films could result in the enhanced resonant transmittance so that an optically thick film becomes semi-transparent. At resonance, the transmittance can be increased by 10^4 times. The transmittance maximum inherits a double-peak structure due to the splitting of SPPs into symmetric and antisymmetric modes. The resonant transmittance increases with a decrease of the system loss; the loss decrease can be accomplished, for example, by cooling the film to the cryogenic temperatures. The resonance amplitude of the SPP field can be larger than the amplitude of the incident field by several orders of magnitude. The optical nonlinearities can become important due to such amplification and, in their turn, result in significant enhancement of EOT. The film can manifest the optical bistability phenomenon at sufficiently large intensities of the incident light. Films with such bistable behavior at the resonant regime can be used, for example, as optical switches. The modulation and, therefore, film's extraordinary transmittance, can be tuned dynamically by using, for example, auxiliary light beams and employing the optical nonlinearity of the film, making use of such films as active optical devices.

Chapter 5

Electromagnetic Properties of Metal-dielectric Crystals

In this chapter, we will consider the electromagnetic properties of bulk metal-dielectric materials irradiated by a high-frequency electromagnetic field under the conditions when the skin effect is strong in metal grains. Two different classes of metal-dielectric systems are analyzed: the percolation composites and artificial electromagnetic crystals [Sievenpiper *et al.*, 1996], [Pendry *et al.*, 1996]. The electromagnetic crystals are three-dimensional periodic structures with metal inclusions embedded in a dielectric host. These crystals are similar to well-known photonic crystals composed of periodic structures of dielectric particles. At high frequencies, when the metal periodic inclusions can sustain plasmon excitations, the electromagnetic crystals can also be referred to as plasmonic crystals.

Three-dimensional metal-dielectric percolation composites and three-dimensional electromagnetic crystals differ very much at the first glance. Still, we show below that the electromagnetic properties of both random composites and electromagnetic crystals can be understood in terms of the effective permittivity and permeability, provided that the wavelength of the incident wave is much larger than an intrinsic scale of the system under consideration. The wavelength inside the metal component can be very small. This is important because the most interesting effects are expected in the limit of the strong skin effect. As we will see, the methods for calculating the effective permittivity and permeability are essentially the same for composites and electromagnetic crystals. Moreover, the results for the effective permittivity and permeability are also, to large extent, similar.

In order to calculate these effective parameters, we follow a simple method suggested in Refs. [Lagarkov *et al.*, 1987], [Vinogradov *et al.*, 1989b], [Lagarkov *et al.*, 1992], [Rousselle *et al.*, 1993], and [Sarychev *et al.*, 2000b]. This approach is similar to [Smith *et al.*, 1999] and to a dualistic approximation by [Maslovski *et al.*, 2002]. More sophisticated approach for the effective parameters of the wire mesh crystals was suggested by [Pokrovsky and Efros, 2002], [Pokrovsky and Efros, 2003], [Efros and Pokrovsky, 2004], [Pokrovsky, 2004]. A general form of material equations in the electrodynamic description of metamaterials was considered by [Vinogradov and Aivazyan, 1999], [Vinogradov, 2002]. Computer simulations and real experiments on the wire mesh crystals can be found in [McGurn and Maradudin, 1993], [Smith *et al.*, 1994], [Sigalas *et al.*, 1995], [Pendry *et al.*, 1998], [Smith *et al.*, 2000], [Bayindir *et al.*, 2001], [Hafner *et al.*, 2005], [Ricci *et al.*, 2005]. Two- dimensional wire mesh structures have been known for a long time in microwave engineering and science, where they were called as *inductive frequency selective surfaces* [Munk, 2000]. Recent interest to the plasmonic crystals was boosted by the possibility to obtain metamaterials with negative dielectric constant [Sievenpiper *et al.*, 1996] and negative refraction index [Shelby *et al.*, 2001], [Smith *et al.*, 2004a] (also see discussion in Sec. 2.5.)

5.1 Metal-dielectric Composites

We consider the propagation of electromagnetic waves in percolation composites consisting of metal and dielectric particles as it is illustrated at Fig. 3.2. The typical scale of inhomogeneity in these composites is known to be represented by a percolation correlation length ξ_p, which is a typical size of metal cluster. The propagation of the EM wave with a wavelength λ less than the percolation correlation length ξ_p may be accompanied by strong scattering [see discussion in Sec. 3.4]. On the other hand, the wave propagation at $\lambda \gg \xi_p$ can be described by Maxwell's equations with the effective permittivity ε_e and effective permeability μ_e. We will discuss below: i) what sense have the effective parameters ε_e and μ_e when the skin effect is strong and ii) how to express the composite effective permittivity and permeability ε_e and μ_e in terms of individual permittivities and permeabilities of the metal and dielectric (ε_m, ε_d, μ_m, μ_d) as well as the size a of metal inclusions.

When we are interested in the balk permittivity ε_e and permeability μ_e, the consideration can be restricted to the optically thin metamaterial of size $\mathcal{L} \ll \lambda/\sqrt{|\varepsilon_e\mu_e|}$, which is still homogeneous from the percolation point of view ($\mathcal{L} \gg \xi_p$). Suppose that the composite is placed inside a resonator (e.g., metal sphere), which has the resonance mode excited at the frequency

ω. It is assumed that the volume V of the resonator is much larger than the characteristic volume of the composite $v \sim \mathcal{L}^3$. Since the composite is described in terms of the effective parameters, the macroscopic Maxwell equations, i.e., the equations averaged over a spatial scale larger than the correlation length ξ_p, can be written in the following form

$$\operatorname{curl} \mathbf{E} = ik\mu\mathbf{H}, \tag{5.1}$$

$$\operatorname{curl} \mathbf{H} = -ik\varepsilon\mathbf{E}, \tag{5.2}$$

where $k = \omega/c$ is a wave vector. Inside the composite, $\varepsilon = \varepsilon_e$ and $\mu = \mu_e$. Outside the composite, $\varepsilon = \mu = 1$. Inside, the electric $\mathbf{E}$ and magnetic $\mathbf{H}$ fields coincide with macroscopic fields $\mathbf{E}_0$ and $\mathbf{H}_0$ in the composite. The fields $\mathbf{E}_1$ and $\mathbf{H}_1$ in the *empty* resonator satisfy the Maxwell equations

$$\operatorname{curl} \mathbf{E}_1 = ik_1\mathbf{H}_1, \tag{5.3}$$

$$\operatorname{curl} \mathbf{H}_1 = -ik_1\mathbf{E}_1, \tag{5.4}$$

where $k_1 = \omega_1/c$ and ω_1 is the resonance frequency of the empty resonator. The shift of the resonance frequency $\Delta\omega = \omega_1 - \omega$ is small $|\Delta\omega|/\omega \ll 1$ since the volume of the resonator is much larger than the composite volume v. The frequency shift $\Delta\omega$, which is proportional to the permittivity of the material placed in the resonator, is widely used for the dielectric measurements (see, e.g., [Chen *et al.*, 2004] and [Hock *et al.*, 2004].)

Let us multiply Eq. (5.1) by $\mathbf{H}_1^*$, and Eq. (5.2) by $-\mathbf{E}_1^*$; the complex conjugated Eqs. (5.3) and (5.4) are multiplied by $\mathbf{H}$ and by $-\mathbf{E}$, respectively. After the multiplication, all the four equations are summed together to receive

$$\begin{aligned}\mathbf{H}_1^* \operatorname{curl} \mathbf{E} - \mathbf{E} \operatorname{curl} \mathbf{H}_1^* + \mathbf{H} \operatorname{curl} \mathbf{E}_1^* - \mathbf{E}_1^* \operatorname{curl} \mathbf{H} \\ = i\left(k\varepsilon - k_1\right)\mathbf{E}\mathbf{E}_1^* + i\left(k\mu - k_1\right)\mathbf{H}\mathbf{H}_1^*.\end{aligned} \tag{5.5}$$

The left hand side (l.h.s.) of Eq. (5.5) can be written as

$$\operatorname{div}\left([\mathbf{H}_1^* \times \mathbf{E}] + [\mathbf{E}_1^* \times \mathbf{H}]\right). \tag{5.6}$$

If Eq. (5.5) is integrated over the resonator volume, the integral in l.h.s. transforms to the surface integral, which vanishes since the tangential component of the electric field is zeroed on the resonator walls. (It is assumed that resonator is made of a perfect conductor). The integral of the r.h.s. of Eq. (5.5) can be rearranged:

$$\int [(\varepsilon - 1)\,\mathbf{E}\cdot\mathbf{E}_1^* + (\mu - 1)\,\mathbf{H}\cdot\mathbf{H}_1^*]\,d\mathbf{r} = \frac{\Delta\omega}{\omega}\int (\mathbf{E}\cdot\mathbf{E}_1^* + \mathbf{H}\cdot\mathbf{H}_1^*)\,d\mathbf{r}. \tag{5.7}$$

The expressions $(\varepsilon - 1)$ and $(\mu - 1)$ in l.h.s. of Eq. (5.7) is nonzero only in the composite. Therefore the integral in l.h.s. of the Eq. (5.7) can be taken just over the volume v, where $\varepsilon = \varepsilon_e$ and $\mu = \mu_e$. Since the size of the composite is much smaller than the wavelength, we can replace the fields $\mathbf{E}_1(\mathbf{r})$ and $\mathbf{H}_1(\mathbf{r})$ by the fields $\mathbf{E}_{10} = \mathbf{E}_1(\mathbf{r}_0)$ and $\mathbf{H}_{10} = \mathbf{H}_1(\mathbf{r}_0)$, where the vector $\mathbf{r}_0$ represents the composite position in the resonator. Recall that $\mathbf{E}_1(\mathbf{r})$ and $\mathbf{H}_1(\mathbf{r})$ are the empty resonator fields. Since the volume v of the composite is much smaller than the volume of the resonator, the maim contribution to the r.h.s. of Eq. (5.7) comes from the distances much larger then the size $\mathcal{L}$ (the composite size). At these distances, the field perturbation due to the composite presence can be neglected so that $\mathbf{E} \approx \mathbf{E}_1$ and $\mathbf{H} \approx \mathbf{H}_1$ in the r.h.s. of integral (5.7). Then the frequency shift estimates from Eq. (5.7) as

$$\frac{\Delta\omega}{\omega} = \frac{\int_v [(\varepsilon_e - 1)\,\mathbf{E}_0(\mathbf{r})\cdot\mathbf{E}_{10}^* + (\mu_e - 1)\,\mathbf{H}_0(\mathbf{r})\cdot\mathbf{H}_{10}^*]\,d\mathbf{r}}{\int \left(|\mathbf{E}_1(\mathbf{r})|^2 + |\mathbf{H}_1(\mathbf{r})|^2\right)d\mathbf{r}}, \tag{5.8}$$

where $\mathbf{E}_0(\mathbf{r})$ and $\mathbf{H}_0(\mathbf{r})$ are the macroscopic fields in the composite. The particular field distribution $\mathbf{E}_1(\mathbf{r})$ and $\mathbf{H}_1(\mathbf{r})$ can be calculated for the empty resonator geometry. The field distributions $\mathbf{E}_0(\mathbf{r})$ and $\mathbf{H}_0(\mathbf{r})$ are unambiguously determined in the composite by the effective parameters ε_e and μ_e and the shape of the composite. Therefore, we can determine the effective parameters ε_e and μ_e by the measurements of the frequency shift $\Delta\omega$ made sampled at different positions of the composite in the resonator.

Suppose that the composite is placed at a maximum of the electric field in the resonator. The magnetic field $\mathbf{H}_{10} = 0$ (corresponding to the maximum of $\mathbf{E}_{10}$), moreover, since the composite is optically thin $\mathcal{L} \ll \lambda/\sqrt{|\varepsilon_e\mu_e|}$, the magnetic field $\mathbf{H}_0$ inside the composite is almost zero also. Therefore, Eq. (5.7) takes the form

$$\frac{\Delta\omega_E}{\omega} = \frac{(\varepsilon_e - 1)\int_v \mathbf{E}_0\cdot\mathbf{E}_{10}^* d\mathbf{r}}{2\int |\mathbf{E}_1|^2\,d\mathbf{r}}, \tag{5.9}$$

where we used the equality $\int |\mathbf{E}_1|^2\, d\mathbf{r} = \int |\mathbf{H}_1|^2\, d\mathbf{r}$ in the denominator, which holds for the fields in a resonator (see, e.g., [Landau *et al.*, 1984], §90). Suppose now that the composite has a spherical shape. Then the macroscopic field inside the sphere $\mathbf{E}_0$ is uniform:

$$\mathbf{E}_0 = \frac{3}{\varepsilon_e + 2}\mathbf{E}_{10} \tag{5.10}$$

(also see the discussion below for Eq. (5.23)). Therefore, the frequency shift,

$$\frac{\Delta\omega_E}{\omega} = \frac{3}{2}\frac{\varepsilon_e - 1}{\varepsilon_e + 2}\frac{|\mathbf{E}_{10}|^2\, v}{\int |\mathbf{E}_1|^2\, d\mathbf{r}}, \tag{5.11}$$

is unambiguously connected to the effective permittivity ε_e and the field distribution in the empty resonator.

If the composite in a form of sphere is placed at a maximum of the magnetic field $\mathbf{H}_{10}$, the corresponding shift in the resonance frequency

$$\frac{\Delta\omega_H}{\omega} = \frac{3}{2}\frac{\mu_e - 1}{\mu_e + 2}\frac{|\mathbf{H}_{10}|^2\, v}{\int |\mathbf{H}_1|^2\, d\mathbf{r}} \tag{5.12}$$

is unambiguously connected to the effective permeability μ_e.

Below, we discuss how to obtain self-consistent equations for the parameters ε_e, μ_e and macroscopic fields $\mathbf{E}_0$ and $\mathbf{H}_0$ in the percolating composites and electromagnetic crystals. We consider a spherical resonator with radius R_0 with the excited electric dipole mode. The coordinate origin coincides with the center of the resonator.

The vector potential $\mathbf{A}$ of the electric dipole mode equals to

$$\mathbf{A} = \mathbf{F}_0\frac{\sin(kr + \chi)}{r}, \tag{5.13}$$

where $r = |\mathbf{r}|$, $\mathbf{r} = \{x, y, z\}$ is radius vector and we choose the vector $\mathbf{F}_0$ to be

$$\mathbf{F}_0 = i\frac{3}{2k^2}\mathbf{E}_1(\mathbf{r} = 0) \equiv i\frac{3}{2k^2}\mathbf{E}_{10}. \tag{5.14}$$

Since the vector potential as well as electric and magnetic fields has no singularities in the resonator, the phase $\chi = 0$ in the empty resonator while the electric field takes the form of

$$\mathbf{E}_1(\mathbf{r}) = \left(k^2 f + \frac{f'}{r}\right)\mathbf{E}_{10} - \left(k^2 f + \frac{3f'}{r}\right)(\mathbf{E}_{10}\cdot\mathbf{n})\,\mathbf{n}, \tag{5.15}$$

where $\mathbf{n} = \mathbf{r}/r$ and $f = 3\sin(kr)/(2k^3 r)$.

It is easy to check that the electric field $\mathbf{E}_1(\mathbf{r}) \approx \mathbf{E}_{10}$ near the center of the empty resonator ($kr \ll 1$, $r \ll R_0$). The tangential component of the electric field vanishes at the internal resonator interface as $[\mathbf{E}_1 \times \mathbf{n}] = 0$ for $r = R_0$. This condition gives resonance frequencies ω_m for the dipole modes

$$\omega_m \simeq \frac{c}{R_0} m\pi \left(1 - \frac{1}{m^2\pi^2} - \frac{5}{3m^4\pi^4} - \frac{73}{15m^6\pi^6}\right), \quad m = 1, 2, 3, \ldots \quad (5.16)$$

We mainly consider the first mode with $m = 1$. The schematic picture of the electric field for $m = 1$ mode is shown in Fig. 5.1(a).

Let us consider now a ball made from the investigated metal-dielectric metamaterial placed in the center of the resonator (see Fig. 5.1(b)). The ball radius $\mathcal{L} \ll R_0$ but we assume that $\mathcal{L} \gg \xi_p$ so that the electric field is determined by the bulk properties of the sample. When the ball is inserted into the resonator, the phase χ does not vanish in Eq. (5.13). Yet, $\chi \ll 1$ since the sample is much smaller than the resonator. Then the resonator field near the center but out of the sample ($r > \mathcal{L}$) has the following form:

$$\mathbf{E}_{out} = \mathbf{E}_{01} - \nabla\varphi_{out}, \quad (5.17)$$

where $\mathbf{E}_{01}$ is the field in the center of the empty resonator. We assume that $kr \ll 1$ and the potential φ_{out}, obtained from the Eq. (5.13), has a dipole form

$$\varphi_{out} = \alpha a^3 \frac{(\mathbf{E}_{01} \cdot \mathbf{r})}{r^3}, \quad (5.18)$$

where the polarizability $\alpha = -(3/2)\chi/(ka)^3$ is proportional to the phase shift χ. The electric field in the resonator with the spherical ball inside is

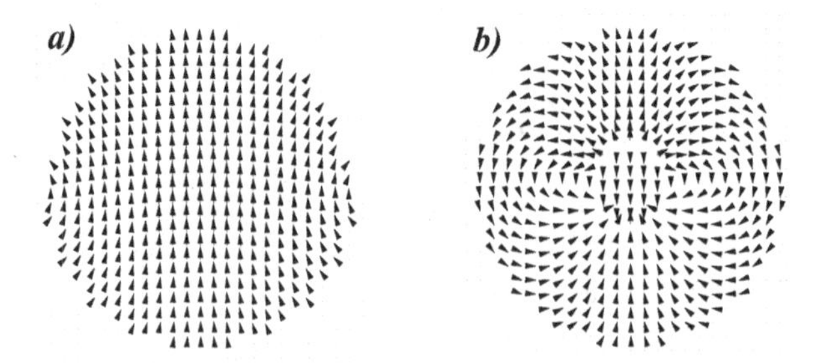

Fig. 5.1 Electric field in electrodipole mode excited in spherical resonator; a — empty resonator, b — spherical particle with permittivity $\varepsilon = -3$ is placed in the center.

shown in Fig. 5.1 for the case when the effective dielectric constant of the material is equal to $\varepsilon_e = -3$.

Due to the boundary conditions, we match the average electric potential $\langle \varphi_{in} \rangle$ in the material with the potential φ_{out} obtaining

$$\langle \varphi_{in} \rangle = (\alpha - 1)(\mathbf{E}_{01} \cdot \mathbf{r}). \tag{5.19}$$

The continuity of the normal component of the electric displacement $\mathbf{D}$ requires that

$$\langle \mathbf{D} \rangle = \langle \varepsilon \mathbf{E}_{in} \rangle = (1 + 2\alpha)\mathbf{E}_{01}, \tag{5.20}$$

where ε is the local, microscopic permittivity in the material. We define the effective permittivity ε_e of the metamaterial as the ratio between average electric displacement $\langle \mathbf{D} \rangle$ and average potential field $\mathbf{E}_0 = -\langle \nabla \varphi_{in} \rangle$:

$$\langle \mathbf{D} \rangle = \varepsilon_e \mathbf{E}_0. \tag{5.21}$$

Substituting this definition into Eq. (5.20) and using Eq. (5.19) we obtain the equation for the polarizability

$$\alpha = \frac{\varepsilon_e - 1}{\varepsilon_e + 1} \tag{5.22}$$

and an equation for the internal field

$$\mathbf{E}_0 = -\langle \nabla \varphi_{in} \rangle = \frac{3}{\varepsilon_e + 2}\mathbf{E}_{01}, \tag{5.23}$$

which coincide with usual expressions for the polarizability and for the internal electric field in a homogeneous material. Therefore, the average electric displacement is proportional to the potential part of the local field averaged over the system and this coefficient is exactly equal to the effective permittivity. The same derivation is true for the magnetic field, which results in the equation for the average magnetic induction $\langle \mathbf{B} \rangle$:

$$\langle \mathbf{B} \rangle = \mu_e \mathbf{H}_0, \tag{5.24}$$

where $\mathbf{H}_0 = -\langle \nabla \psi_{in} \rangle$ is the average potential part of the local field magnetic field. Equations (5.21) and (5.24) replace the usual constitutive equations $\langle \mathbf{D} \rangle = \varepsilon_e \langle \mathbf{E} \rangle$ and $\langle \mathbf{B} \rangle = \mu_e \langle \mathbf{H} \rangle$, which hold in the quasistatic case.

Note that the electric field $\mathbf{E}(\mathbf{r})$ inside the metamaterial is not potential in general. It can be represented as

$$\mathbf{E}(\mathbf{r}) = -\nabla \varphi_{in}(\mathbf{r}) + 4\pi \mathbf{L}(\mathbf{r}), \tag{5.25}$$

where $\mathbf{L}(\mathbf{r})$ is the solenoidal field arising from the eddy currents in the metal grains. The field $\mathbf{L}(\mathbf{r})$ exists in the metal inclusions and decays on ξ_p scale outside of the metal. That is why we have matched the potential part of the internal electric field $-\nabla\varphi_{in}$ with the outside field given by Eq. (5.17).

Let consider a "microscopical" reason for the frequency shift of the resonator when the metamaterial is placed in there. The field change when the spherical composite is placed in the resonator is determined by superposition of the fields scattered from the composite individual metal and dielectric particles. The interaction between the particles can be taken into account in the self-consistent approximation known as the effective medium theory [Bergman and Stroud, 1992], [Bruggeman, 1935] (see discussion in Sec. 2.2). In this theory, the interaction of a given metal or a dielectric particle with the rest of the system can be found by replacing the latter by a homogeneous medium with the effective parameters ε_e and μ_e. Assuming that the composite inclusions are spherical in shape, the electric fields $\mathbf{E}_{in,m}$ and $\mathbf{E}_{out,m}$, excited by electric field $\mathbf{E}_0$, are calculated inside and outside of the metal inclusion of the size a (see [Landau *et al.*, 1984], §59; [Lagarkov *et al.*, 1992] and [Vinogradov *et al.*, 1989b]):

$$\mathbf{E}_{in,m}(\mathbf{r}) = \mathbf{E}_{in,m0} + 4\pi\mathbf{L}(\mathbf{r}), \tag{5.26}$$

where

$$\mathbf{E}_{in,m0} = \frac{3\varepsilon_e}{2\varepsilon_e + \tilde{\varepsilon}_m}\mathbf{E}_0, \tag{5.27}$$

$\tilde{\varepsilon}_m$ is the renormalized permittivity of the metal defined as

$$\tilde{\varepsilon}_m = \varepsilon_m \frac{2F(k_m a)}{1 - F(k_m a)}, \qquad F(x) = \frac{1}{x^2} - \frac{\cot(x)}{x}, \tag{5.28}$$

and $k = \omega/c$, $k_m = k\sqrt{\varepsilon_m \mu_m}$. The skin (penetration) depth δ is defined as $\delta = 1/\operatorname{Im} k_m$. When the metal conductivity σ_m is a real quantity (this is true at microwave- and radio- frequencies) and $\mu_m = 1$, the skin depth $\delta = c/\sqrt{2\pi\sigma_m\omega}$. In the Cartesian coordinate system with the "z"-axis directed along the field $\mathbf{E}_0$, the local solenoidal field $\mathbf{L}$ in Eq. (5.26) is determined by the equation

$$\operatorname{curl}\mathbf{L}(\mathbf{r}) = \frac{1}{4\pi}\operatorname{curl}\mathbf{E}_{in,m}(\mathbf{r}) = \frac{ik}{4\pi}\mathbf{B}_E, \tag{5.29}$$

where the loop field

$$\mathbf{B}_E = -3iE_0 \frac{ak\varepsilon_m\varepsilon_e \sin(k_m r)F(k_m r)}{(2\varepsilon_e + \tilde{\varepsilon}_m)\sin(k_m a)\left(F(k_m a) - 1\right)}\{\frac{y}{r}, -\frac{x}{r}, 0\} \tag{5.30}$$

is the circular magnetic induction generating in the metal particle by the electric current. Therefore the electric field inside of the metal consists of uniform potential part $\mathbf{E}_{in,m0}$ and the solenoidal part $\mathbf{L}(\mathbf{r})$ that depends on the coordinate. The field outside of the metal particle equals to

$$\mathbf{E}_{out,m} = \mathbf{E}_0 + a^3 \frac{\varepsilon_e - \tilde{\varepsilon}_m}{2\varepsilon_e + \tilde{\varepsilon}_m} \nabla \left(\frac{\mathbf{E}_0 \cdot \mathbf{r}}{r^3} \right), \tag{5.31}$$

i.e., it is the potential field. The wavelength in the dielectric inclusion relates to an incident field wavelength by a well-known material relation: $\lambda_d = \lambda/\sqrt{\varepsilon_d}$ and is assumed to be much larger than the dielectric grain size: $\lambda_d \gg a$. Then the electric fields inside and outside a dielectric particle are given by the well known equations (see e.g. [Landau *et al.*, 1984], §8)

$$\mathbf{E}_{in,d} = \mathbf{E}_0 \frac{3\varepsilon_e}{2\varepsilon_e + \varepsilon_d} \tag{5.32}$$

and

$$\mathbf{E}_{out,d} = \mathbf{E}_0 + a^3 \frac{\varepsilon_e - \varepsilon_d}{2\varepsilon_e + \varepsilon_d} \nabla \left(\frac{\mathbf{E}_0 \cdot \mathbf{r}}{r^3} \right). \tag{5.33}$$

Similar equations can be obtained for the magnetic field excited by uniform magnetic field $\mathbf{H}_0$ inside and outside of the metal and the dielectric particle:

$$\mathbf{H}_{in,m} = \mathbf{H}_{in,m0} + 4\pi\mathbf{M}, \tag{5.34}$$

where

$$\mathbf{H}_{in,m0} = \frac{3\mu_e}{2\mu_e + \tilde{\mu}_m}\mathbf{H}_0, \tag{5.35}$$

and the renormalized metal permeability $\tilde{\mu}_m$ equals to

$$\tilde{\mu}_m = \mu_m \frac{2F(k_m a)}{1 - F(k_m a)}. \tag{5.36}$$

The function F is still defined in Eq. (5.28). Note that the renormalized metal permeability $\tilde{\mu}_m$ is not equal to one, even if metal is non-magnetic and the seed permeability $\mu_m = 1$. The solenoidal magnetic field $\mathbf{M}$ in Eq. (5.34) satisfies the equations, which are similar to Eqs. (5.29) and (5.30):

$$\operatorname{curl}\mathbf{M} = \frac{1}{4\pi}\operatorname{curl}\mathbf{H}_{in,m} = -\frac{ik}{4\pi}\mathbf{D}_H, \tag{5.37}$$

$$\mathbf{D}_H = 3iH_0\frac{ak\mu_m\mu_e\sin(k_m r)F(k_m r)}{(2\mu_e+\tilde{\mu}_m)\sin(k_m a)\,(F(k_m a)-1)}\{\frac{y}{r},-\frac{x}{r},0\}, \tag{5.38}$$

where $\mathbf{D}_H$ is the circular electric displacement vector induced in the metal particle by the high-frequency magnetic field $\mathbf{H}_0$. The displacement vector $\mathbf{D}_H$ can be written as $\mathbf{D}_H = i\frac{4\pi}{\omega}\mathbf{j}$, where the eddy currents $\mathbf{j}$ are known as Foucault currents. The field $\mathbf{H}_{in,m0}$ is a potential part, while $\mathbf{M}$ as a rotational (solenoid) part of the local magnetic field. The magnetic field outside the metal particle is an irrotational (curl-free) field, which equals to

$$\mathbf{H}_{out,m} = \mathbf{H}_0 + a^3\frac{\mu_e-\tilde{\mu}_m}{2\mu_e+\tilde{\mu}_m}\nabla\left(\frac{\mathbf{H}_0\cdot\mathbf{r}}{r^3}\right). \tag{5.39}$$

If the dielectric component of the composite is non-magnetic, i.e., the dielectric permeability $\mu_d = 1$, then the magnetic fields inside and outside of a dielectric particle are equal to

$$\mathbf{H}_{in,d} = \mathbf{H}_0\frac{3\mu_e}{2\mu_e+1} \tag{5.40}$$

and

$$\mathbf{H}_{out,d} = \mathbf{H}_0 + a^3\frac{\mu_e-1}{2\mu_e+1}\nabla\left(\frac{\mathbf{H}_0\cdot\mathbf{r}}{r^3}\right) \tag{5.41}$$

respectively.

The effective parameters ε_e and μ_e are determined by the self-consistent condition that the scattered fields averaged over all the inclusions should vanish, i.e. $\langle\mathbf{E}_{out}\rangle = p\mathbf{E}_{out,m} + (1-p)\mathbf{E}_{out,d} = \mathbf{E}_0$ and $\langle\mathbf{H}_{out}\rangle = p\mathbf{H}_{out,m} + (1-p)\mathbf{H}_{out,d} = \mathbf{H}_0$, where $\langle\cdots\rangle$ stands for the volume averaging. Substitution the scattered electric and magnetic fields from Eqs. (5.31), (5.33) and (5.39), (5.41) into the above relations result in the following system of equations

$$p\frac{\varepsilon_e-\tilde{\varepsilon}_m}{2\varepsilon_e+\tilde{\varepsilon}_m} + (1-p)\frac{\varepsilon_e-\varepsilon_d}{2\varepsilon_e+\varepsilon_d} = 0, \tag{5.42}$$

$$p\frac{\mu_e - \tilde{\mu}_m}{2\mu_e + \tilde{\mu}_m} + (1-p)\frac{\mu_e - 1}{2\mu_e + 1} = 0, \tag{5.43}$$

that gives the effective permittivity ε_e and permeability μ_e in terms of the metal concentration p and $\tilde{\varepsilon}_m$, ε_d, and $\tilde{\mu}_m$.

The equations (5.42) and (5.43) are similar to the equations of the traditional effective medium theory [Bergman and Stroud, 1992]. The eddy currents, confined to the metal grains, result in renormalization of the permittivity and permeability: the metal permittivity ε_m and permeability μ_m are replaced by $\tilde{\varepsilon}_m$ and $\tilde{\mu}_m$, which are given by Eqs. (5.28) and (5.36), respectively. This has a substantial effect on the frequency dependence of the effective parameters. For example, it is commonly accepted that the effective conductivity of a composite ($\sigma_e = -i\omega\varepsilon_e/4\pi$) is dispersion-free, when the conductivity of metal component σ_m is frequency independent and is large in comparison with the applied frequency, $\sigma_m \gg \omega$ (which is typical for the microwave and far-infrared ranges). The traditional effective medium theory predicts $\sigma_e = \sigma_m(3p-1)/2$ for the metal concentration p sufficiently above the percolation threshold. Equation (5.42) gives the same result for the effective conductivity σ_e but the metal conductivity is renormalized according to Eq. (5.28), which results in the following formula $\sigma_e = \sigma_m F(x)(3p-1)/[1-F(x)]$, $x = a\sqrt{2\pi\sigma_m\omega}/c$. From this expression, we see that the effective conductivity has a dispersion behavior when the skin effect is important in the metal inclusions. In the limit of very strong skin effect $\delta \ll a$, the renormalized metal permittivity $\tilde{\varepsilon}_m$ and the permeability $\tilde{\mu}_m$ are approximated by

$$\tilde{\varepsilon}_m \approx 2i\varepsilon_m/(k_m a) = 2i\sqrt{\varepsilon_m}/(ka), \tag{5.44}$$

and by

$$\tilde{\mu}_m \approx 2i/(k_m a), \tag{5.45}$$

as it follows from Eqs. (5.28) and (5.36). Accordingly, the effective conductivity decreases with frequency as $\sigma_e \propto \sigma_m/(k_m a) \sim \sigma_m(\delta/a) \sim (c/a)\sqrt{\sigma_m/\omega}$. Another interesting result is that the percolation composites exhibit magnetic properties, even those properties are absent in each component (the metal and the dielectric), i.e. $\mu_m = \mu_d = 1$. In this case, the real part μ_e' of the effective permeability is less than unity and decreases with frequency. The imaginary part μ_e'' of the effective permeability has its maximum at the frequencies when the skin depth $\delta(\omega) \sim a$.

We show now that the effective parameters ε_e and μ_e determine propagation of the electromagnetic wave in the metal-dielectric composites. In the effective medium approximation, used here, the average electric field equals to

$$\langle \mathbf{E} \rangle = p\mathbf{E}_{in,m0} + 4\pi \langle \mathbf{L} \rangle + (1-p)\mathbf{E}_{in,d}, \tag{5.46}$$

where Eq. (5.26) was used. When Eqs. (5.27) and (5.32) are substituted in Eqs. (5.46) and (5.42) is taken into account, the Eq. (5.46) simplifies to

$$\langle \mathbf{E} \rangle = \mathbf{E}_0 + 4\pi \langle \mathbf{L} \rangle, \tag{5.47}$$

where $\langle \cdots \rangle$ denotes again the volume average. Therefore, the curl-free part of the local field, being averaged over the volume, gives the field $\mathbf{E}_0$, while the second term in Eq. (5.47) results from the skin effect in the metal grains. The above consideration being applied to the average magnetic field $\langle \mathbf{H} \rangle$ [see Eqs. (5.34), (5.35) and (5.40)] results in

$$\langle \mathbf{H} \rangle = p\mathbf{H}_{in,m} + (1-p)\mathbf{H}_{in,d} = \mathbf{H}_0 + 4\pi \langle \mathbf{M} \rangle, \tag{5.48}$$

where the solenoidal field $\mathbf{M}$ in the metal grains is given by Eq. (5.37) (the field $\mathbf{M}$ equals to zero in the dielectric grains). Again, the average irrotational (curl-free) part of the local magnetic field gives $\mathbf{H}_0$, while the term $4\pi \langle \mathbf{M} \rangle$ represents the average rotational (curl-type) magnetic field.

Equations (5.47) and (5.48) can be considered as a definitions for the fields $\mathbf{E}_0$ and $\mathbf{H}_0$. Indeed, if the local fields $\mathbf{E}(\mathbf{r})$ and $\mathbf{H}(\mathbf{r})$ are known in the composite, the fields $\mathbf{L}(\mathbf{r})$, $\mathbf{M}(\mathbf{r})$ can be found from Eqs. (5.29), (5.37) and, ultimately, the fields $\mathbf{E}_0$ and $\mathbf{H}_0$ from the Eqs. (5.47) and (5.48).

We then proceed with derivation of the equations for the macroscopic electromagnetism in metal-dielectric composites. In the above consideration, an optically thin composite sample has been placed either at the maximum of the electric or the magnetic field. After that, the effective parameters ε_e and μ_e were determined.

Consider now an em wave spread out in a bulk metal-dielectric composite. To obtain the necessary macroscopic equations for the composite behavior, the "microscopic" Maxwell equations are averaged over the scale $\mathcal{L}$, which is larger than ξ_p but much smaller than the "internal" wavelength $\lambda_e = \lambda/\sqrt{|\varepsilon_e \mu_e|}$, $\xi_p < \mathcal{L} \ll \lambda_e$. The macroscopic electric $\mathbf{E}_0$ and magnetic $\mathbf{H}_0$ fields are still defined by Eqs. (5.47) and (5.48), but both fields are nonzero in the volume $v \sim \mathcal{L}^3$. Equation (5.21) gives the average electric displacement excited by the electric field $\mathbf{E}_0$, but magnetic field $\mathbf{H}_0$ also excites eddy electric currents (Foucault currents). Adding of the electric displacement $\mathbf{D}_H$ given by Eq. (5.37) to the average displacement given by Eq. (5.21) now represents the full electric displacement

$$\langle \mathbf{D} \rangle_f = \varepsilon_e \mathbf{E}_0 + \frac{i4\pi}{k} \langle \operatorname{curl} \mathbf{M} \rangle . \tag{5.49}$$

Note that the second term in Eq. (5.49) disappears when the skin effect vanishes, i.e., when $|k_m|\, a \to 0$. Similarly, the average full magnetic induction $\langle \mathbf{B} \rangle_f$ equals to

$$\langle \mathbf{B} \rangle_f = \mu_e \mathbf{H}_0 - \frac{i4\pi}{k} \langle \operatorname{curl} \mathbf{L} \rangle , \tag{5.50}$$

where the vector $\mathbf{L}$ is given by Eq. (5.29). Now the Maxwell equations are averaged over the macroscopic volume $v \sim \mathcal{L}^3$ for the sample, centered at the point $\mathbf{r}$. The Maxwell equations rewritten in the frequency domain give

$$\langle \operatorname{curl} \mathbf{E} \rangle = ik \langle \mathbf{B} \rangle_f = ik\mu_e \mathbf{H}_0 + 4\pi \langle \operatorname{curl} \mathbf{L} \rangle , \tag{5.51}$$

$$\langle \operatorname{curl} \mathbf{H} \rangle = -ik \langle \mathbf{D} \rangle_f = -ik\varepsilon_e \mathbf{E}_0 + 4\pi \langle \operatorname{curl} \mathbf{M} \rangle , \tag{5.52}$$

where the respective Eqs. (5.49) and (5.50) are substituted for the electric and magnetic inductances. The order of the curl operation and the volume average in Eqs. (5.51) and (5.52) can be interchanged as it commonly done in the macroscopic theory based on Maxwell equations (see [Jackson, 1998] Ch. 6, Sec. 6.6). Operation $\langle \operatorname{curl} \mathbf{E} \rangle$ can be written as $\langle \operatorname{curl} \mathbf{E} \rangle = \operatorname{curl} [\langle \mathbf{E} \rangle (\mathbf{r})]$, where the differentiation in l.h.s. is performed over the position of the radius-vector $\mathbf{r}$ of the volume v. Note that the fields $\mathbf{E}_0$ and $\mathbf{H}_0$ defined by Eqs. (5.47) and (5.48) are also functions of $\mathbf{r}$. Then the Maxwell equations (5.51) and (5.52) acquire the form

$$\operatorname{curl} \mathbf{E}_0(\mathbf{r}) = ik\mu_e \mathbf{H}_0(\mathbf{r}), \tag{5.53}$$

$$\operatorname{curl} \mathbf{H}_0(\mathbf{r}) = -ik\varepsilon_e \mathbf{E}_0(\mathbf{r}), \tag{5.54}$$

i.e., they have the form typical for the macroscopic electromagnetism, describing the propagation of electromagnetic waves in the composite media.

Equations (5.53) and (5.54) are self-consistent: the fields $\mathbf{E}_0(\mathbf{r})$ and $\mathbf{H}_0(\mathbf{r})$ are indeed the potential fields on the scales $\mathcal{L} \ll \lambda_e$, where λ_e is the wavelength in the metamaterial. These equations describe the EM wave propagation in the metamaterials. For example, they can be used to find the wave propagation in "macroscopically inhomogeneous" composite, where the effective permittivity $\varepsilon_e(\mathbf{r})$ and permeability $\mu_e(\mathbf{r})$ are non-uniform. Thus, Eqs. (5.53) and (5.54), as it is noted in the beginning of the chapter, indeed coincide with Eqs. (5.1) and (5.2) for the meta-material composite placed in a resonator.

It is important that all the quantities in Eqs. (5.21), (5.24), (5.53) and (5.54) are well defined and do not depend on the assumptions made in the course of their derivation. The vector $\mathbf{M}$ in Eq. (5.49) can be determined as a magnetic moment of the eddy currents per unit volume, so that

$$\langle \mathbf{M} \rangle = -i\frac{k}{8\pi v}\int [\mathbf{r}\times\mathbf{D}_H]\, d\mathbf{r} = \frac{1}{2cv}\int [\mathbf{r}\times\mathbf{j}_H]\, d\mathbf{r}, \tag{5.55}$$

where the integration is performed over the macroscopic volume v. This definition of vector $\mathbf{M}$ is in an agreement with Eq. (5.37), except it is not required that the eddy currents $\mathbf{j}_H$ are all the same in the metal particles. In a similar way, the vector $\mathbf{L}$ is defined as the spatial density of the "electric moments of the magnetic eddy currents":

$$\langle \mathbf{L} \rangle = i\frac{k}{8\pi v}\int [\mathbf{r}\times\mathbf{B}_E]\, d\mathbf{r}, \tag{5.56}$$

where the integration is taken over the volume v and $\mathbf{B}_E = (-i/k)\,\mathrm{curl}\,\mathbf{E}$, with $\mathbf{E}$ being the local electric field. Note that the vector $\langle \mathbf{L} \rangle$ has no direct analog in the classical electrodynamics since there is no such thing as circular magnetic currents in the atoms and the molecules. After definition of the vectors $\mathbf{L}$ and $\mathbf{M}$, equations (5.21), (5.24), (5.53) and (5.54) form a complete system of equations that determine the effective parameters and the electromagnetic wave propagation in metal-dielectric composite media. Various approximations such as the effective medium theory can still be very useful in the actual calculations of the effective parameters. We do not consider here rather special metal inclusions, which support chiral or quadrupole modes. The excellent discussion of the modes beyond the dipole approximation can be found in [Vinogradov and Aivazyan, 1999] and [Vinogradov, 2002].

5.2 Electromagnetic Crystals

In this subsection the method, discussed above, is used to calculate the effective permittivity and permeability of metal-dielectric crystals, also known as the electromagnetic crystals. In these artificial crystals, the metal component is assembled in a periodic lattice fashion. Two limiting cases of the electromagnetic crystals are considered: a cubic lattice of unconnected metal spheres and a three-dimensional conducting wire mesh configured into a cubic lattice.

5.2.1 *Cubic lattice of metal spheres*

The local electromagnetic fields and effective parameters are considered for the system of the metal spheres with radius a, which are embedded into a dielectric host at sites of a cubic lattice with the period $\mathcal{L} > 2a$. The wavelength λ is assumed to be much larger than the lattice period $\mathcal{L}$. Consider first the electric field distribution in the lattice cell centered at a metal sphere. The electric field $\mathbf{E}_{out}$ outside the sphere can be expanded in multipole series; for simplicity, the dipole approximation is used [Nicorovici *et al.*, 1995a], [Nicorovici *et al.*, 1995b], and [McPhedran *et al.*, 1997], which holds in the limit $\mathcal{L} \gg a$. In this approximation, the outside field has constant and dipole components only, namely

$$\mathbf{E}_{out}(\mathbf{r}) = \mathbf{E}_1 + \alpha a^3 \nabla \left(\frac{\mathbf{E}_1 \cdot \mathbf{r}}{r^3} \right), \qquad (5.57)$$

where $\mathbf{E}_1$ is an electric field vector aligned with the electric field of the incident wave and α is an unknown coefficient. Provided that the external field $\mathbf{E}_{out}$ is given, the electric field $\mathbf{E}_{in}$ inside the metal grain can be found unambiguously by solving the Maxwell equations at the boundary conditions $\mathbf{E}_{in} \times \mathbf{n} = \mathbf{E}_{out} \times \mathbf{n}$ and $\varepsilon_m \mathbf{E}_{in} \cdot \mathbf{n} = \varepsilon_d \mathbf{E}_{out} \cdot \mathbf{n}$ are imposed at the metal surface ($\mathbf{n} = \mathbf{r}/r$ is the unit vector directed outward the metal sphere). Thus the internal field $\mathbf{E}_{in}$ is given by Eqs. (5.26) and (5.27), where the field $\mathbf{E}_0$ is replaced by $\mathbf{E}_1$. The pre-factor α is also found by matching the fields $\mathbf{E}_{in}$ and $\mathbf{E}_{out}$ at the surface of the metal grain: This gives the result $\alpha = (\varepsilon_d - \tilde{\varepsilon}_m) / (2\varepsilon_d + \tilde{\varepsilon}_m)$, which coincides formally with Eq. (5.31), with the renormalized metal permittivity $\tilde{\varepsilon}_m$ given by Eq. (5.28).

When the local electric field $\mathbf{E}(\mathbf{r})$ is known for one lattice cell, the effective permittivity of the whole electromagnetic crystal can be found following the procedure described above. The average electric field is calculated as

$$\langle \mathbf{E} \rangle = \frac{1}{v} \int \mathbf{E}(\mathbf{r})\, d\mathbf{r} = \frac{1}{v} \left(\int \mathbf{E}_{in}(\mathbf{r})\, d\mathbf{r} + \int \mathbf{E}_{out}(\mathbf{r})\, d\mathbf{r} \right), \qquad (5.58)$$

where the first integral limits are taken over the volume $v = \mathcal{L}^3$ of the lattice cell, and the fields $\mathbf{E}_{in}$ and $\mathbf{E}_{out}$ are integrated inside and outside of the metal grain respectively. Note that in the considered dipole approximation, when $\mathbf{E}_{out}$ has the form of Eq. (5.57), the integration of the dipole term in Eq. (5.57) results in zero. The second integral in Eq. (5.58) gives $v^{-1}\int \mathbf{E}_{out}(\mathbf{r})\, d\mathbf{r} = \mathbf{E}_1 (1-p)$, where $p = (4/3)\,\pi a^3/v$ is the volume concentration of the metal (a filling factor). The average potential part of the field then equals to

$$\mathbf{E}_0 = \langle \mathbf{E} \rangle - 4\pi \langle \mathbf{L} \rangle = \frac{1}{v}\left(\int \mathbf{E}(\mathbf{r})\, d\mathbf{r} - \frac{1}{2}\int [\mathbf{r}\times \operatorname{curl} \mathbf{E}(\mathbf{r})]\, d\mathbf{r}\right), \quad \text{(III)}$$

which gives

$$\mathbf{E}_0 = [p\, 3\varepsilon_d / (2\varepsilon_d + \tilde{\varepsilon}_m) + (1-p)]\, \mathbf{E}_1, \quad (5.59)$$

for the considered metal-sphere crystal. Permittivity $\tilde{\varepsilon}_m$ is given by Eq. (5.28).

The average electric displacement is then equal to

$$\langle \mathbf{D} \rangle = \left[p \frac{3\varepsilon_d \tilde{\varepsilon}_m}{2\varepsilon_d + \tilde{\varepsilon}_m} + (1-p)\,\varepsilon_d \right] \mathbf{E}_1, \quad (5.60)$$

and the effective permittivity ε_e is defined as

$$\varepsilon_e = \frac{\langle D \rangle}{E_0} = \varepsilon_d \frac{(1+2p)\,\tilde{\varepsilon}_m + 2\,(1-p)\,\varepsilon_d}{(1-p)\,\tilde{\varepsilon}_m + (p+2)\,\varepsilon_d}. \quad (5.61)$$

The above homogenization procedure can be easily repeated for the magnetic field to define the effective permeability μ_e for the cubic crystal composed of metal spheres:

$$\mu_e = \frac{(1+2p)\,\tilde{\mu}_m + 2\,(1-p)}{(1-p)\,\tilde{\mu}_m + (p+2)}\,. \quad (5.62)$$

The renormalized metal permeability $\tilde{\mu}_m$ is given by Eq. (5.36), and we have continued to assume for simplicity that neither a dielectric host nor metal spheres have magnetic properties, i.e. $\mu_d = \mu_m = 1$.

In the quasistatic case, when the skin effect is negligible, Eqs. (5.61) and (5.62) give the well known Maxwell-Garnet formulae for the effective parameters. It has been demonstrated that the Maxwell-Garnet approximation, which emerges from the dipole approximation, gives very accurate results for the effective properties of various metal-dielectrics periodic composites, even at large filling factors $p \lesssim 0.5$ [Sheng, 1995], [Nicorovici *et al.*, 1995a], [Nicorovici *et al.*, 1995b], and [McPhedran *et al.*, 1997]. Then it is reasonable to conjecture that in a non-quasistatic case, Eqs. (5.61) and (5.62) will still hold for the above concentration range of $p's$. For larger filling factors p, the local fields and effective parameters of the electromagnetic crystals can be defined using the Rayleigh technique developed by McPhedran and co-workers in the above cited works. Again, provided that the internal field has been found, the effective parameters ε_e and μ_e can be calculated from the Eqs. (5.21) and (5.24). The multipole corrections to the effective permittivity and permeability are considered in Refs. [Vinogradov and Aivazyan, 1999] and [Vinogradov, 2002].

Consider now the strong skin effect when $\operatorname{Im}(k_m)\, a \to +\infty$, that is the electric and magnetic fields fall exponentially in the metal grain, being confined to the skin depth $\delta = 1/\operatorname{Im}(k_m)$. We then recall that the skin depth is equal to $\delta = c/\sqrt{2\pi\sigma_m\omega}$ and the wavevector $k_m = (1+i)/\delta$ for the positive values of the metal conductivity σ_m (which is typical for the radio-, microwave- and far-infrared frequencies ω in most metals). In optical and infrared spectral ranges, a typical metal permittivity ε_m is negative and the wavevector $k_m \simeq i\sqrt{|\varepsilon_m|}k$. As it follows from Eqs. (5.44) and (5.45), the renormalized permittivity $|\tilde{\varepsilon}_m| \to \infty$ while renormalized permeability $|\tilde{\mu}_m| \to 0$ when $|\varepsilon_m| \to \infty$. Substitution of these estimates in Eqs. (5.61) and (5.62) results in simple expressions for the effective parameters of the metal-sphere crystal in the case of strong skin effect:

$$\varepsilon_e = \varepsilon_d \frac{1+2p}{1-p}, \tag{5.63}$$

and

$$\mu_e = 2\frac{1-p}{p+2}. \tag{5.64}$$

The obtained ε_e and μ_e do not depend on the metal properties at all. The effective refractive index $n = \sqrt{\varepsilon_e\mu_e} = \sqrt{\varepsilon_d 2\,(1+2p)\,/\,(p+2)}$ is close to unity for almost all the filling factors. The effective surface impedance $\zeta = \sqrt{\mu_e/\varepsilon_e}$ determines reflection at the system interface. It follows from Eqs. (5.63) and (5.64) that the effective surface impedance ζ can be estimated as

$$\zeta = (1-p)\sqrt{\frac{2}{\varepsilon_d\,(1+2p)\,(p+2)}}. \tag{5.65}$$

It is also independent of the metal properties and almost linearly decreases with increasing the filling factor p.

The normal reflectance R at the metal-vacuum interface equals to

$$R \simeq \frac{\zeta - 1}{\zeta + 1} = \frac{n-1}{n+1} + \frac{9pn}{2(1+n)^2}, \tag{5.66}$$

where $n = \sqrt{\varepsilon_d}$ is the refractive index for a dielectric host. The reflectance R is real and it increases almost linearly until the filling factor takes the value $p_c \approx 0.524$, which corresponds to the case of the close-packed metal spheres embedded into the cubic lattice. It follows from Eqs. (5.63)-(5.64) that losses in the crystal are negligible. Electromagnetic field "squeezes" between the metal grains so that the electromagnetic crystal is essentially transparent. Note that the dipole approximation used for obtaining Eqs. (5.63)–(5.64) does not hold when the filling factor p approaches the close-packing limit $p_c \approx 0.524$ since the direct contact between the metal spheres becomes important.

5.2.2 *A wire-mesh electromagnetic crystal*

We consider now the electromagnetic properties of a three-dimensional metal wire mesh configured into the cubic lattice. The conducting wires are embedded into a dielectric medium with permittivity ε_d. The constructed electromagnetic crystal can be thought of as the opposite limit of the non-connected metal spheres considered above.

It is worthwhile to consider first a two-dimensional array of metal cylinders assembled into a square lattice with period $\mathcal{L}$. The direction of the incident electromagnetic wave is thought to be perpendicular to the cylinder axis of symmetry. Then there are two basic polarizations implied: a TE (transverse electric) and a TM (transverse magnetic) polarization. The most interesting results are obtained for the TM polarization when the electric field is parallel to the cylinders while the magnetic field is perpendicular. In the long-wavelength limit considered throughout the chapter (when the wavelength λ is much larger than the size $\mathcal{L}$ of the lattice cell), the local electric field can be found in the dipole approximation following the reasoning of the above sub-section. In the dipole approximation, the local electric field has the circular symmetry inside the wire:

$$\mathbf{E}(\mathbf{r}) = \mathbf{E}_1 J_0(k_m r), \qquad r < a;$$

For the rest of the cell, it becomes

$$\mathbf{E}(\mathbf{r}) = \mathbf{E}_1 \left[J_0(k_m a) - k_m a J_1(k_m a) \ln\left(\frac{r}{a}\right)\right], \qquad (5.67)$$

$$r > a\,, \quad |y| < \mathcal{L}/2, \quad |x| < \mathcal{L}/2;$$

where $\mathbf{r} = \{x, y\}$ is the two-dimensional vector in the plane perpendicular to the cylinder, $\mathbf{E}_1 = \{0, 0, E_1\}$ is a vector, which is aligned with the cylinders and is proportional to the amplitude of the external electric field; $J_0(k_m a)$ and $J_1(k_m a)$ are Bessel functions of the zeroth and the first order, respectively.

The average curl-free electric field $\mathbf{E}_0$ is extracted from the field $\mathbf{E}(\mathbf{r})$ by using the following equation:

$$\begin{aligned} \mathbf{E}_0 &= \langle \mathbf{E}(\mathbf{r})\rangle - \langle \mathbf{L}(\mathbf{r})\rangle \\ &= \frac{1}{v}\int \mathbf{E}(\mathbf{r})\, d\mathbf{r} - \frac{1}{2v}\int [\mathbf{r}\times \operatorname{curl}\mathbf{E}(\mathbf{r})]\, d\mathbf{r} \\ &= \mathbf{E}_1 \left\{ J_0(k_m a) + \frac{1}{4} k_m a J_1(k_m a)\left[4 - \pi + 2\ln\left(\frac{2a^2}{\mathcal{L}^2}\right)\right]\right\}, \end{aligned} \qquad (5.68)$$

where the integration is performed over the elementary cell area $v = \mathcal{L}^2$. From the symmetry of the crystal, it follows that the solenoidal part $\mathbf{L}(\mathbf{r})$ of the electric field $\mathbf{E}(\mathbf{r})$ vanishes on average at the cell boundary. Then we can unambiguously extract the average solenoidal field from the equation $\langle \mathbf{L}(\mathbf{r})\rangle = v^{-1}\int [\mathbf{r}\times \operatorname{curl}\mathbf{E}(\mathbf{r})]\, d\mathbf{r}/2$.

The electric displacement averaged over the cell is given by

$$\begin{aligned} \langle \mathbf{D}\rangle = \mathbf{E}_1 \Bigg\{ & \varepsilon_d (1-p) J_0(k_m a) \\ & + J_1(k_m a)\left[\frac{2k_m p}{ak^2} - \frac{\varepsilon_d k_m a}{4}\left(2p - 6 + \pi - 2\ln\left(\frac{2a^2}{\mathcal{L}^2}\right)\right)\right]\Bigg\}, \end{aligned} \qquad (5.69)$$

where the filling factor $p = \pi a^2/\mathcal{L}^2$. The effective permittivity, defined in the standard way, $\varepsilon_e = \langle D\rangle / E_0$ is then equals to

$$\varepsilon_e = \varepsilon_d - \frac{2a^2\left[4\pi\varepsilon_d - (4\pi + \mathcal{L}^2 k^2 \varepsilon_d)\,\tilde{\varepsilon}_m\right]}{\mathcal{L}^2\left[8 + a^2 k^2 \left(4 - \pi - 2\ln(\mathcal{L}^2/2a^2)\right)\tilde{\varepsilon}_m\right]}, \qquad (5.70)$$

where the renormalized metal permittivity $\tilde{\varepsilon}_m$ for the TM polarization is given by the following expression:

$$\tilde{\varepsilon}_m = \varepsilon_m \frac{2J_1(k_m a)}{J_0(k_m a)}. \qquad (5.71)$$

It is interesting to note that this renormalized metal permittivity $\tilde{\varepsilon}_m$ coincides with the renormalized permittivity for the conducting-stick composites considered by [Lagarkov *et al.*, 1992].

For the important case of large metal permittivity ($|\varepsilon_m| \gg 1$, $a|k_m| > 1$), Eq. (5.70) simplifies to the expression

$$\varepsilon_e = \varepsilon_d - \frac{2\left(4\pi + \mathcal{L}^2 k^2 \varepsilon_d\right)}{\mathcal{L}^2 k^2 \left(\pi - 4 + 2\ln(\mathcal{L}^2/2a^2)\right)}, \tag{5.72}$$

which does not depend on the metal conductivity at all. The effective permittivity takes very simple form for the very thin good conductors. We assume that the radius a is small enough so that $\ln(\mathcal{L}/a) \gg 1$ while the conductivity is large so that the strong skin effect condition $a|k_m| \gg 1$ is fulfilled. In this limit, we obtain from Eq. (5.72) that

$$\varepsilon_e = \varepsilon_d - \frac{2\pi}{k^2 \mathcal{L}^2 \log(\mathcal{L}/a)} = \varepsilon_d - \frac{\omega_p^2}{\omega^2}, \tag{5.73}$$

where the "plasma frequency" is given by

$$\omega_p^2 = \frac{2c^2 p}{a^2 \ln\left(\frac{\mathcal{L}}{a}\right)} = \frac{2\pi c^2}{\mathcal{L}^2 \ln\left(\frac{\mathcal{L}}{a}\right)}. \tag{5.74}$$

It follows from Eq. (5.72) that the effective permittivity becomes negative for the frequency ω smaller than the renormalized plasma frequency $\tilde{\omega}_p = c\sqrt{2\pi}/\left(\mathcal{L}\sqrt{\varepsilon_d}\sqrt{\ln(\mathcal{L}/a) + \pi/4 - 3/2 - \ln 2/2}\right) \approx \omega_p/\sqrt{\varepsilon_d}$. Therefore the incident TM electromagnetic wave decays exponentially in the electromagnetic crystal. The negative values for the effective permittivity in metal-dielectric composites containing conducting sticks were first predicted theoretically by [Lagarkov *et al.*, 1992], [Lagarkov and Sarychev, 1996] and then obtained experimentally [Lagarkov *et al.*, 1997a], [Lagarkov *et al.*, 1998]. The cut-off frequency in Eq. (5.74) was first suggested by [Pendry *et al.*, 1996], who related it to the effective plasma frequency of the electron gas in the electromagnetic crystal.

We consider now the effective permeability μ_e of the mesh crystal. We still consider the TM polarization when the external magnetic field is perpendicular to the main symmetry axis of the cylinders. The problem of finding the local magnetic field in an elementary cell is actually a two-dimensional analog of the magnetic field distribution for the elementary cell in the metal-sphere crystals, which was considered earlier.

The following expression can be written for the magnetic field outside the cylinder [cf. Eq. (5.57)] in the dipole approximation:

$$\mathbf{H}_{out}(\mathbf{r}) = \mathbf{H}_1 + \alpha a^2 \nabla \left(\frac{\mathbf{H}_1 \cdot \mathbf{r}}{r^2} \right), \tag{5.75}$$

where $\mathbf{H}_1$ is a vector aligned with the external magnetic field, α is a magnetic "polarizability", $\mathbf{r}$ is two-dimensional vector in the plane perpendicular to the cylinder axis. The magnetic field inside the cylinder $\mathbf{H}_{in}$, which matches the outside field $\mathbf{H}_{out}$, can be written as [Landau *et al.*, 1984](§ 59):

$$\mathbf{H}_{in}(\mathbf{r}) = \beta \, \mathrm{curl}\,\mathrm{curl}\left[J_0(k_m r)\,\mathbf{H}_1\right], \tag{5.76}$$

where β is another coefficient and $J_0(k_m r)$ is the Bessel function of the zeroth order. The coefficients α and β can be found from the boundary conditions $\mathbf{H}_{in} = \mathbf{H}_{out}$ imposed at the surface of the cylinder. If we recall that we consider the crystal with no intrinsic magnetic properties $(\mu_m = \mu_d = 1)$, we obtain

$$\alpha = \frac{1 - \tilde{\mu}_m}{1 + \tilde{\mu}_m}, \tag{5.77}$$

and

$$\beta = \frac{4}{(1 + \tilde{\mu}_m)\, k_m^2 J_0(k_m a)\left[2 - F_1(k_m a)\right]}, \tag{5.78}$$

where the renormalized metal permeability $\tilde{\mu}_m$ and the function F_1 are given by

$$\tilde{\mu}_m = F_1(k_m a) / \left[2 - F_1(k_m a)\right], \tag{5.79}$$

and

$$F_1(k_m a) = \frac{2 J_1(k_m a)}{k_m a J_0(k_m a)} \tag{5.80}$$

respectively.

If the magnetic field is known in the lattice cell, the effective permeability of the wire mesh crystal can be found from the similar procedure that was outlined above: The magnetic field, averaged over the cell, equals to

$$\langle \mathbf{H} \rangle = \frac{1}{\mathcal{L}^2}\left(\int_{r<a} \mathbf{H}_{in}\, d\mathbf{r} + \int_{r>a} \mathbf{H}_{out}\, d\mathbf{r} \right) = \mathbf{H}_1 \left[\frac{2p\tilde{\mu}_m}{(1 + \tilde{\mu}_m)} + (1 - p) \right], \tag{5.81}$$

where $p = \pi a^2/\mathcal{L}^2$ is the filling factor. The average potential field is given by

$$\mathbf{H}_0 = \langle \mathbf{H} \rangle - \frac{1}{2\mathcal{L}^2} \int [\mathbf{r} \times \operatorname{curl} \mathbf{H}(\mathbf{r})] \, d\mathbf{r} = \mathbf{H}_1 \left[\frac{2p}{(1+\tilde{\mu}_m)} + (1-p) \right], \quad (5.82)$$

where $\operatorname{curl} \mathbf{H}(\mathbf{r})$ vanishes outside the metal cylinder (see Eqs. (5.76) and (5.75)). The average magnetic induction $\langle \mathbf{B} \rangle$ coincides with average magnetic field $\langle \mathbf{B} \rangle = \langle \mathbf{H} \rangle$ since we consider nonmagnetic crystal ($\mu_m = \mu_d = 1$). Then the effective permeability of the mesh crystal equals to

$$\mu_{eTM} \equiv 1 + 4\pi\chi_{TM} = \frac{|\langle \mathbf{B} \rangle|}{|\mathbf{H}_0|} = \frac{|\langle \mathbf{H} \rangle|}{|\mathbf{H}_0|} = 1 - \frac{2p(1-\tilde{\mu}_m)}{1+p+\tilde{\mu}_m(1-p)}. \quad (5.83)$$

When the skin effect is strong, i.e., $|k_m| \, a \gg 1$, the renormalized metal permeability $\tilde{\mu}_m$ goes to zero and the effective permeability takes the following an asymptotic form

$$\mu_{eTM} = 1 - \frac{2p}{1+p}, \quad (5.84)$$

which is independent of the metal properties. Therefore, the mesh crystal has a diamagnetic response $|\mu_e| < 1$. Note that specially organized electromagnetic crystal can also have a giant paramagnetic response as it was discussed in Sec. 2.5.

We repeat the above consideration for the TE polarization when the magnetic field is parallel to the wires, obtaining

$$\mu_{eTE} \equiv 1 + 4\pi\chi_{TE} = pF_1(k_m a) + (1-p), \quad (5.85)$$

where the function $F_1(k_m a)$ is given by Eq. (5.80). For $|k_m| \, a \gg 1$, the permeability $\mu_{eTE} = 1-p$, which means that the magnetic field is a constant between the wires, but is zero inside the wires, which are occupying fraction p of the space. The potential part of the field is the constant everywhere.

Consider now a three-dimensional conducting wire mesh configured in the cubic lattice. Let suppose, for simplicity, that the wires are sufficiently thin and, therefore, the filling factor p is small. Then the effect of the intersections of the wires can be neglected since these effects give the corrections on the order of $p^2 \ll 1$. The component of the electric field, which is parallel to the wires, gives the major impact on the dielectric response. We will not consider here the plasmon resonance, which takes place when $\tilde{\varepsilon}'_m = -\varepsilon_d$. In the absence of the resonances and when the filling factor $p \ll 1$, the effective permittivity of the cubic mesh crystal is approximated by Eq. (5.70). The effective permeability equals to $\mu_e = 1 + 8\pi\chi_{TM} + 4\pi\chi_{TE}$, as it follows from the cubic symmetry, where the filling factor changes from p to $p/3$, which corresponds to three-metal cylinders in the lattice cell.

First, we consider the permittivity of the mesh crystal in the microwave frequency range assuming that the metal conductivity σ_m is real. The real $\varepsilon_e'(\lambda)$ and the imaginary $\varepsilon_e''(\lambda)$ parts of the effective permittivity are shown in Figs. 5.2 and 5.3. The behavior of $\varepsilon_e(\lambda)$ changes dramatically when the ratio of the skin depth to the wire radius $\delta/a = c/(a\sqrt{2\pi\omega\sigma_m})$ decreases. The real part of the permittivity increases almost hundred times when the ratio δ/a decreases from 0.8 to 0.3. The loss is very significant when $\delta/a > 0.2$. Under this condition the imaginary part ε_e'' of the effective permittivity is larger than the real part, $\varepsilon_e'' > \varepsilon_e'$. On the other hand, the permittivity ε_e is almost real when the skin effect is strong. It becomes negative and large when the frequency decreases below the cut-off frequency. This means that the crystal reflects incident EM waves completely and the loss is negligible. The simple Eq. (5.73) approximates the effective permeability well when skin effect is strong $\delta/a < 0.1$ (a dashed line in Fig. 5.2). To understand the physical meaning of the negative permittivity, let take into account

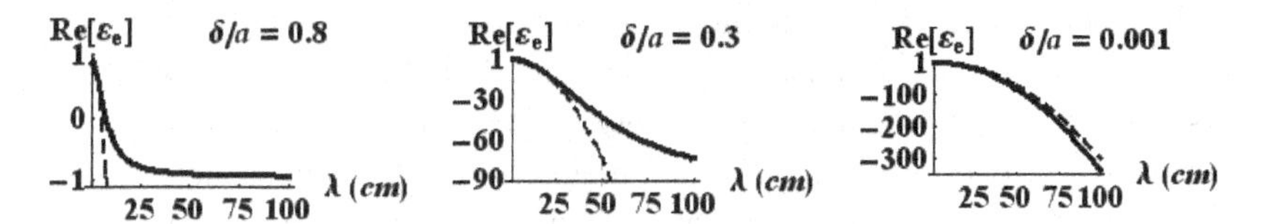

Fig. 5.2 Real part of effective dielectric constant $\varepsilon_e(\lambda)$ in cubic lattice of metal wires. Period of lattice $\mathcal{L}$ equals to $\mathcal{L} = 1\,cm$, diameter of wire $2a = 0.1\,mm$. Figures correspond to different magnitudes of the skin effect $\delta/a = \left(c/\sqrt{2\pi\sigma\omega}\right)/a$ at wavelength $\lambda = 1\,cm$.

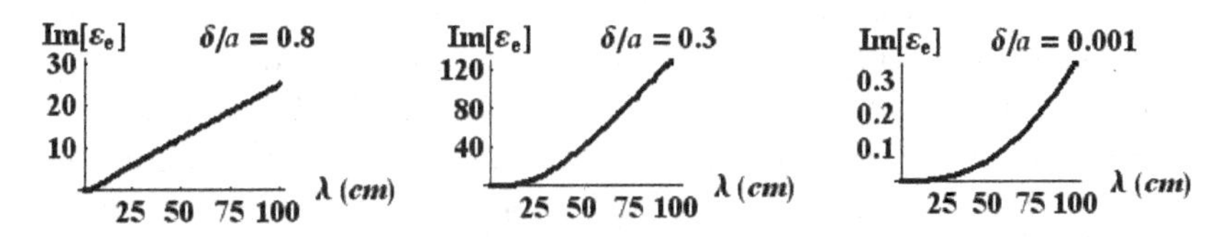

Fig. 5.3 Imaginary part of effective dielectric constant $\varepsilon_e(\lambda)$ in cubic lattice of metal wires. Period of lattice $\mathcal{L}$ equals to $\mathcal{L} = 1\,cm$, diameter of the wire $2a = 0.1\,mm$. Figures correspond to different magnitudes of the skin effect $\delta/a = \left(c/\sqrt{2\pi\sigma\omega}\right)/a$ at wavelength $\lambda = 1\,cm$.

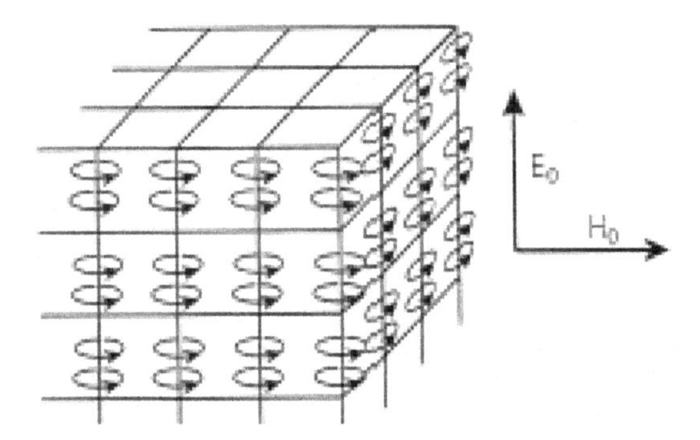

Fig. 5.4 Circular magnetic field excited in the cubic mesh crystal by external electric field.

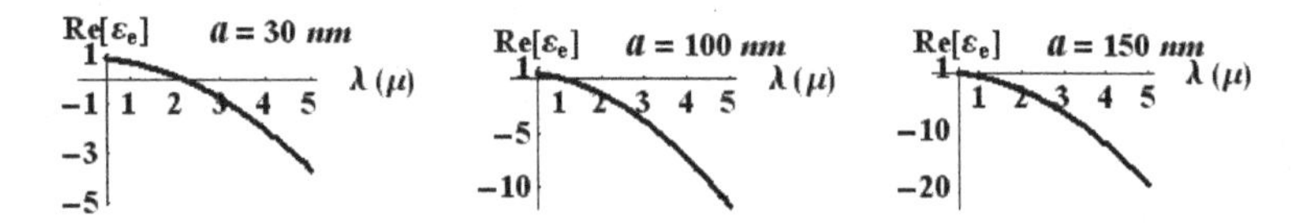

Fig. 5.5 Real part of effective dielectric constant $\varepsilon_e(\lambda)$ in silver mesh crystal. Period of lattice equals to $\mathcal{L} = 500$ nm, radius of wire $a = 30,\ 100,\ 150$ nm.

that the electric field of an incident wave excites not only the current in the metal wires but also the circular magnetic induction $\mathbf{B}_E$ around the wires as it is sketched in Fig. 5.4. Thus, the energy of the incident wave reversibly converts to the energy of the circular magnetic field, which concentrates around the wires. The magnetic field, in turn, generates an electric field, which is phase-shifted by π with respect to the external electric field. The average electric field is directed in the opposite direction with respect to the external field, when the secondary field is larger than the primary electric field (which occurs at the strong skin effect). The effective permittivity becomes negative in this case.

In Figs. 5.5 and 5.6, we show an optical permittivity of silver mesh crystal. Variation of the (nano)wire diameter and the lattice period allow to fit the negative effective permittivity in the optical range. It is interesting to note that the loss remains nearly the same when we increase the wire radius from 30 nm to 100 nm. Effectively, this is the result of the optical skin effect.

To summarize, the above equations describe the "macroscopic" electromagnetism in metal-dielectric media. The equations hold on scales much

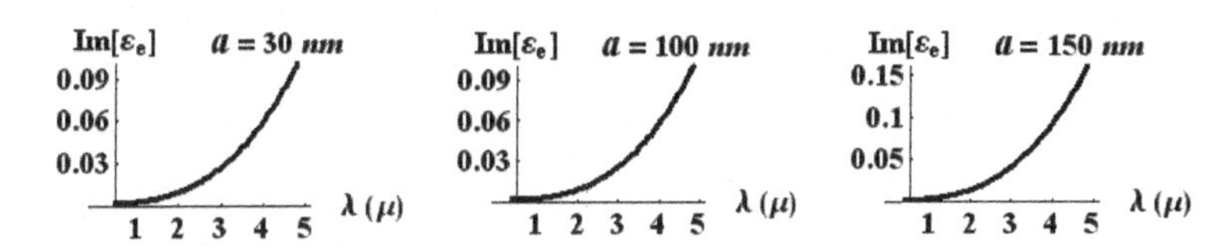

Fig. 5.6 Imaginary part of effective dielectric constant $\varepsilon_e(\lambda)$ in silver mesh crystal. Period of lattice equals to $\mathcal{L} = 500$ nm, radius of wire $a = 30,\ 100,\ 150$ nm.

larger than the spatial scale of inhomogeneity, e.g., the size of a metal grain. In the derivation of the macroscopic equations, the eddy currents are taken into account. These currents are excited in the metal grains or the (nano)wires by the high-frequency magnetic field. The eddy currents of the magnetic induction are induced by a high-frequency electric field. The latter has no analogy in the classical electrodynamics since an electric field does not generate the magnetic induction in the atoms and the molecules. The presented theory offers the macroscopic Maxwell equations, which describe the wave propagation in metal-dielectric media that include the effective permittivity and permeability.

The theory also provides the unambiguous procedure for the calculation of the effective parameters. In the case of periodic metal-dielectric structures, known as electromagnetic crystals, the explicit equations are obtained for the effective permittivity and permeability. Thus the cubic lattice of thin conducting wires appears to have a negative permittivity with negligible loss when the skin effect is strong. The negative values of the permittivity result from the eddy currents of the magnetic induction induced by the high-frequency electric field inside and around the metal wires. Such electromagnetic crystals have properties, which are similar to the bulk metal in the optical and near-infrared spectral ranges, e.g., the surface plasmons can be excited at the boundaries and at defects inside the crystal. Another interesting high-frequency property of the electromagnetic crystals is the effective magnetism, which can be observed in composite systems where neither of the components (the metal and the dielectric) has the inherent magnetism.

Bibliography

Abrahams, E., P. Anderson, D. Licciardello, and T. Ramakrishnan: 1979. *Phys. Rev. Lett.* **42**, 673.
Agranovich, M. V. and D. L. Mills (eds.): 1982, *Surface Polaritons*. Amsterdam: North-Holland.
Aharony, A.: 1987. *Phys. Rev. Lett.* **58**, 2726.
Aharony, A.: 1994. *Physica A* **205**, 330.
Aharony, A., R. Blumenfeld, and A. Harris: 1993. *Phys. Rev. B* **47**, 5756.
Aizpurua, J., P. Hanarp, D. S. Sutherland, M. Kall, G. W. Bryant, and F. J. G. de Abajo: 2003. *Phys. Rev. Lett.* **90**, 057401–1.
Aktsipetrov, O., O. Keller, K. Pedersen, A. Nikulin, N. Novikova, and A. Fedyanin: 1993. *Phys. Let. A* **179**, 149.
Alu, A. and N. Engheta: 2004. *IEEE T. Microw. Theory* **52**, 199.
Anderson, P.: 1958. *Phys. Rev.* **109**, 1492.
Antonov, A., V. Batenin, A. Vinogradov, A. Kalachev, A. Kulakov, A. Lagarkov, A. Matitsin, L. Panina, K. Rozanov, A. Sarychev, and Y. Smychkovich: 1990, *Electric and Magnetic Properties of the Percolating Systems*. Moscow: Institute for High Temperatures, Academy of Science.
Ashcroft, N. W. and N. D. Mermin: 1976, *Solid State Physics*. NY: Holt, Rinehart and Winston.
Astilean, S., P. Lalanne, and M. Palamaru: 2000. *Optics Comm.* **175**, 265.
Avrutsky, I., Y. Zhao, and V. Kochergin: 2000. *Optics Lett.* **25**, 595.
Aydin, K., K. Guven, M. Kafesaki, L. Zhang, C. M. Soukoulis, and E. Ozbay: 2004. *Optic Letters* **29**, 2623.
Balanis, C.: 2005, *Antenna Theory: Analysis and Design*. NY: John Wiley.
Balberg, I.: 1985. *Phys. Rev. B* **31**, 4053.
Balberg, I.: 1986. *Phys. Rev. B* **30**, 3618.
Balberg, I.: 1987. *Philos. Mag.* **56**, 1991.
Balberg, I.: 1991. *Phys. Rev. B* **31**, 1287.
Balberg, I., N. Binenbaum, and C. Anderson: 1983. *Phys. Rev. Lett.* **51**, 1605.
Balberg, I., N. Binenbaum, and N. Wagner: 1984. *Phys. Rev. Lett.* **52**, 1465.
Ballesteros, H., L. Fernandez, V. Martin-Mayor, A. Sudupe, G. Parisi, and J. Ruiz-Lorenzo: 1999. *J. Phys. A* **32**, 1.

Bardhan, K.: 1997. *Physica A* **241**, 267.
Bardhan, K. and R. Chakrabarty: 1994. *Phys. Rev. Lett.* **72**, 1068.
Barnes, W., A. Dereux, and T. Ebbesen: 2003. *Nature* **424**, 824.
Barnes, W., W. Murray, J. Dintinger, E. Devaux, and T. Ebbesen: 2004. *Phys. Rev. Lett.* **92**, 107401.
Barnes, W., T. Preist, S. Kitson, and J. Sambles: 1996. *Phys. Rev. B* **54**, 6227.
Barnes, W., T. Preist, S. Kitson, J. Sambles, N. Cotter, and D. Nash: 1995. *Phys. Rev. B* **51**, 11164.
Barrera, R., J. Giraldo, and W. Mochan: 1993. *Phys. Rev. B* **47**, 8528.
Baskin, E. M., M. V. Entin, A. K. Sarychev, and A. A. Snarskii: 1997. *Physica A* **242**, 49.
Bayindir, M., E. Cubukcu, I. Bulu, T. Tut, E. Ozbay, and C. Soukoulis: 2001. *Phys. Rev. B* **64**, 195113.
Bennett, P., V. Albanis, Y. Svirko, and N. Zheludev: 1999. *Opt. Lett.* **24**, 1373.
Bergman, D.: 1989. *Phys. Rev. B* **39**, 4598.
Bergman, D.: 1991. In: G. D. Maso and G. Dell'Antinio (eds.): *Composite Media and Homogenization Theory.* Boston, p. 67, Brikhauser.
Bergman, D., E. Duering, and M. Murat: 1990. *J. Stat. Phys.* **58**, 1.
Bergman, D. and Y. Imry: 1977. *Phys. Rev. Lett.* **39**, 1222.
Bergman, D., O. Levy, and D. Stroud: 1994. *Phys. Rev. B* **49**, 129.
Bergman, D. and D. Stroud: 1992. *Solid State Phys.* **46**, 147.
Bernasconi, J.: 1978. *Phys. Rev. B* **18**, 2185.
Bethe, H. A.: 1944. *Phys. Rev.* **66**, 163.
Blacher, S., F. Brouers, P. Gadenne, and J. Lafait: 1993. *J. Appl. Phys.* **74**, 207.
Blacher, S., F. Brouers, and A. K. Sarychev: 1995, 'Multifractality of Giant Field Fluctuations in Semicontinuous Metal Films'. In: M. Novak (ed.): *Fractals in the Natural and Applied Sciences.* London, Chapman and Hall.
Bloembergen, N., R. Chang, S. Jha, and C. Lee: 1968. *Phys. Rev.* **174**, 813.
Boardman, A. (ed.): 1982, *Electromagnetic Surface Modes.* NY: John Wiley.
Boliver, A.: 1992. *J. Phys. A: Math. Gen.* **25**, 1021.
Borisov, A., F. G. de Abajo, and S. Shabanov: 2005. *Phys. Rev. B* **71**, 075408.
Botten, L. C. and R. C. McPhedran: 1985. *Optica Acta* **32**, 595.
Boyd, R. W.: 1992, *Nonlinear Optics.* NY: Academic Press.
Bozhevolnyi, S. and V. Coello: 2001. *Phys. Rev. B* **64**, 115414.
Bozhevolnyi, S., J. Erland, K. Leosson, P. Skovgaard, and J. M. Hvam: 2001. *Phys. Rev. Lett.* **86**, 3008.
Bozhevolnyi, S., V. Markel, V. Coello, W. Kim, and V. Shalaev: 1998. *Phys. Rev. B* **58**, 11441.
Bozhevolnyi, S. and F. Pudonin: 1997. *Phys. Rev. Lett.* **78**, 2823.
Bravo-Abad, J., A. Degiron, F. Przybilla, C. Gernet, F. Garcia-Vidal, L. Martin-Moreno, and T. Ebbesen: 2006. *Nature Physics* **2**, 120.
Breit, M., V. Podolskiy, S. Gresillon, G. von Plessen, J. Feldmann, J. Rivoal, P. Gadenne, A. Sarychev, and V. Shalaev: 2001. *Phys. Rev. B* **64**, 125106.
Brouers, F.: 1986. *J. Phys. C* **19**, 7183.
Brouers, F., S. Blacher, A. N. Lagarkov, A. K. Sarychev, P. Gadenne, and V. M. Shalaev: 1997a. *Phys. Rev. B* **55**, 13234.

Brouers, F., S. Blacher, and A. K. Sarychev: 1998. *Phys. Rev. B* **58**, 15897.

Brouers, F., A. K. Sarychev, S. Blacher, and O. Lothaire: 1997b. *Physica A* **241**, 146.

Bruggeman, D. A. G.: 1935. *Ann. Phys.* **24**, 636.

Butenko, A., V. Shalaev, and M. Stockman: 1988. *Sov. Phys. JETP* **67**, 60.

Byshkin, M. and A. Turkin: 2005. *J. Physics A* **38**, 5057.

Caloz, C., A. Lai, and T. Itoh: 2005. *New J. Phys.* **7**, 167.

Caloz, C., C.-J. Lee, D. Smith, J. Pendry, and T. Itoh: 2004, 'Existence and Properties of Microwave Surface Plasmons at the Interface Between a Right-Handed and a Left-Handed Media'. In: *IEEE-AP-S USNC/URSI National Radio Science Meeting.* IEEE.

Cao, H. and A. Nahata: 2004. *Optic Express* **12**, 1004.

Cao, Q. and P. Lalanne: 2002. *Phys. Rev. Lett.* **88**, 057403.

Chaikin, P. and T. Lubensky: 1995, *Principles of Condensed Matter Physics.* Cambridge: Cambridge Univ. Press.

Chang, R. and T. Furtak (eds.): 1982, *Surface Enhance Raman Scattering.* NY: Plenum Press.

Chang, S.-H., S. Gray, and G. Schatz: 2005. *Optic Express* **13**, 3150.

Chang, T.-E., A. Kisliuk, S. Rhodes, W. Brittain, and A. Sokolov: 2006. *Polymer* **47**, 7740.

Chen, C., A. de Castro, and Y. Shen: 1981. *Phys. Rev. Lett.* **46**, 145.

Chen, L., C. Ong, C. Neoa, V. Varadan, and V. Varadan: 2004, *Microwave Electronics: Measurement and Materials Characterization.* NY: John Wiley.

Clerc, J., V. Podolskiy, and A. Sarychev: 2000. *Euro. Phys. Journ. B* **15**, 507.

Clerc, J. P., G. Giraud, and J. M. Luck: 1990. *Adv. Phys.* **39**, 191.

Cohen, R. W., G. D. Cody, M. D. Coutts, and B. Abeles: 1973. *Phys. Rev. B* **8**, 3689.

Coyle, S., M. Netti, J. Baumberg, M. Ghanem, P. Birkin, and P. Bartlett: 2001. *Phys. Rev. Lett.* **87**, 176801.

Darmanyan, S. and A. Zayats: 2003. *Phys. Rev. B* **67**, 035424.

de Abajo, F. G., R. Gomez-Medina, and J. Saenz: 2005a. *Phys. Rev. E* **72**, 016608.

de Abajo, F. G. and J. Saenz: 2005. *Phys. Rev. Lett.* **95**, 233901.

de Abajo, F. J. G., G. Gomez-Santos, L. Blanco, A. Borisov, and S. Shabanov: 2005b. *Phys. Rev. Lett.* **95**, 067403.

de Arcangelis, L., S. Redner, and H. Herrmann: 1985. *J. de Physique Lett.* **46**, L585.

de Gennes, P. G.: 1976. *J. Phys. Paris Lett.* **37**, L1.

Debrus, S., J. Lafaitand, M. May, N. Pincon, D. Prot, C. Sella, and J. Venturini: 2000. *J. Appl. Phys.* **88**, 4469.

Debye, P.: 1912. *Phys. Z.* **9**, 341.

Degiron, A. and T. Ebbesen: 2004. *Optic Express* **12**, 3694.

Depardieu, G., P. Frioni, and S. Berthier: 1994. *Physica A* **207**, 110.

Depine, R. A. and A. Lakhtakia: 2004. *Phys. Rev. E* **69**, 057602.

Dolling, G., C. Enkrich, M. Wegener, J. Zhou, and C. Soukoulis: 2005. *Optics Lett.* **30**, 3198.

Drachev, V., V. Shalaev, and A. Sarychev: 2006, 'Raman Imaging and Sensing Apparatus Employing Nanoantennas'. *U.S. Patent* **No 6,985,223**.

Dragila, R., B. Luther-Davies, and S. Vukovic: 1985. *Phys. Rev. Lett.* **55**, 1117.

Draine, B.: 1988. *Astrophys. J.* **333**, 848.

Ducourtieux, S., S. Gresillon, A. Boccara, J. Rivoal, X. Quelin, P. Gadenne, V. Drachev, W. Bragg, V. Safonov, V. Podolskiy, Z. Ying, R. Armstrong, and V. Shalaev: 2000. *J. Nonlinear Opt. Phys. Mater.* **9**, 105.

Ducourtieux, S., V. Podolskiy, S. Gresillon, P. Gadenne, A. Boccara, J. Rivoal, W. Bragg, V. Safonov, V. Drachev, Z. Ying, A. Sarychev, and V. Shalaev: 2001. *Phys. Rev. B* **64**, 165403.

Duxbury, P. and P. Leath: 1994. *Phys. Rev. B* **49**, 12676.

Duxbury, P., P. Leath, and P. Beale: 1987. *Phys. Rev. B* **36**, 367.

Dykhne, A., A. Sarychev, and V. Shalaev: 2003. *Phys. Rev. B* **67**, 195402.

Dykhne, A. M.: 1971. *Sov. Phys. JETP* **32**, 348.

Ebbesen, T. W., H. J. Lezec, H. F. Ghaemi, T. Thio, and P. A. Wolff: 1998. *Nature* **391**, 667.

Efros, A. and B. Shklovskii: 1976. *Phys. Status Solidi B* **76**, 475.

Efros, A. L. and A. L. Pokrovsky: 2004. *Solid St. Comm.* **129**, 643.

Enkrich, C., M. Wegener, S. Linden, S. Burger, L. Zschiedrich, F. Schmidt, J. Zhou, T. Koschny, and C. Soukoulis: 2005. *Phys. Rev. Lett.* **95**, 203901.

Felderhof, B.: 1989. *Phys. Rev. B* **39**, 5669.

Flytzanis, C.: 1996, 'Nonlinear Optics'. In: AIP (ed.): *Encyclopedia of Applied Physics*. New York, p. 487, VHC Publisher.

Flytzanis, C., F. Hache, M. Klein, D. Ricard, and P. Roussignol: 1992. *Prog. Opt.* **29**, 2539.

Foteinopoulou, S. and C. M. Soukoulis: 2003. *Phys. Rev. B* **67**, 235107.

Frank, D. J. and C. J. Lobb: 1988. *Phys. Rev. B* **37**, 302.

Freilikher, V. D., E. Kanzieper, and A. A. Maradudin: 1997. *Phys. Rep.* **288**, 127.

Gadenne, P., A. Beghadi, and J. Lafait: 1988. *Optics Comm.* **65**, 17.

Gadenne, P., F. Brouers, V. M. Shalaev, and A. K. Sarychev: 1998. *J. Opt. Soc. Am. B* **15**, 68.

Gadenne, P., X. Quelin, S. Ducourtieux, S. Gresillon, L. Aigouy, J. Rivoal, V. Shalaev, and A. Sarychev: 2000. *Physica B* **279**, 52.

Gadenne, P. and J. Rivoal: 2002, *Surface-Plasmon-Enhanced Nonlinearities in Percolating 2-D Metal-Dielectric Films: Calculation of the Localized Giant Field and their Observation in SNOM*, Vol. 82. Springer Verlag.

Gadenne, P., Y. Yagil, and G. Deutscher: 1989. *J. Appl. Phys.* **66**, 3019.

Garanov, V., A. Kalachev, A. Karimov, A. Lagarkov, S. Matytsin, A. Pakhomov, B. Peregood, A. Sarychev, A. Vinogradov, and A. Virnic: 1991. *J. Phys. C: Condens. Matter* **3**, 3367.

Garcia-Vidal, F., L. Martin-Moreno, H. Lezec, and T. Ebbesen: 2003. *Appl. Phys. Lett.* **83**, 4500.

Garnett, J. C. M.: 1904. *Philos. Trans. Roy. Soc.* **203**, 385.

Genchev, Z. and D. Dosev: 2004. *J. Exp. Theor. Phys.* **99**, 1129.

Genov, D., A. Sarychev, and V. Shalaev: 2003a. *Phys. Rev. E* **67**, 056611.

Genov, D., A. Sarychev, and V. Shalaev: 2003b, 'Local Field Statistic and Plasmon Localization in Random Metal-Dielectric Films'. In: S. Skipetrov and B. V. Tiggelen (eds.): *Wave Scattering in Complex Media: From Theory to Applications*. Dordrecht, Netherlands, p. 139, Kluwer Acad.

Genov, D., A. Sarychev, V. Shalaev, and A. Wei: 2004. *Nano Letters* **4**, 167401.

Genov, D., V. Shalaev, and A. Sarychev: 2005. *Phys. Rev. B* **72**, 113102.

Ghaemi, H., T. Thio, D. Grupp, T. Ebbesen, and H. Lezec: 1998. *Phys. Rev. B* **58**, 6779.

Gibson, U. and R. Buhrman: 1983. *Phys. Rev. B* **27**, 5046.

Gippius, N., S. Tikhodeev, and T. Ishihara: 2005. *Phys. Rev. B* **72**, 045138.

Govyadinov, A. A. and V. A. Podolskiy: 2006. *Phys. Rev. B* **73**, 115108.

Grannan, D. M., J. C. Garland, and D. B. Tanner: 1981. *Phys. Rev. Lett.* **46**, 375.

Granqvist, C. and O. Hunderi: 1977. *Phys. Rev. B* **16**, 3513.

Granqvist, C. and O. Hunderi: 1978. *Phys. Rev. B* **18**, 1554.

Gresillon, S., L. Aigouy, A. C. Boccara, J. C. Rivoal, X. Quelin, C. Desmarest, P. Gadenne, V. A. Shubin, A. K. Sarychev, and V. M. Shalaev: 1999a. *Phys. Rev. Lett.* **82**, 4520.

Gresillon, S., J. C. Rivoal, P. Gadenne, X. Quelin, V. M. Shalaev, and A. K. Sarychev: 1999b. *Phys. Stat. Sol. A* **175**, 337.

Grigorenko, A., A. Geim, H. Gleeson, Y. Zhang, A. Firsov, I. Khrushchev, and J. Petrovic: 2005. *Nature* **438**, 335.

Grimes, C., E. Dickey, C. Mungle, K. Ong, and D. Qian: 2001. *J. Appl. Phys.* **90**, 4134.

Grimes, C., C. Mungle, D. Kouzoudis, S. Fang, and P. Eklund: 2000. *Chem. Phys. Lett.* **319**, 460.

Grupp, D., H. Lezec, T. Ebbesen, K. Pellerin, and T. Thio: 2000. *Appl. Phys. Lett.* **77**, 1569.

Hafner, C., X. Cui, and R. Vahldieck: 2005. *J. Comp. Theor. Nanoscience* **2**, 240.

Hallen, E.: 1962, *Electromagnetic Theory*. London: Chapman and Hall.

Hecht, B., H. Bielefeldt, L. Novotny, Y. Inouye, and D. Pohl: 1996. *Phys. Rev. Lett.* **77**, 1889.

Herrmann, H. and S. Roux (eds.): 1990, *Statistical Models for the Fracture of Disordered Media*. Amsterdam: Elsevier Science, North-Holland.

Hibbins, A., B. Evans, and J. Sambles: 2005. *Science* **308**, 670.

Hibbins, A. and J. Sambles: 2002. *Appl. Phys. Lett.* **81**, 4661.

Hibbins, A., J. Sambles, and C. Lawrence: 1999. *J. Appl. Phys.* **86**, 1797.

Hilfer, R.: 1992. *Phys. Rev. B* **45**, 7115.

Hilfer, R.: 1993. *Physica A* **194**, 406.

Hilke, M.: 1994. *J. Phys. A: Math. Gen.* **27**, 4773.

Hock, L., O. Kim, S. Matitsin, and G. Beng (eds.): 2004, 'Proceedings of the Symposium F Electromagnetic Materials: Suntec, Singapore, 7-12 December 2003'. World Scientific.

Hofstetter, E. and M. Schreiber: 1993. *Phys. Rev. B* **48**, 16979.

Hongxing, X., J. Aizpurua, M. Käll, and P. Apell: 2000. *Phys. Rev. E* **62**, 4318.

Houck, A., J. Brock, and I. Chuang: 2003. *Phys. Rev. Lett.* **90**, 137401.
Hui, P.: 1996, 'Nonlinearity and Breakdown in Soft Condensed Matter'. In: *Lecture Notes in Physics*, Vol. 437 of *Lecture Notes in Physics*. Berlin, Springer Verlag.
Hui, P., P. Cheung, and Y. Kwong: 1997. *Physica A* **241**, 301. and references therein.
Hui, P., P. Cheung, and D. Stroud: 1998. *J. Appl. Phys.* **84**, 3451.
Hui, P. and D. Stroud: 1994. *Phys. Rev. B* **49**, 11729.
Hui, P. and D. Stroud: 1997. *J. Appl. Phys.* **82**, 4740.
Hui, P., C. Xu, and D. Stroud: 2004a. *Phys. Rev. B* **69**, 014202.
Hui, P., C. Xu, and D. Stroud: 2004b. *Phys. Rev. B* **69**, 014202.
Huinink, H., L. Pel, and M. Michels: 2003. *Phys. Rev. E* **68**, 056114.
Ilyinsky, A. S., G. Y. Slepyan, and A. Y. Slepyan: 1993, *Propagation, Scattering and Dissipation of Electromagnetic Waves*, Vol. 36 of *IEE Electromagnetic Waves Series*. London: Peregrinus.
Jackson, J.: 1998, *Classical Electrodynamics*. NY: John Wiley, 3rd edition.
Jacob, Z., L. V. Alekseyev, and E. Narimanov: 2006. *Opt. Express* **14**, 8247.
Johnson, B. and R. Christy: 1972. *Phys. Rev. B* **6**, 4370.
Kadanoff, L., W. Gotze, D. Hamblen, R. Hecht, E. Lewis, V. Palciaus, M. Rayl, J. Swift, D. Aspnes, and J. Kane: 1967, 'Static Phenomena Near Critical Points - Theory and Experiment'. *Rev. Mod. Phys.* **39**, 395.
Kalachev, A., I. Kukolev, S. Matytsin, K. Rozanov, and A. Sarychev: 1991. *Soviet J. Nondestruct. Test.* **27**, 427.
Kats, A. and A. Nikitin: 2004. *Phys. Rev. B* **70**, 235412.
Katsarakis, . N., T. Koschny, M. Kafesaki, E. N. Economy, and C. M. Soukoulis: 2004. *Appl. Phys. Lett.* **84**, 2943.
Kawarabayashi, T., B. Kramer, and T. Ohtsuki: 1998. *Phys. Rev. B* **57**, 11842.
Kawata, S. (ed.): 2001, *Near-Field Optics and Surface Plasmon Polaritons*. Springer.
Keblinski, P. and F. Cleri: 2004. *Phys. Rev. B* **69**, 184201.
Khang, B., G. Bartrouni, S. Redner, L. de Arcangelis, and H. Herrmann: 1988. *Phys. Rev. B* **37**, 7625.
Kildishev, A., W. Cai, U. Chettiar, H.-K. Yuan, A. Sarychev, V. Drachev, and V. Shalaev: 2006. *JOSA B* **23**, 423.
Kim, T. J., T. Thio, T. W. Ebbesen, D. E. Grupp, and H. J. Lezec: 1999. *Opt. Lett.* **24**, 256.
Kirkpatrik, S.: 1973. *Rev. Mod. Phys.* **45**, 574.
Kneipp, K., H. Kneipp, I. Itzkan, R. Dasari, and M. Feld: 1999. *Chem. Rev.* **99**, 2957.
Kneipp, K., H. Kneipp, I. Itzkan, R. Dasari, and M. Feld: 2002, 'Single Molecules Detection Using Near Infrared Surface-Enhanced Raman Scattering'. In: R.Rigler, M. Orrit, and T. Basche (eds.): *Single Molecule Spectroscopy, Nobel Conference Lectures*. Heidelberg, Germany, p. 144, Springer Verlag.
Kneipp, K., Y. Wang, H. Kneipp, L. Perelman, I. Itzkan, R. Dasari, and M. Feld: 1997. *Phys. Rev. Lett.* **78**, 1667.
Kolesnikov, A., A. Lagarkov, L. Novogrudskiy, S. Matitsin, K. Rozanov, and

A. Sarychev: 1991. In: J. Emerson and J. Torkelson (eds.): *Optical and Electrical Properties of Polymers*, Vol. 214 of *MRS Symposia Proceedings*. Pittsburgh, p. 119, Material Research Society.

Koschny, T., M. Kafesaki, E. N. Economo, and C. Soukoulis: 2004. *Phys. Rev. Lett.* **93**, 107402–1.

Kramer, B. and A. MacKinnon: 1993, 'Localization: Theory and Experiment'. *Rep. Prog. Phys.* **56**, 1469.

Kreibig, U. and M. Volmer: 1995, *Optical Properties of Metal Clusters*. Berlin: Springer-Verlag.

Krenn, J., A. Dereux, J. Weeber, E. Bourillot, Y. Lacroute, and J. Goudonnet: 1999. *Phys. Rev. Lett.* **82**, 2590.

Krishnan, A., T. Thio, T. Kim, H. Lezec, T. Ebbesen, P. Wolff, J. Pendry, L. Martin-Moreno, and F. Garcia-Vidal: 2001. *Opt. Comm.* **200**, 1.

Krug, J., G. Wang, S. Emory, and S. Nie: 1999. *J. Am. Chem. Soc.* **121**, 9208.

Kuang, L. and H. Simon: 1995. *Phys. Lett. A* **197**, 257.

Kukhlevsky, S., M. Mechler, L. Csapo, K. Janssens, and O. Samek: 2005. *Phys. Rev. B* **72**, 165421.

Lagarkov, A., S. Matitsin, K. Rozanov, and A. Sarychev: 1998. *J. Appl. Phys.* **84**, 3806.

Lagarkov, A., S. Matytsin, K. Rozanov, and A. Sarychev: 1997a. *Physica A* **241**, 58.

Lagarkov, A. and A. Sarychev: 1996. *Phys. Rev. B* **53**, 6318.

Lagarkov, A., A. Sarychev, Y. Smychkovich, and A. Vinogradov: 1992. *J. Electromagn. Waves Appl.* **6**, 1159.

Lagarkov, A., V. Semenenko, V. Chistyaev, D. Ryabov, S. Tretyakov, and C. Simovski: 1997b. *Electromagnetics* **17**, 213.

Lagarkov, A. N. and V. N. Kissel: 2004. *Phys. Rev. Lett.* **92**, 077401–2.

Lagarkov, A. N., L. V. Panina, and A. Sarychev: 1987. *Sov. Phys. JETP* **66**, 123.

Lagarkov, A. N., K. N. Rozanov, A. K. Sarychev, and A. N. Simonov: 1997c. *Physica A* **241**, 199.

Laks, B., D. Mills, and A. Maradudin: 1981. *Phys. Rev. B* **23**, 4965.

Lalanne, P., J. Hugonin, S. Astilean, M. Palamaru, and K. Moller: 2000. *J. Opt. A.* **2**, 48.

Lalanne, P., J. Rodier, and J. Hugonin: 2005. *J. Opt. A* **7**, 422.

Landau, L. D., E. M. Lifshitz, and L. P. Pitaevskii: 1984, *Electromagnetics of Continuous Media*. Oxford: Pergamon, 2nd edition.

Landauer, R.: 1978. In: J. G. D. Tanner (ed.): *Electrical Transport and Optical Properties of Inhomogeneous Media*, Vol. 40 of *AIP. Conf. Proc.* New York, p. 2, AIP.

Laroche, M., R. Carminati, and J.-J. Greffet: 2005. *Phys. Rev. B* **71**, 155113.

Lee, K. and Q.-H. Park: 2005. *Phys. Rev. Lett.* **95**, 103902.

Leveque, G. and O. Martin: 2006. *Optics Express* **14**, 9973.

Levy, O. and D. Bergman: 1993. *J. Phys. C: Condens. Matter* **5**, 7095.

Levy, O. and D. Bergman: 1994. *Physica A* **207**, 157.

Levy, O., D. Bergman, and D. Stroud: 1995. *Phys. Rev. E* **52**, 3184.

Levy-Nathansohn and D. J. Bergman: 1997a. *Phys. Rev. B* **55**, 5425.
Levy-Nathansohn and D. J. Bergman: 1997b. *Physica A* **241**, 166.
Lezec, H., A. Degiron, E. Devaux, R. Linke, L. Martin-Moreno, F. Garcia-Vidal, and T. Ebbesen: 2002. *Science* **297**, 820.
Lezec, H., J. Dionne, and H. Atwater: 2007. *Science* **316**, 431.
Lezec, H. and T. Thio: 2004. *Optic Express* **12**, 3629.
Liao, H., R. Xiao, H. Wang, K. Wong, and G. Wong: 1998. *Appl. Phys. Lett.* **72**, 1817.
Liebsch, A. and W. Schaich: 1989. *Phys. Rev. B* **40**, 5401.
Lifshits, I. M., S. A. Gredeskul, and L. A. Pastur: 1988, *Introduction to the Theory of Disordered Systems.* NY: John Wiley.
Linde, S., C. Enkrich, M. Wegener, J. Zhou, T. Koschny, and C. M. Soukoulis: 2004. *Science* **306**, 1351.
Liu, Z., H. Lee, Y. Xiong, C. Sun, and X. Zhang: 2007. *Science* **315**, 1686.
Lockyear, M., A. Hibbins, and J. Sambles: 2004. *Appl. Phys. Lett.* **84**, 2040.
Lopez-Rios, T., D. Mendoza, F. Garcia-Vidal, J. Sanchez-Dehesa, and B. Pannetier: 1998. *Phys. Rev. Lett.* **81**, 665.
Lyon, L., C. Keating, A. Fox, B. Baker, L. He, S. Nicewarner, S. Mulvaney, and M. Natan: 1998. *Anal. Chem.* **70**, R341.
Ma, H., R. Xiao, and P. Sheng: 1998. *J. Opt. Soc. Am. B* **15**, 1022.
Madrazo, A. and M. Nieto-Vesperinas: 1997. *Appl. Phys. Lett.* **70**, 31.
Mahmoud, F.: 1991, *AND Electromagnetic Waveguides: Theory and Applicatio.* London: Peregrinus.
Makhnovskiy, D., L. Panina, D. Mapps, and A. Sarychev: 2001. *Phys. Rev. B.* **64**, 134205.
Markel, V. and T. George (eds.): 2000, *Optics of Nanostructured Materials.* NY: John Wiley.
Marques, R., F. Medina, and R. Rafii-El-Idrissi: 2002. *Phys. Rev. B* **65**, 144440.
Martin-Moreno, L., F. Garcia-Vidal, H. Lezec, A. Degiron, and T. Ebbesen: 2003. *Phys. Rev. Lett.* **90**, 167401.
Martin-Moreno, L., F. Garcia-Vidal, H. Lezec, K. Pellerin, T. Thio, J. Pendry, and T. Ebbesen: 2001. *Phys. Rev. Lett.* **86**, 1114.
Martins, P. and J. Plascak: 2003. *Phys. Rev. E* **67**, 046119.
Maslovski, S., S. Tretyakov, and P. Belov: 2002. *Microwave and Optical Technology Let.* **35**, 47.
McGurn, A. and A. Maradudin: 1993. *Phys. Rev. B* **48**, 17576.
McGurn, A., A. Maradudin, and V. Celli: 1985. *Phys. Rev. B* **31**, 4866.
McGurn, A. R. and A. A. Maradudin: 1985. *Phys. Rev. B* **31**, 4866.
McPhedran, R. and G. Milton: 1981. *Appl. Phys. A* **26**, 207.
McPhedran, R., N. Nicorovici, and L. Botten: 1997. *J. Electromag. Waves Appl.* **11**, 981.
Meakin, P.: 1991. *Science* **252**, 226.
Milton, G.: 1981. *J. Appl. Phys.* **52**, 5286.
Mittra, R., C. Chan, and T. Cwik: 1988. *Proc. IEEE* **76**, 1593.
Monette, L., M. Anderson, and G. Grest: 1994. *J. Appl. Phys.* **75**, 1155.
Moresco, F., M. Rocca, T. Hildebrandt, and M. Henzler: 1999. *Phys. Rev. Lett.*

83, 2238.

Moskovits, M.: 1985. *Rev. Mod. Phys.* **57**, 783.

Moskovits, M.: 2004. private communication.

Moukarzel, C. and P. Duxbury: 1995. *Phys. Rev. Lett.* **75**, 4055.

Munk, B.: 2000, *Frequency Selective Surfaces. Theory and Design.* NY: John Wiley.

Murphy, R., M. Yeganeh, K. Song, and E. Plummer: 1989. *Phys. Rev. Lett* **63**, 318.

Nandi, U. and K. Bardhan: 1995. *Europhys. Lett.* **31**, 101.

Nettelblab, S. and G. Niklasson: 1994. *Solid State Commun.* **90**, 201.

Newman, M. and R. Ziff: 2001. *Phys. Rev. E* **64**, 016706.

Nicorovici, N. A., R. C. McPhedran, and L. C. Botten: 1995a. *Phys. Rev. Lett.* **75**, 1507.

Nicorovici, N. A., R. C. McPhedran, and L. C. Botten: 1995b. *Phys. Rev. E* **52**, 1135.

Nie, S. and S. Emory: 1997. *Science* **275**, 1102.

Niklasson, G. A. and C. G. Granquist: 1984. *J. Appl. Phys.* **55**, 3382.

Nishiyama, Y.: 2006. *Phys. Rev. E* **73**, 016114.

Noh, T. W., P. H. Song, S.-I. Lee, D. C. Harris, J. R. Gaines, and J. C. Garland: 1992. *Phys. Rev. B* **46**, 4212.

Nolte, D.: 1999. *J. Appl Phys.* **85**, 6259.

O'Brien, S., D. McPeake, S. Ramakrishna, and J. Pendry: 2004. *Phys. Rev. B* **69**, 241101.

Olbright, G., N. Peyghambarian, S. Koch, and L. Banyai: 1987. *Opt. Lett.* **12**, 413.

Palik, E. D. (ed.): 1985, *Hand Book of Optical Constants of Solids.* NY: Academic Press.

Panina, L., A. Grigorenko, and D. Makhnovskiy: 2002. *Phys. Rev. B* **66**, 155411.

Panina, L., A. Lagarkov, A. Sarychev, Y. Smychkovich, and A. Vinogradov: 1990. In: G. Cody, T. Geballe, and P. Sheng (eds.): *Physical Phenomena in Granular Materials*, Vol. 195 of *MRS Symposia Proceedings.* Pittsburgh, p. 275, Material Research Society.

Parage, F., M. Doria, and O. Buisson: 1998. *Phys. Rev. B* **58**, R8921.

Parazzoli, C., R. Greegor, K. Li, B. Koltenbah, and M. Tanielian: 2003. *Phys. Rev. Lett.* **90**, 107401.

Park, S., N. Hur, S. Guha, and S. Cheong: 2004. *Phys. Rev. Lett.* **92**, 167206.

Pendry, J.: 2000. *Phys. Rev. Lett.* **85**, 3966.

Pendry, J.: 2004a. *Science* **306**, 1353.

Pendry, J., A. Holden, D. Robbins, and W. Stewart: 1998. *J. Phys.: Cond. Mat.* **10**, 4785.

Pendry, J., A. Holden, D. Robbins, and W. Stewart: 1999. *IEEE T. Microw. Theory* **47**, 2075.

Pendry, J., A. Holden, W. Stewart, and I. Youngs: 1996. *Phys. Rev. Lett.* **76**, 4773.

Pendry, J., L. Martin-Moreno, and F. Garcia-Vidal: 2004. *Science* **305**, 847.

Pendry, J. and D. Smith: 2004. *Physics Today* **57**, 37.

Pendry, J. B.: 2003. *Opt. Express* **11**, 755.
Pendry, J. B. and S. A. Ramakrishna: 2002. *J. Phys. Condens. Matter* **14**, 8463.
Pendry, S. R. J.: 2004b. *Phys. Rev. B* **69**, 115115.
Peng, C.: 2001. *Appl. Optics* **40**, 3922.
Podolskiy, V., A. Sarychev, and V. Shalaev: 2002a. *J. Nonlinear Opt. Phys. and Mat.* **11**, 65.
Podolskiy, V., A. Sarychev, and V. Shalaev: 2002b. *Laser Physics* **12**, 292.
Podolskiy, V., A. Sarychev, and V. Shalaev: 2003. *Optic Express* **11**, 735.
Pokrovsky, A.: 2004. *Phys. Rev. B* **69**, 195108.
Pokrovsky, A. and A. Efros: 2002. *Phys. Rev. Lett.* **89**, 093901.
Pokrovsky, A. L. and A. L. Efros: 2003. *Physica B* **338**, 196.
Poliakov, E., V. Shalaev, V. Markel, and R. Botet: 1996. *Opt. Lett.* **21**, 1628.
Popov, E., M. Nevière, S. Enoch, and R. Reinisch: 2000. *Phys. Rev. B* **62**, 16100.
Porto, J., F. Garcia-Vidal, and J. Pendry: 1999. *Phys. Rev. Lett.* **83**, 2845.
Porto, J. A., L. Martín-Moreno, and F. García-Vidal: 2004. *Phys. Rev. B* **70**, 081402.
Purcell, E. and C. Pennypacker: 1973. *Astrophys. J.* **405**, 705.
Qu, D. and D. Grischkowsky: 2004. *Phys. Rev. Lett.* **93**, 196804.
Raether, H.: 1988, *Surface Plasmons on Smooth and Rough Surfaces and on Gratings*. Berlin: Springer-Verlag.
Rayleigh, L.: 1896, *The Theory of Sound*. London: MacMillan.
Rayleigh, L.: 1920. *Philos. Mag.* **39**, 225.
Reider, G. and T. Heinz: 1995, 'Second-Order Nonlinear Optical Effects at Surfaces and Interfaces: Recent Advances'. In: P. Halevi (ed.): *Photonic Probes of Surfaces: Electromagnetic Waves*. Amsterdam, Elsevier.
Reynolds, P. J., W. Klein, and H. E. Stanley: 1977. *J. Phys. C* **10**, L167.
Ricci, M., N. Orloff, and S. Anlage: 2005. *Appl. Phys. Lett.* **87**, 034102.
Rivas, J., C. Schotsch, P. Bolivar, and H. Kurz: 2003. *PRB* **68**, 201306.
Rousselle, D., A. Berthault, O. Acher, J. Bouchaud, and P. Zerah: 1993. *J. Appl. Phys.* **74**, 475.
Roussignol, R., D. Ricard, K. Rustagi, and C. Flytzanis: 1985. *Opt. Commun.* **55**, 1413.
Salandrino, A. and N. Engheta: 2006. *Phys. Rev. B* **74**, 075103.
Salomon, L., F. Grillot, A. Zayats, and F. Fornel: 2001. *Phys. Rev. Lett.* **86**, 1110.
Sanches-Gil, J. A., A. A. Maradudin, J. Q. Lu, V. D. Freilikher, M. Pustilnik, and I. Yurkevich: 1994. *Phys. Rev. B* **50**, 15 353.
Sarychev, A., V. Drachev, H. Yuan, V. Podolskiy, and V. Shalaev: 2003. In: *Nanotubes and Nanowires*. p. 93, SPIE Proc. 5219.
Sarychev, A., V. Podolskiy, A. Dykhne, and V. Shalaev: 2002. *IEEE Journal of Quantum Electronics* **8**, 956.
Sarychev, A. and V. Shalaev: 2002, 'Theory of Nonlinear Optical Response in Metal-Dielectric Composites, in Optical Properties of Random Nanostructures'. In: V. Shalaev (ed.): *Topics in Applied Physics*, Vol. 82. p. 231, Springer Verlag.
Sarychev, A. and V. Shalaev: 2003, 'Optical Properties of Metal–Dielectric Film'.

In: W. Weiglhofer and A. Lakhtakia (eds.): *Introduction to Complex Mediums for Optics and Electromagnetics*. WA, USA, SPIE, Bellingham.
Sarychev, A. and V. Shalaev: 2004. In: *Light and Complexity*. p. 137, SPIE Proc. 5508.
Sarychev, A., V. Shubin, and V. Shalaev: 2000a. *Physica B* **279**, 87.
Sarychev, A. and Y. Smychkovich: 1990. In: G. Cody, T. Geballe, and P. Sheng (eds.): *Physical Phenomena in Granular Materials*, Vol. 195 of *MRS Symposia Proceedings*. Pittsburgh, p. 279, Material Research Society.
Sarychev, A. and G. Tartakovsky: 2006. In: G. Dewar, M. McCall, M. Noginov, and N. Zheludev (eds.): *Complex Photonic Media*. p. 0A, SPIE.
Sarychev, A. and G. Tartakovsky: 2007. *Phys. Rev. B* **75**, 085436.
Sarychev, A. and A. Vinogradov: 1981. *J. Phys. C* **14**, L487.
Sarychev, A. and A. Vinogradov: 1983. *Sov. Phys., JETP* **58**, 665.
Sarychev, A. K.: 1977. *Sov. Phys. JETP* **72**, 524.
Sarychev, A. K., D. J. Bergman, and Y. Yagil: 1995. *Phys. Rev. B* **51**, 5366.
Sarychev, A. K. and F. Brouers: 1994. *Phys. Rev. Lett.* **73**, 2895.
Sarychev, A. K., R. McPhedran, and V. M. Shalaev: 2000b. *Phys. Rev. B* **62**, 8531.
Sarychev, A. K. and V. M. Shalaev: 2000. *Phys. Rep.* **335**, 275.
Sarychev, A. K. and V. M. Shalaev: 2001, 'Field Distribution, Anderson Localization, and Optical Phenomena in Random Metal-Dielectric Films'. In: V. Markel and T. George (eds.): *Optics of Nanostructered Materials*. NY, p. 227, John Wiley.
Sarychev, A. K., V. A. Shubin, and V. M. Shalaev: 1999a. *Phys. Rev. E* **59**, 7239.
Sarychev, A. K., V. A. Shubin, and V. M. Shalaev: 1999b. *Physica A* **266**, 115.
Sarychev, A. K., V. A. Shubin, and V. M. Shalaev: 1999c. *Phys. Rev. B* **60**, 16389.
Schill, R. and J. S. Seshadri: 1988. *J. Appl. Phys.* **64**, 6530.
Schill, R. and J. S. Seshadri: 1989. *J. Appl. Phys.* **65**, 4420.
Schroter, U. and D. Heitmann: 1998. *Phys. Rev. B* **58**, 15419.
Schroter, U. and D. Heitmann: 1999. *Phys. Rev. B* **60**, 4992.
Seal, K., D. Genov, A. Sarychev, H. Noh, V. Shalaev, Z. Ying, X. Zhang, and H. Cao: 2006. *Phys. Rev. Lett.* **97**, 206103.
Seal, K., M. Nelson, Z. Ying, D. Genov, A. Sarychev, and V. Shalaev: 2002. *J. Mod. Optics* **49**, 2423.
Seal, K., M. Nelson, Z. Ying, D. Genov, A. Sarychev, and V. Shalaev: 2003. *Phys. Rev. B* **67**, 035318.
Seal, K., A. Sarychev, H. Noh, D. Genov, A. Yamilov, V. Shalaev, Z. Ying, and H. . Cao: 2005. *Phys. Rev. Lett.* **94**, 226101.
Selcuk, S., K. Woo, D. Tanner, A. Hebard, A. Borisov, and S. Shabanov: 2006. *Phys. Rev. Lett.* **97**, 067403.
Semin, D. J., A. Lo, S. E. Roak, R. T. Skodje, and K. L. Rowlen: 1996. *J. Chem. Phys.* **105**, 5542.
Seshadri, S.: 1986. *J. Appl. Phys.* **60**, 1514.
Shalaev, V.: 1996. *Phys. Reports* **272**, 61.

Shalaev, V. (ed.): 2002, 'Optical Properties of Nanostructured Random Media', Topics in Applied Physics. NY: Springer.
Shalaev, V., R. Botet, and R. Jullien: 1991. *Phys. Rev. B* **44**, 12216.
Shalaev, V., W. Cai, U. Chettiar, H.-K. Yuan, A. Sarychev, V. Drachev, and A. Kildishev: 2005a. *Opt. Lett.* **30**, 3356.
Shalaev, V., W. Cai, U. Chettiar, H.-K. Yuan, A. Sarychev, V. Drachev, and A. Kildishev: 2006. *Laser Phys. Lett.* **3**, 49.
Shalaev, V., V. Podolskiy, and A. Sarychev: 2001, 'Local Fields and Optical Properties of Metal-Dielectric Films'. In: A. Lakhtakia, W. Weiglhofer, and I. Hodgkinson (eds.): *Complex Mediums II*, Vol. 4467 of *SPIE Proc.* p. 207.
Shalaev, V., V. Podolskiy, and A. Sarychev: 2002, 'Plasmonic Nanophotonics: Manipulating light and sensing molecules'. In: A. Lakhtakia and G. D. M. McCall (eds.): *Complex Mediums III*, Vol. 4806 of *SPIE Proc.* p. 33.
Shalaev, V. and M. Stockman: 1987. *Sov. Phys. JETP* **65**, 287.
Shalaev, V. M.: 2000, *Nonlinear Optics of Random Media*, Vol. 158 of *Springer Tracts in Modern Physics*. Berlin: Springer.
Shalaev, V. M., W. Cai, U. Chettiar, H.-K. Yuan, A. K. Sarychev, V. P. Drachev, and A. V. Kildishev: 2005b. *arXiv:physics/0504091* **Apr. 13**.
Shalaev, V. M., V. A. Markel, E. Y. Poliakov, R. L. Armstrong, V. P. Safonov, and A. K. Sarychev: 1998. *Nonlin. Optic. Phys. Mat.* **7**, 131.
Shalaev, V. M., E. Y. Poliakov, V. A. Markel, V. P. Safonov, and A. K. Sarychev: 1997. *Fractals* **5**, 63.
Shalaev, V. M. and A. K. Sarychev: 1998. *Phys. Rev. B* **57**, 13265.
Shelby, R. A., D. R. Smith, and S. Schultz: 2001. *Science* **292**, 77.
Shen, Y.: 1984, *The Principles of Nonlinear Optics*. NY: John Wiley.
Sheng, P.: 1980. *Phys. Rev. Lett.* **45**, 60.
Sheng, P.: 1995, *Introduction to Wave Scattering, Localization, and Mesoscopic Phenomena*. San Diego: Academic Press.
Shi, X. and L. Hesselink: 2004. *J. Optical. Soc. Am.* **21**, 1305.
Shin, F., W. Tsui, and Y. Yeung: 1990. *J. Mater. Sci. Lett.* **9**, 1002.
Shin, F. and Y. Yeung: 1988. *J. Mater. Sci. Lett.* **7**, 1066.
Shklovskii, B. and A. Efros: 1984, *Electronic Properties of Doped Semiconductors*, Springer Series on Solid State Physics. Berlin: Springer-Verlag.
Shubin, V., A. Sarychev, J. Clerc, and V. Shalaev: 2000. *Phys. Rev. B* **62**, 11230.
Shvets, G.: 2003. *Phys. Rev. B* **67**, 035109.
Shvets, G. and Y. A. Urzhumov: 2004. *Phys. Rev. Lett.* **93**, 243902.
Shvets, G. and Y. A. Urzhumov: 2005. *J. Opt. A: Pure Appl. Opt.* **7**, S23.
Sievenpiper, D.: 1999, 'High-Impedance Electromagnetic Surfaces'. Ph.D. thesis, University of California LA.
Sievenpiper, D., M. Sickmiller, and E. Yablonovitch: 1996. *Phys. Rev. Lett.* **76**, 2480.
Sievenpiper, D., E. Yablonovitch, J. Winn, S. Fan, P. Villeneuve, and J. Joannopoulos: 1998. *Phys. Rev. Lett.* **80**, 2829.
Sievenpiper, D., L. Zhang, R. Broas, N. Alexpolous, and E. Yablonovitch: 1999. *IIEEE Trans. Microwave Theory Tech.* **47**, 2059.

Sigalas, M., C. Chan, K. Ho, and C. Soukoulis: 1995. *Phys. Rev. B* **52**, 11 744.
Skal, A. and B. I. Shklovskii: 1975. *Sov. Phys Semicond.* **8**, 1089.
Smith, D. and N. Kroll: 2000. *Phys. Rev. Lett.* **85**, 2933.
Smith, D., W. Padilla, D. Vier, S. Nemat-Nasser, and S. Shultz: 2000. *Phys. Rev. Lett.* **84**, 4184.
Smith, D., J. Pendry, and M. Wiltshire: 2004a. *Science* **305**, 788.
Smith, D., S. Schultz, N. Kroll, M. Sigals, K. Ho, and C. Soukoulis: 1994. *Appl. Phys. Lett.* **65**, 645.
Smith, D., D. Vier, W. Padilla, S. Nemat-Nasser, and S. Schultz: 1999. *Appl. Phys. Lett.* **75**, 1425.
Smith, D. R., P. Rye, D. Vier, A. Starr, J. Mock, and T. Perram: 2004b. *IEICE T. Electronics* **E87C**, 359.
Smith, D. R. and D. Schurig: 2003. *Phys. Rev. Lett.* **90**, 077405.
Smolyaninov, I., Y.-J. Hung, and C. Davis: 2005. *Appl. Phys. Lett.* **87**, 041101.
Sobnack, M., W. Tan, N. Wanstall, T. Preist, and J. Sambles: 1998. *Phys. Rev. Lett.* **80**, 5667.
Sonnichsen, C., A. Duch, G. Steininger, M. Koch, G. V. Plessen, and J. Feldmann: 2000. *Appl. Phys. Lett.* **76**, 140.
Stanley, H.: 1981, *Introduction to Phase Transition and Critical Phenomena.* London: Oxford Press.
Stauffer, D. and A. Aharony: 1994, *Introduction to Percolation Theory.* London: Taylor and Francis, 2nd edition.
Stinchcombe, R. and B. Watson: 1976. *J. Phys C* **9**, 3221.
Stockman, M. and D. Bergman: 2003. *Phys. Rev. Lett.* **90**, 027402.
Stockman, M., S. Faleev, and D. Bergman: 2001. *Phys. Rev. Lett.* **87**, 167401.
Stockman, M. I.: 1997a. *Phys. Rev. E* **56**, 6494.
Stockman, M. I.: 1997b. *Phys. Rev. Lett.* **79**, 4562.
Stockman, M. I., L. N. Pandey, L. S. Muratov, and T. F. George: 1994. *Phys. Rev. Lett.* **72**, 2486.
Straley, J.: 1977. *Phys. Rev. B* **15**, 5733.
Straley, J. P.: 1976. *J. Phys. C* **9**, 783.
Stratton, J. A.: 1935. *Proc. Natl. Acad. Sci. USA* **21**, 51.
Stroud, D.: 1975. *Phys. Rev. B* **12**, 3368.
Stroud, D.: 1998. *Super-lattice and Microstructures* **23**, 567.
Stroud, D. and P. Hui: 1988. *Phys. Rev. B* **37**, 8719.
Stroud, D. and X. Zhang: 1994. *Physica A* **207**, 55.
Suckling, J., A. Hibbins, M. Lockyear, T. Preist, J. Sambles, and C. Lawrence: 2004. *Phys. Rev. Lett.* **92**, 147401.
Sugawara, Y., T. Kelf, and J. Baumberg: 2006. *Phys. Rev. Lett.* **97**, 266808.
Tadayoni, M. and N. Dand: 1991. *Appl. Spec.* **45**, 1613.
Takakura, Y.: 2001. *Phys. Rev. Lett.* **86**, 5601.
Tan, W., T. Preist, and J. Sambles: 2000. *Phys. Rev. B* **80**, 11134.
Theocaris, P.: 1987, *The Concept of Mesophase in Composites*, Springer Series in Polymer: Properties and Applications. Springer-Verlag.
Thio, T.: 2006. *American Scientist* **94**, 40.
Thio, T., H. Ghaemi, H. Lezec, P. Wolff, and T. Ebbesen: 1999. *J. Opt. Soc.*

Am. B **16**, 1743.

Thio, T., H. Lezec, T. Ebbesen, K. Pellerin, G. Lewen, A. Nahata, and R. Linke: 2002. *Nanotechnology* **13**, 429.

Thio, T., K. Pellerin, R. Linke, H. Lezec, and T. Ebbesen: 2001. *Opt. Lett.* **26**, 1972.

Tominaga, J., J. Kim, H. Fuji, D. Buchel, T. Kikukawa, L. Men, H. Fukuda, A. Sato, T. Nakano, A. Tachibana, Y. Yamakawa, M. Kumagai, T. Fukaya, and N. Atoda: 2001. *Japan. J. Appl. Phys.* **40**, 1831.

Tomita, Y. and Y. Okabe: 2002. *J. Phys. Soc. Japan* **71**, 1570.

Tortet, L., J. R. Gavarri, J. Musso, G. Nihoul, J. P. Clerc, A. N. Lagarkov, and A. K. Sarychev: 1998. *Phys. Rev. B* **58**, 5390.

Treacy, M.: 1999. *Appl. Phys. Lett.* **75**, 606.

Treacy, M.: 2002. *Phys. Rev. B* **66**, 195105.

Tsang, T.: 1995. *Phys. Rev. A* **52**, 4116.

Tsang, T.: 1996. *Opt. Lett.* **21**, 245.

Tsui, F. S. W. and Y. Yeung: 1988. *J. Mater. Sci. Lett.* **8**, 1383.

Vainshtein, L.: 1969, *The Theory of Diffraction and the Factorization Method (Generalized Wiener-Hopf Technique).* Boulder Colorado: Golem Press.

Vainshtein, L.: 1988, *Electromagnetic Waves.* Moscow: Radio and Telecommunications, 2nd edition.

Veselago, V.: 1968. *Soviet Physics Uspekhi* **10**, 509.

Vigoureux, J.: 2001. *Optics Comm.* **198**, 257.

Vinogradov, A.: 2002, 'On the Form of Constitutive Equations in Electrodynamics'. *Physics - Uspekhi (Uspekhi Fizicheskikh Nauk, Russian Academy of Sciences)* **45**, 331.

Vinogradov, A. and A. Aivazyan: 1999. *Phys. Rev. E* **60**, 987.

Vinogradov, A., A. Goldenshtein, and A. Sarychev: 1989a. *Sov. Phys. Techn. Phys.* **34**, 125.

Vinogradov, A., A. Karimov, A. Kunavin, A. Lagarkov, A. Sarychev, and N. Stember: 1984a. *Sov. Phys. Dokl.* **29**, 214.

Vinogradov, A., A. Karimov, and A. Sarychev: 1988. *Sov. Phys. JETP* **67**, 2129.

Vinogradov, A., A. Lagarkov, and A. Sarychev: 1984b. *Sov. Phys., JETP Lett.* **40**, 1083.

Vinogradov, A., L. Panina, and A. Sarychev: 1989b. *Sov. Phys. Dokl.* **34**, 530.

Vinogradov, A., A. Virnic, V. Garanov, A. Kalachev, S. Matytsin, I. Oklahoma, A. Pakhomov, and A. Sarychev: 1992. *Sov. Phys., Techn. Phys.* **37**, 506.

Wan, W., H. Lee, P. Hui, and K. Yu: 1996. *Phys. Rev. B* **54**, 3946.

Wannemacher, R.: 2001. *Optic Comm.* **195**, 107.

Watts, R., T. Preist, and J. Sambles: 1997a. *Phys. Rev. Lett.* **79**, 3978.

Watts, R., T. Preist, and J. Sambles: 1997b. *Phys. Rev. Lett.* **79**, 3978.

Wei, A.: 2003. In: R. Shenhar and V. Rotello (eds.): *Nanoparticles: Scaffolds and Building Blocks.* N.Y., p. 143, Kluver Academic.

Weiglhofer, W.: 2003, 'Constitutive Characterization of Simple and Complex Mediums'. In: W. Weiglhofer and A. Lakhtakia (eds.): *Introduction to Complex Mediums for Optics and Electromagnetics.* WA, USA, SPIE, Bellingham.

Wiltshire, M., J. Hajnal, J. Pendry, D. Edwards, and C. Stevens: 2003. *Optic Express* **11**, 709.

Wood, R.: 1902. *Proc. R. Soc. London* **A18**, 269.

Wurtz, G., R. Pollard, and A. Zayats: 2006. *Phys. Rev. Lett.* **97**, 057402.

Xiong, Y. ., S. Shen, and X. Xie: 2001. *Phys. Rev. B* **63**, 140418.

Xu, H., E. Bjerneld, M. Käll, and L. Börjesson: 1999. *Phys. Rev. Lett.* **83**, 4357.

Yagil, Y., P. Gadenne, C. Julien, and G. Deutscher: 1992. *Phys. Rev. B* **46**, 2503.

Yagil, Y., M. Yosefin, D. J. Bergman, G. Deutscher, and P. Gadenne: 1991. *Phys. Rev. B* **43**, 11342.

Yen, T., W. Padilla, N. Fang, D. Vier, D. Smith, J. Pendry, D. Basov, and X. Zhang: 2004. *Science* **303**, 1494.

Yoshida, K., Y. Sano, and Y. T. Y: 1993. *Physica C* **206**, 127.

Zayats, A., I. Smolyaninov, and A. Maradudin: 2005. *Phys. Rep.* **408**, 131.

Zeng, X., P. Hui, D. Bergman, and D. Stroud: 1989. *Physica A* **157**, 10970.

Zhang, S., W. Fan, N. Panoiu, K. Malloy, R. Osgood, and S. Brueck: 2005. *Phys. Rev. Lett.* **95**, 137404.

Zhang, X. and D. Stroud: 1994. *Phys. Rev. B* **49**, 944.

Zuev, C. and A. Sidorenko: 1985a. *Theor. Mat. Fiz.* **62**, 253.

Zuev, C. and A. Sidorenko: 1985b. *Theor. Mat. Fiz.* **62**, 76.